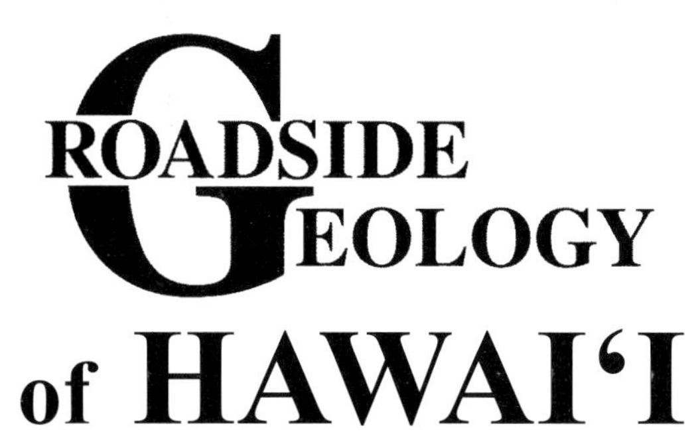

# of HAWAI'I

Second Edition

RICK HAZLETT
CHERYL GANSECKI
STEVE LUNDBLAD

2022
Mountain Press Publishing Company
Missoula, Montana

First Printing, August 2022

Geologic maps were refined by Chelsea Feeney (www.cmcfeeney.com)
Photographs by authors unless otherwise credited.

Library of Congress Cataloging-in-Publication Data

Names: Hazlett, Richard W., author. | Gansecki, Cheryl, author. | Lundblad, Steve, 1963- author.
Title: Roadside geology of Hawai'i / Rick Hazlett, Cheryl Gansecki, Steve Lundblad ; illustrated by Chelsea M. Feeney.
Description: Second edition. | Missoula, Montana : Mountain Press Publishing Company, 2022. | Series: Roadside geology | Revised edition of: Roadside geology of Hawai'i / Richard W. Hazlett, Donald W. Hyndman. 1996. | Includes bibliographical references and index. | Summary: "This completely revised, full-color second edition of Roadside Geology of Hawai'i details the evolution of this volcanic island chain, from its first tumultuous appearance above the sea to ongoing eruptions, including the 2018 eruption from Kīlauea. — Provided by publisher.
Identifiers: LCCN 2022022496 | ISBN 9780878427116 (paperback)
Subjects: LCSH: Geology—Hawaii—Guidebooks. | Hawaii—Guidebooks.
Classification: LCC QE349.H3 H39 1996 | DDC 559.69—dc23/eng20220723
LC record available at https://lccn.loc.gov/2022022496

PRINTED IN THE UNITED STATES

P.O. Box 2399 • Missoula, MT 59806 • 406-728-1900
800-234-5308 • info@mtnpress.com
www.mountain-press.com

TO THE MEMORY OF

HAROLD T. STEARNS, 1900–1986,

PIONEERING HAWAIʻI GEOLOGIST

Stearns in front of the family home in Pāhala in 1924. —Hawaiian Collection Archive, Edwin H. Moʻokini Library, University of Hawaiʻi, Hilo

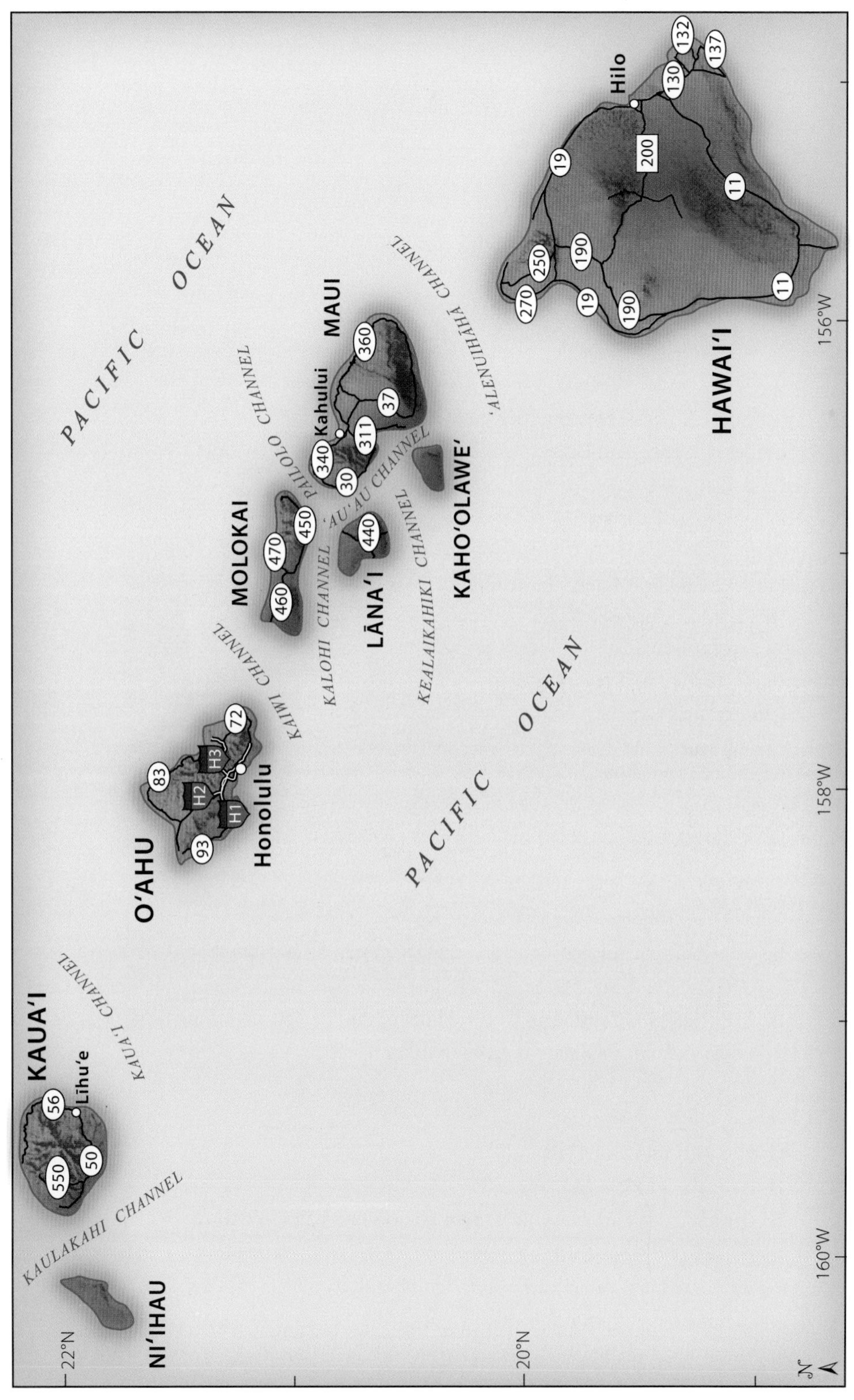

Roads and sections of Roadside Geology of Hawai'i.

CONTENTS

Huge lava flow entering the ocean near Cape Kumukahi, island of Hawai'i. Between early June and late August 2018, this flow added 875 acres of new land to the island. —Photo courtesy US Geological Survey

PREFACE AND ACKNOWLEDGMENTS

We wrote this book for people who would like to learn more about Hawai'i. Most of what is known about Hawaiian geology is widely scattered through the professional literature, much of which is difficult to locate and even more difficult to read. Because almost all published geologic research is buried in institutional libraries or pay-for-access websites, we feel that the results should be made more easily available, so the public can enjoy the wonderful geologic history of Hawai'i.

The book begins with a general introduction to the geology of Hawai'i. Following chapters cover each of the inhabited and easily accessible main islands. Each chapter begins with a general discussion of the rocks of that island, then proceeds to a series of road guides with local details.

Like all scientists, geologists use more than their fair share of jargon, a "shorthand" which makes what they do and say seem deeper and more mysterious than it really is. Most of the rocks and ideas that geologists investigate are understandable in an everyday sense. We did our best to eliminate as much jargon as we could, but some words, especially names of the rocks, have no vernacular equivalents. You cannot write about geology without discussing processes and rocks, so a glossary is provided at the back of the book.

We thank the following people for their help in producing this book: Donald Hyndman and David Alt of the University of Montana contributed much to the book's first-edition content and, with Jane Taylor, Kathleen Ort, and Jennifer Carey, improved its readability. Geology student Raelyn Pisco-Eckert assisted with typing, figure production, and data collection. Chelsea Feeney assisted handsomely in the preparation of geologic maps. Mike Zoeller (US Geological Survey Hawaiian Volcano Observatory) provided photography and field support for work on Lāna'i, and Darcy Bevens (Center for the Study of Active Volcanoes) on Molokai. Their contributions were essential. Ken Hon (US Geological Survey Hawaiian Volcano Observatory) gave important field assistance and advice on several occasions. Nancy Lundblad undertook close editing of revised drafts: an arduous chore. Don Swanson and Frank Trusdell of the Hawaiian Volcano Observatory were sources of key information and insights. The UHH Geology Department also supported this project with office space and computing services. Finally, to the good folks at various local booksellers and the US National Park Service, a big *mahalo*!

A NOTE ON THE HAWAIIAN LANGUAGE

The Hawaiians did not have a written language. Early missionaries introduced the Latin alphabet to Hawaiians to enable Bible reading and education beginning in the 1820s, and during the time of Kamehameha's Kingdom (1795–1893) the Hawaiian people developed a high degree of literacy. Hawaiian language newspapers were widely read 150 years ago. Some finer points of pronunciation and meaning in Hawaiian words continue to be discussed up to the moment, assisted by grammatical notations unfamiliar to English-only speakers. Though just a few thousand people still use it as a primary means of communication, Hawaiian is very much a living language.

Immersed in island geography, virtually every resident and visitor to Hawai'i routinely reads and speaks Hawaiian words. The names of many place names tell a story and are compounds of common Hawaiian nouns that even non-Hawaiian speakers learn to recognize; for instance, Uēkahuna, which means "wailing [u(w)e-] priest [-kahuna]," is a sacred precinct overlooking the active craters at the summit of Kīlauea Volcano.

The following guidelines will assist readers who are seeking to pronounce Hawaiian words for the first time. The written Hawaiian language is characterized by clusters of vowel combinations and a small number of consonants. To English ears the language seems fluid, almost even musical. Two diacritical marks, explained below, affect pronunciation and are important to understand. They also affect a word's meaning—for example, lanai (stiff-backed) and lānai (porch, veranda), and ono (a type of fish) and 'ono (delicious).

Hawaiian has eight constants (h, k, l, m, n, p, w, and '). The first seven are pronounced roughly the same as in English. The sound represented by an okina, an upside-down apostrophe ('), is a glottal stop or break in pronunciation. For example, in English the double vowel oo (as in soon) is pronounced "ew," whereas the Hawaiian word "o'o" is pronounced "oh-oh," with the vowel repeated. In general, double vowels (the same vowel twice in a row) do not exist in Hawaiian words without an okina separating them.

Some words, as in 'ono, begin with okinas. In fact, the word *okina* is technically spelled 'okina. The presence of an okina in this position usually signifies that another consonant once was placed there, but it fell out of use as the language evolved. The okina here represents a missing silent letter; it has no impact on word pronunciation in this case.

Unlike English, sounds of vowels are consistent in Hawaiian. Two vowels, e and i, are pronounced differently than they usually are in English: e sounds close to a long a, and i sounds like a long e. Vowels with a kahakō (or macron)—a bar over them—are given greater than normal stress and duration. Pronounce Hawaiian vowels as follows:

a as in lava	mahalo (thank you)
e like the "a" in bay	hele (go)

i as in feet	Waikīkī (Wye-kee-kee)
o as in hole	aloha
u as in boot	hula

Combinations of vowels (diphthongs) include:

ai, pronounced "eye"	kai (sea)
au, pronounced "ow," as in "cow"	lau (leaf, frond)
oe, pronounced "oy" as in "boy"	oe (prolonged sound or chanting)
ei, pronounced "ay" as in "hay"	lei (flower garland)

Determining where to place stress in a Hawaiian word can be baffling for someone not used to hearing Hawaiian words pronounced. For shorter words, the next-to-last syllable is often stressed (wahine, aloha, O'ahu). Also, a syllable with a diphthong in a short word is often stressed, as in makai, meaning "toward the sea."

A couple of Hawaiian words you will encounter frequently in this book are descriptive terms for two different kinds of lava: 'a'ā (ah-AH), sharp, rough, cindery lava, and pāhoehoe (PAH-hoy-hoy), smooth, ropy lava.

The following are some other translated Hawaiian words you may encounter as you explore local roads and trails in Hawai'i:

Ahu "AH-who"	stack of stones, typically marking a trail
Ala "AH-lah"	way, route
'Āina "EYE-nah"	area of land, earth
Iki "EE-kee"	little
Kīpuka	patch of older land or forest surrounded by younger lavas
Lua "LOO-ah"	pit, small crater
Makai "Mah-KYE"	toward the sea
Mauka "MOW-kah"	toward the mountains, or interior (of an island)
Mauna "MOW-nah"	mountain
Pali "PAH-lee"	cliff
Pu'u "POO-oo"	hill
Wai "W-EYE"	freshwater course, stream

A final note on the spelling of place names: there has been a recent trend to combine separate words in many Hawaiian place names into a single name. So, you may see Mauna Kea written as Maunakea or Pu'u Loa as Pu'uloa. We have mostly kept the older format currently found on signs and maps, but be aware that may change.

ISLAND	GEOLOGIC UNITS		AGE (years ago)
KAUA'I	WAIMEA CANYON BASALT	Nāpali Member	5.5 to 4 million
		Hā'upu Member	4.4 to 4.3 million
		Makaweli Member	4 to 3.5 million
		Olokele Member	4 million
	KŌLOA VOLCANICS	Kōloa Volcanics	3.65 million to 150,000
O'AHU	WAI'ANAE VOLCANO (Wai'anae Volcanics)	Lualualei Member	3.9 to 3.55 million
		Kamaile'unu Member	3.55 to 3.06 million
		Pālehua Member	3.0 million
		Kolekole Member	3.0 to 2.8 million
	KO'OLAU VOLCANO	Ko'olau Basalt	3.2 to 1.8 million
		Honolulu Volcanics	1.0 million to 40,000
MOLOKAI	WEST MOLOKAI VOLCANO	West Molokai Volcanics	2.0 to 1.8 million
		Wai'eli Volcanics	1.8 million
	EAST MOLOKAI VOLCANO (East Molokai Volcanics)	upper member	1.75 to 1.52 million
		lower member	1.52 to 1.31 million
	KALAUPAPA VOLCANO	Kalaupapa Volcanics	570,000 to 340,000
LĀNA'I	LĀNA'I BASALT		1.5 to 1.3 million
MAUI	WEST MAUI VOLCANO	Wailuku Basalt	2.1 to 1.3 million
		Honolua Volcanics	1.3 to 1.1 million
		Lahaina Volcanics	600,000 to 300,000
	HALEAKALĀ VOLCANO	Honomanū Basalt	1.1 to 0.9 million
		Kula Volcanics	900,000 to 150,000
		Hāna Volcanics	140,000 to recent
HAWAI'I ISLAND	KOHALA VOLCANO	Pololū Volcanics	770,000 to 300,000
		Hāwī Volcanics	250,000 to 60,000
	MAUNA KEA VOLCANO	Hāmākua Volcanics	300,000 to 65,000
		Laupāhoehoe Volcanics	65,000 to recent
	HUALĀLAI VOLCANO	Hualālai Volcanics	130,000 to recent
		Wa'awa'a Trachyte Member	114,000 to 92,000
	MAUNA LOA VOLCANO	Nīnole Basalt	200,000 to 100,000
		Kahuku Basalt	100,000 to 30,000
		Ka'ū Basalt	30,000 to recent
	KĪLAUEA VOLCANO	Hilina Basalt	100,000 to 16,000
		Puna Basalt	16,000 to recent

Approximate ages of major rock formations seen along Hawaiian roadsides, organized by island and volcano. "Recent" eruptions occurred within the past few thousand years, could erupt again in the near geological future, or both. —Modified from Sherrod and others, 2021

The Hawaiian Hot Spot

Polynesians probably first reached the Hawaiian Islands sometime in the thirteenth century. These early Hawaiians developed a rich oral tradition explaining the origin of islands. The goddess Namaka represented the sea, and her sister Pele was the Goddess of Fire whom Hawaiians came to associate with eruptions of lava and the creation of new land. These two supernatural beings fought jealously with one another, with Namaka getting the upper hand—at first.

Pele initially arrived on Kaua'i, where she used her sacred digging stick Paoa to establish a new home—a volcano. But she could not escape Namaka until she reached her final home, Halema'uma'u Crater on the island of Hawai'i (Big Island) far to the southeast. As a metaphor, this shows that the Hawaiians understood that their islands were geologically youngest in the southeast, and that the islands to the northwest were ancient volcanoes eroding back into the sea.

By the early nineteenth century, scientists realized that the Hawaiian Islands were simply the summits of very large, submerged volcanoes rising above sea level. James Dana, who traveled to Hawai'i with the US Wilkes Exploration Expedition in 1840–1841, began developing a modern theory for their origin. He thought that a large gash, or propagating crack, was slowly opening in the ocean floor toward the southeast. Volcanic eruptions healed the tear in its wake, but its southeastern tip, now under the Big Island, remained open, permitting lava to erupt and build new land. He was

Steam rises where lava enters the ocean during the 2018 eruption of Kīlauea Volcano on the Big Island.

convinced that the Big Island's volcanoes would someday die and that a new island would rise in the ocean even farther to the southeast. He turned out to be right about that when in 1955 a new Hawaiian volcano, Kamaʻehuakanaloa (formerly known as Lōʻihi), was discovered underwater 25 miles south of the Big Island. By this time, though, his propagating-crack explanation was no longer persuasive. In 1946, Harold T. Stearns, one of the first to map the geology of the islands, noted that there was simply no evidence of a great fracture running the length of the island chain.

To understand the formation of Hawaiʻi, we must first understand the Earth's structure. The outermost rocky layer of the Earth is called the crust. The boundary between it and the mantle below is so sharp that earthquake waves reflect from it and echo back to the surface. Geologists measure the thickness of the crust by timing the echo. The crust beneath continents is between 15 and 40 miles thick, made mostly of granite, schist, granulite, and gneiss, lighter rocks consisting largely of the minerals feldspar and quartz, speckled with darker micas and amphiboles. The Hawaiian Islands have no continental crust; they are strictly oceanic. Thus, Hawaiʻi is the only truly oceanic state in the United States.

Oceanic crust of the seafloor consists almost entirely of gabbro capped with basalt flows. The crust is generally about 5 miles thick, but much denser and heavier than its continental counterpart, which explains why the ocean floor rests at a lower level than the continents. Were it not for this density difference, almost all of Earth's surface today would be underwater! The average depth to the ocean floor in the Hawaiian region is about 16,000 feet.

Beneath the crust lies the mantle, reaching all the way to Earth's core. The upper part of the mantle is mainly a rock called peridotite, which consists chiefly of olivine and pyroxene crystals. Hawaiian peridotites also contain small amounts of other minerals. Spinel, an oxide of magnesium and aluminum, occurs as tiny octahedral crystals of various colors in peridotite that forms at depths of less than 30 miles. Peridotite from greater depths contains beautiful dark-red garnets.

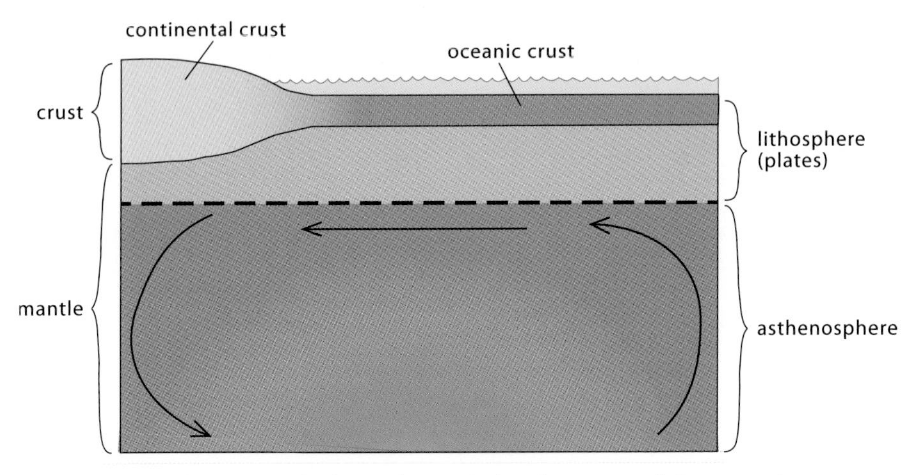

Major layering inside the Earth is critical to the process of plate tectonics.

In the 1950s, revolutionary new technology allowed people to study the ocean floor as never before, and evidence emerged from the magnetization of marine lavas that the world's crust and uppermost mantle are divided into a set of giant shifting pieces, or tectonic plates. (*Tekton* is the Greek word for "builder"—and plate tectonics is the process of Earth's shifting crustal plates.) We have learned that the Hawaiian Islands lie in the middle of the gigantic Pacific Plate, which is gliding toward the northwest at a rate of 30 to 35 feet every century. That's not a blinding speed in human terms, but a century means nothing to a rock.

The Pacific Plate consists of oceanic crust and a layer of peridotite at the top of the mantle—a rigid layer called the lithosphere. The mantle beneath, the asthenosphere, is so hot that the rock there flows plastically, but it is not molten for the most part. It may be hard to think of solid rock as flowing, but the same process can be observed with ice in glaciers. Convection currents in the mantle are driven by a combination of hotter material rising; cooler, denser material sinking; and the pull of downgoing slabs in subduction zones. These currents cause the overriding lithospheric plates to move.

In 1963, Canadian geophysicist J. Tuzo Wilson proposed that the Hawaiian Islands formed as the Pacific Plate moved for millions of years across a mantle hot spot fixed in position below the lithosphere. Wilson argued that so much upwelling heat is channeled into the hot spot that the mantle partly melts, forming magma—underground molten rock. Less dense than the surrounding solid rock, the magma rises buoyantly and ultimately erupts to build new land.

Wilson explained that one island after another has formed right where the island of Hawai'i is today, only to be rafted piggyback to the northwest on the drifting plate. Cut off from their hot spot source, each aging volcano has died, sinking back into the ocean after a period of 5 to 10 million years.

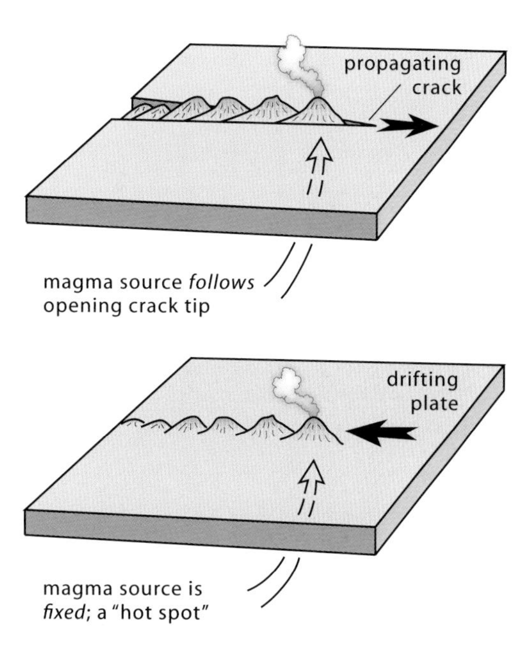

The upper model for the origin of the Hawaiian Islands was proven incorrect. The lower model, the hot spot theory, is now widely accepted.

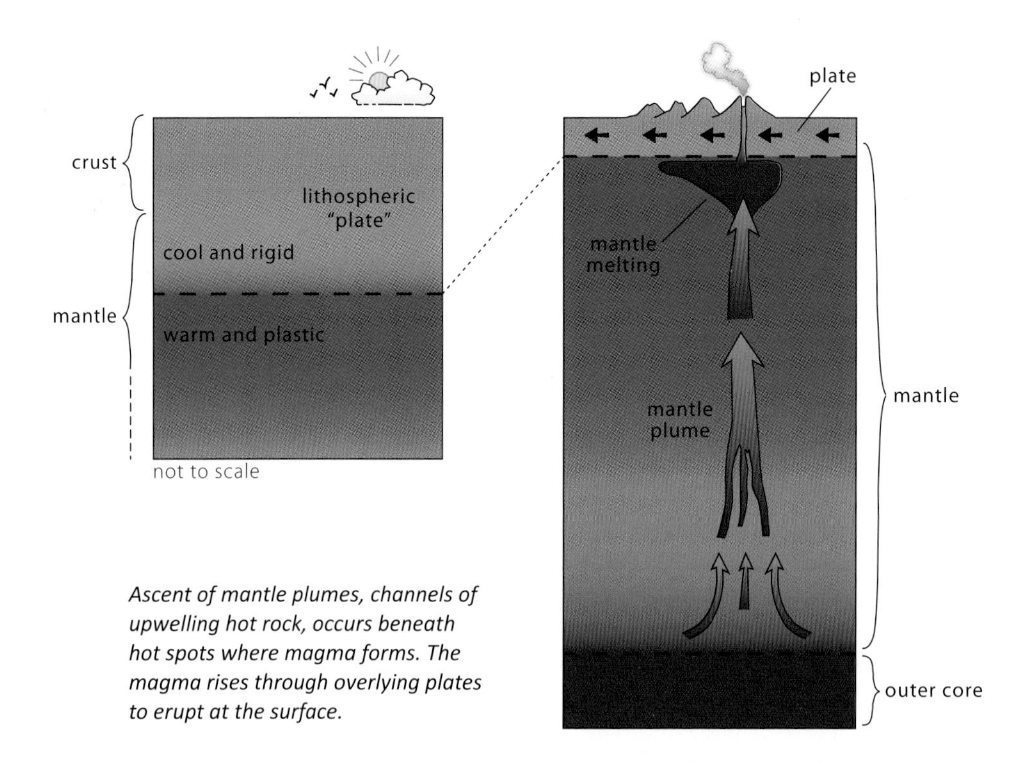

Ascent of mantle plumes, channels of upwelling hot rock, occurs beneath hot spots where magma forms. The magma rises through overlying plates to erupt at the surface.

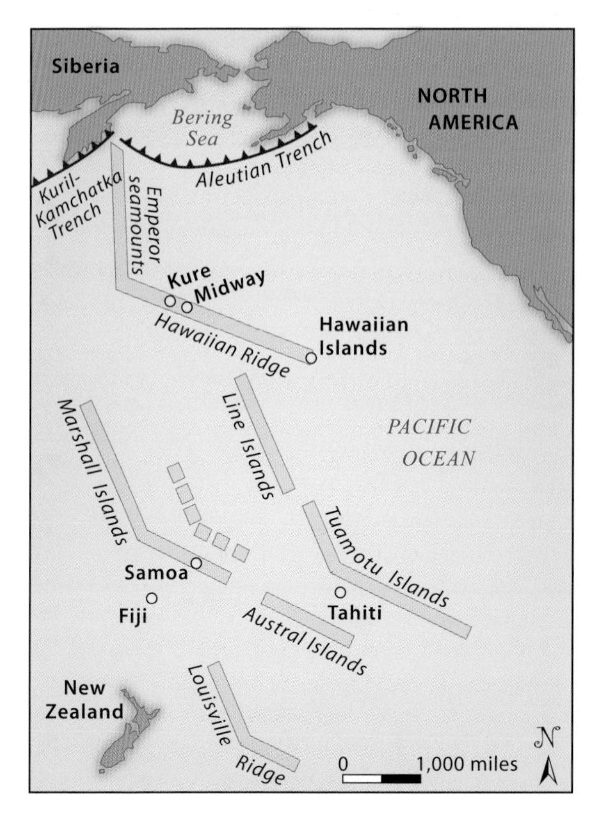

Multiple hot spots exist beneath the Pacific Plate, giving rise to a set of island chains and seamounts (undersea mountains) parallel to one another. Each shows aging of volcanoes to the northwest, underscoring how stable these deep-seated hot spots have been over tens of millions of years.

While this idea has been modified and refined since Wilson's day, his essential proposal about how Hawai'i came into being is widely accepted by scientists today. Moreover, dozens of other hot spots around the world are now known. The Hawaiian hot spot is one of the biggest, most active, and possibly the oldest. It is roughly centered beneath three of Hawai'i's most active volcanoes—Kīlauea, Mauna Loa, and Kama'ehuakanaloa (Lō'ihi), though it seems to uplift a larger area up to 500 miles across known as the Hawaiian swell. Seismic imaging shows its deep roots farther to the southeast and likely rising from near the core-mantle boundary; the shallow plume seems to be dragged to the northwest along with the Pacific Plate. No one knows exactly why hot spots form, or why some are short-lived and others last for many millions of years. They are certainly very efficient pathways for heat to escape Earth's deep interior, and our planet is continuously trying to cool itself off.

The Hawaiian-Emperor Seamounts

The eight major Hawaiian Islands constitute only about 10 percent of the 3,600-mile-long hot spot–related volcanic chain that stretches off to the northwest almost entirely submerged. Whereas 17 volcanoes compose the main islands from Hawai'i to Ni'ihau, 112 others form the rest of the chain. Each long-dead underwater volcano likely formed an island similar to the ones existing today, albeit millions of years ago.

The islands-that-once-were have vanished into the sea on their long tectonic journey for several reasons. Erosion by waves, streams, and landsliding eats away at the land bit by bit until nothing remains. Also important to the islands' demise is subsidence of the Pacific Plate while it drifts northwestward, cools, and settles into the mantle. The fate of many islands is to be pulled underwater as their plate foundation sinks, even before they completely erode down to sea level.

This island-drowning is generally postponed for millions of years, though, so long as each aging island remains within subtropical waters—less than about 30 degrees latitude. In warm seas, corals build fringing reefs that protect eroding hinterland from wave erosion. As the last ancient lava outcrops ultimately drop out of sight, the corals grow across the entire footprint of the island, forming sandy shoals or atolls, which are circular reef systems enclosing broad, shallow lagoons. You may think of this process as the biological echo of disappearing volcanoes.

Midway Island (also called Pihemanu or Kauihelani) is one example. This 7-mile-wide atoll was volcanically active between 30 and 27.5 million years ago, probably near where the Big Island is today. Massive coral reefs and light-colored sands form a cap over the ancient volcano around a quarter of a mile thick.

Eventually the Pacific Plate carries the islands into cooler ocean waters at latitudes greater than 30 degrees, and coral growth can no longer keep pace with tectonic subsidence. The islands completely disappear, morphing into submerged, flat-topped seamounts called guyots (pronounced GEE-ohs). The oldest known possible Hawaiian Islands, near the western Bering Sea, are now submerged by nearly 1 mile. Kure Atoll, a seabird rookery and emergency landing strip, is the last bit of land to the northwest of Hawai'i; it's about 1,400 miles up the chain from Honolulu.

From Hawai'i to just a few hundred miles past Kure, the Hawaiian range keeps a northwesterly trend. It then takes an abrupt 60-degree turn to the north, corresponding to a geologic change that took place around 50 million years ago possibly due to the Pacific Plate changing direction and/or southward drift of the hot spot plume. The long-dead volcanoes to the north belong to the Emperor seamount group. Geographers have named most big seamounts in this group after famous Japanese emperors.

The northernmost seamount, Meiji, is also the oldest—82 million years, a time when dinosaurs still romped on distant continents. It lies perched at the edge of the Kuril-Kamchatka Trench, a giant marine canyon seam through which the Pacific Plate slowly plunges back into the Earth beneath the overriding tectonic plate. If there are any older once-Hawaiian volcanoes, as seems likely, they were subducted or scraped off and incorporated along the edge of the plate. Unfortunately, we can't tell for sure that this has happened, given the remoteness of this region, ocean depth, and cover by marine sediments. Nevertheless, the Kuril-Kamchatka Trench is the graveyard of Hawaiian volcanoes.

HAWAI'I'S COMMON ROCK TYPES

Hawai'i is formed almost entirely of the products of volcanic eruptions, primarily lava flows of basalt, a dense, dark-gray rock. High concentrations of iron and magnesium in the lava give basalt its dark color and heaviness. Stacks of basalt flows give Hawaiian landscapes a layered appearance where erosion or faulting has exposed the flows in mountainsides and crater walls.

Several kinds of basalt exist, though only a geochemist or petrologist (a geologist who studies the compositions of rocks) may be able to distinguish them. The most common type is tholeiitic basalt, or just tholeiite (THO-lee-ite), named for Tholey, the district in Germany where it was first identified and described.

Tholeiitic basalts commonly include small, scattered crystals of glassy green olivine, also known as the gemstone peridot. Occasionally you may see white crystals of the feldspar plagioclase, too, which commonly have an elongate or tabular shape. Scattered crystals generally compose no more than a few percent of the total volume of a typical basalt. Everything else is formless groundmass and vesicles, holes formed from gas bubbles in the lava.

Less common are alkalic basalts, named because they contain a high concentration of the alkali elements, principally sodium in Hawai'i. In addition to olivine and plagioclase, you may discover crystals of chunky black pyroxene or even tabular hornblende in the lava. If the basalt is ultra-alkalic—very highly enriched in sodium—plagioclase may not appear at all but is replaced by the rare minerals melilite and nepheline.

Petrologists have described a series of different alkalic basalt types according to variations in crystal and chemical composition, including alkali basalt, hawaiites, basanites, ankaramites, benmoreites, mugearites, and nephelinites. In most places we ignore these categories in this guide. We simply refer to them all as alkalic basalts.

Trachyte is another sodium-rich volcanic rock that also contains much more silica (SiO_2) than basalt. Whereas the active lava flows of tholeiitic and alkalic basalts tend to be runny and thin, trachyte lava flows are much more viscous (stiffer), issuing as thick, pasty flows. This reflects their high silica content—60 to 65 percent of total composition compared to only 48 to 50 percent for most basalts. Silica is the material that goes into ordinary window glass, but these lavas in most samples show little if any glassiness except when freshly erupted. Glassy trachyte obsidian exists in Hawai'i, but can be found only at one location on the Big Island.

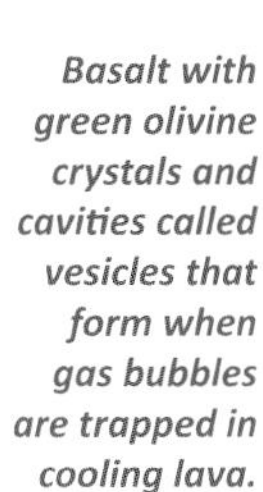

Basalt with green olivine crystals and cavities called vesicles that form when gas bubbles are trapped in cooling lava.

Greenish olivine and black pyroxene crystals in an alkalic basaltic rock known as an ankaramite, from Mauna Kea.

Not all molten basalt rising into Hawaiian volcanoes necessarily erupts. It may instead stagnate and completely crystallize underground. If this happens, the resulting rock type is called gabbro. The densest of all rock types in Hawai'i, gabbro can be detected underground in sensitive gravity surveys. Erosion may expose gabbro bodies in the slopes of aging Hawaiian volcanoes, though they are uncommon at the surface.

Not all Hawaiian rocks are volcanic in origin. Uplifted terraces of ancient reefs, and the cemented sands of older dunes and beaches, form limestone. Cemented dune and beach sand forms sandstone if the sand is composed of quartz, but in Hawai'i, most of the sand is calcareous—composed of tiny pieces of coral—so it consolidates into limestone. Many limestone beds contain the fossils of sea creatures, plant and tree roots, even small animals, particularly birds. Unlike Hawaiian lavas, limestone is a light-colored rock, beige to white in appearance. Like lava, it can form fantastically jagged surfaces, especially where rainwater partly dissolves it. As you might expect, it is commonly found at the shoreline or nearby.

THE LOA AND KEA TRENDS

In the 1840s, James Dwight Dana made an observation that continues to intrigue geologists. He noticed that the summits of Hawaiian volcanoes toward the southeastern end of the chain line up in parallel, arcuate lines. He named these alignments the Loa and Kea Trends, based upon the names of their two most prominent volcanoes, Mauna Loa and Mauna Kea. Compositionally, the lavas erupted in these two trends are chemically distinctive in terms of certain elements that occur only in very small concentrations. These differences show that the composition of the mantle that melts beneath Hawai'i varies—ever so slightly—even across short distances.

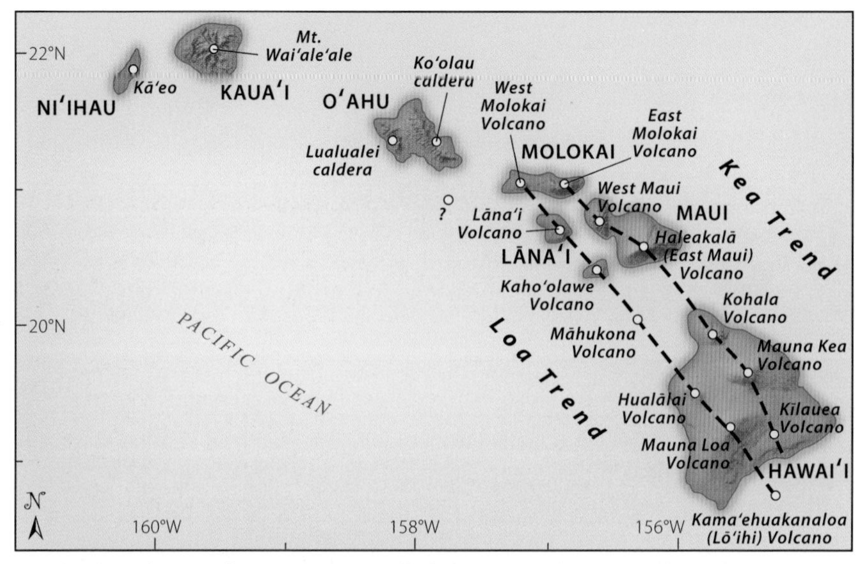

Hawaiian volcanoes line up in arcs called the Loa and Kea Trends in the youngest Hawaiian Islands. The older islands to the northwest do not show the same trends. Geologists do not know what the arcs signify.

Why do these two trends exist? One idea is that as the younger islands of Hawai'i—the Big Island, Maui, Lāna'i, and Molokai—have grown over the past few million years, they have pressed down the lithosphere, causing it to warp and split, like a board that is bent too much. This split reduces pressure in the mantle, allowing it to produce the magma that builds two parallel lines of volcanoes. When smaller volumes of lava are erupted from the hot spot, creating isolated islands with single volcanoes, this pattern cannot form. Another idea is that a slight change in plate direction about 3 million years ago caused melts from different parts of the tilted mantle plume to reach the surface and build separate, chemically distinct, volcanoes.

That the lithosphere between the large southeastern Hawaiian Islands is sagging under their enormous weights is certainly happening. The submerged bases of these younger, heavier islands lie as much as 3,000 feet below the average ocean depth found elsewhere in the region. The resulting depression, called the Hawaiian Deep, is about 40 miles wide and is enclosed by an outer upwarp called the Hawaiian Arch, also a result of flexing lithosphere. On the southeastern coast of the Big Island the rate of subsidence is as much as 4 to 7 inches every decade. The island continues to grow larger in this area, though, because lava flows build new land faster than it can disappear underwater.

To the northwest of the youngest islands, these deep-sea features disappear. The islands are smaller, and they are essentially floating in gravitational balance with the underlying mantle, like ice cubes coming to rest in a glass of water. Sediment eroded from the nearby islands has filled the deeps. Meanwhile, the rate of magma production from the Hawaiian hot spot has increased over the past few million years.

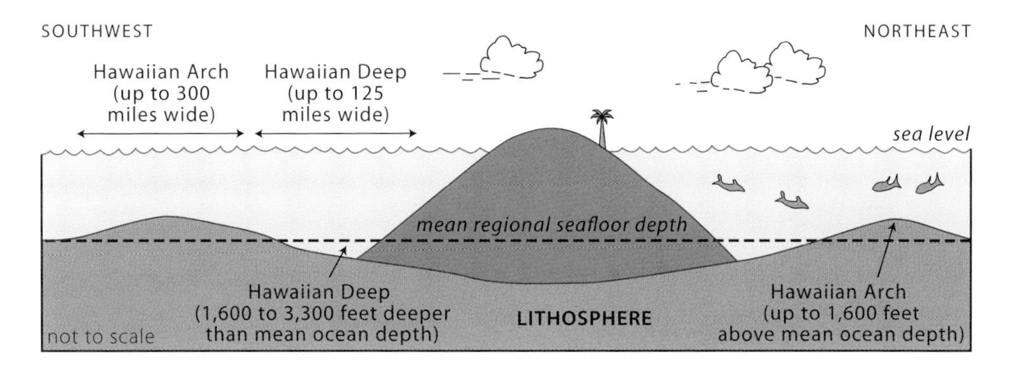

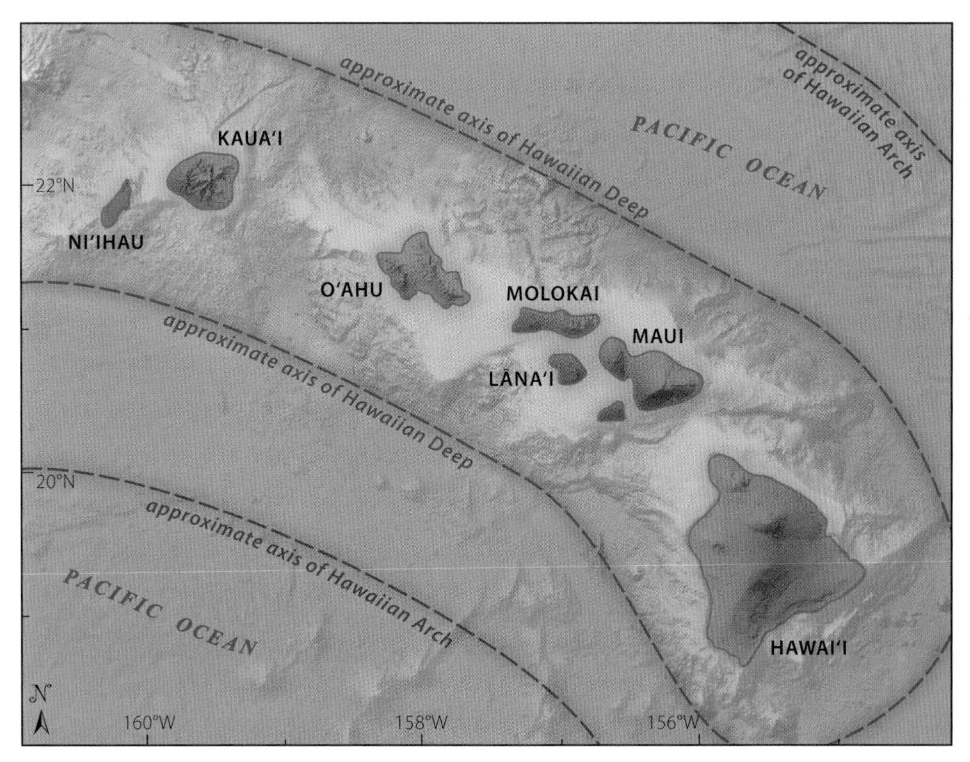

A young Hawaiian volcano depresses the lithosphere below, causing it to warp like an overloaded bookshelf. The warps around the base of the volcano explain the Hawaiian Deep and Hawaiian Arch, seafloor structures enclosing the youngest Hawaiian Islands.

Detailed studies of the earthquakes generated by Hawaiian magma show that it can well up continuously or ascend in batches taking a few weeks or months to reach the surface. As a big volcano evolves, magma begins accumulating a few miles beneath the summit to form a giant underground reservoir—a magma chamber. Many factors cause this to happen, including rapid cooling from escaping steam and gases, and decreasing buoyancy. A high rate of magma recharge, or resupply, is also important to maintain large, shallow magma reservoirs. The magma in the chamber will pressurize until it escapes through weak zones in the volcano and erupts, driven by expanding gases.

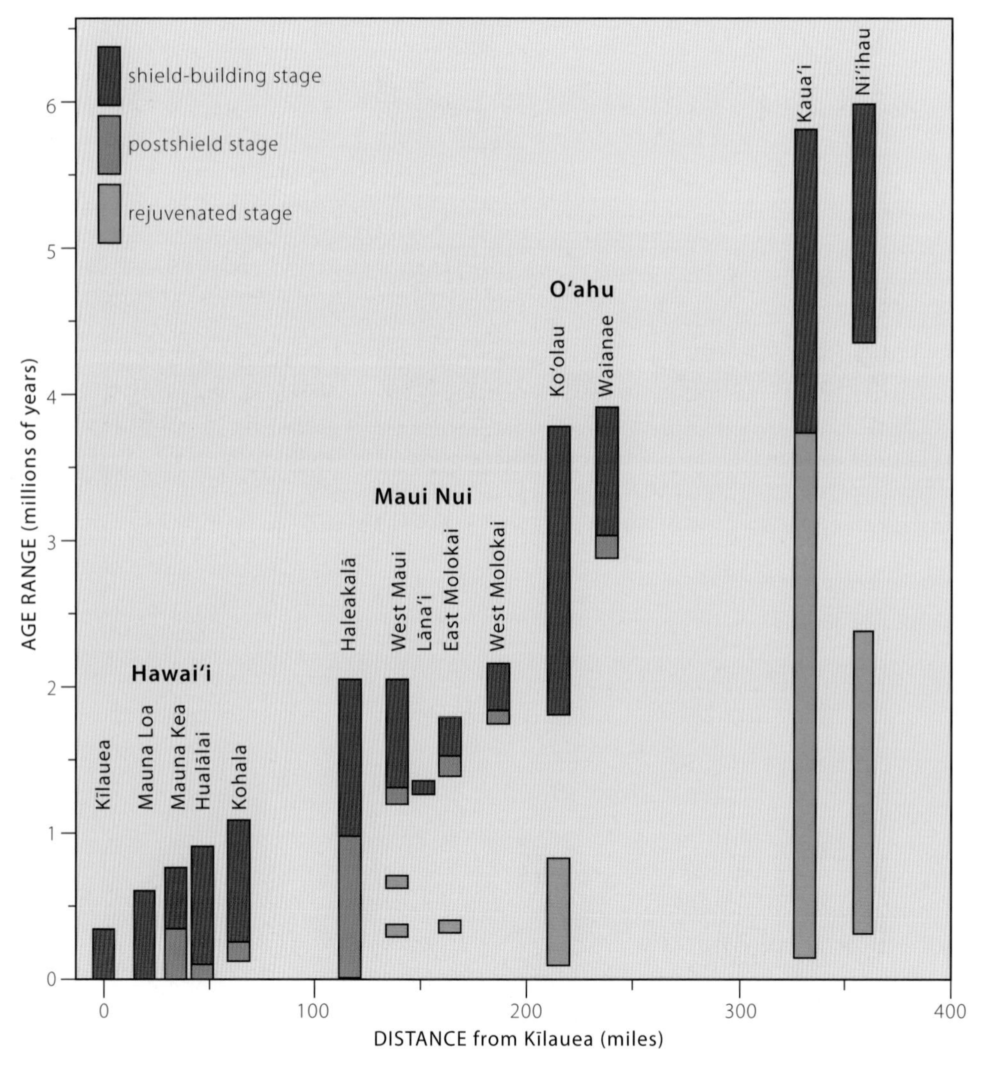

Ages of lavas erupted in the Hawaiian Islands increase toward the northwest from the hot spot, presently positioned beneath Kīlauea and the southern part of Hawai'i Island. The different-colored bars represent lavas erupted during the main stages of volcano development. The volcanoes all become extinct as they drift away from the hot spot at the zero point.
—Data fom Clague and Sherrod, 2014

STAGES OF VOLCANO DEVELOPMENT

Just as a human life can be described through various life stages—young, mature, old—so, too, can we describe a life cycle for Hawaiian volcanoes. Not every Hawaiian volcano necessarily goes through all the stages, but it is a well-documented pattern for most.

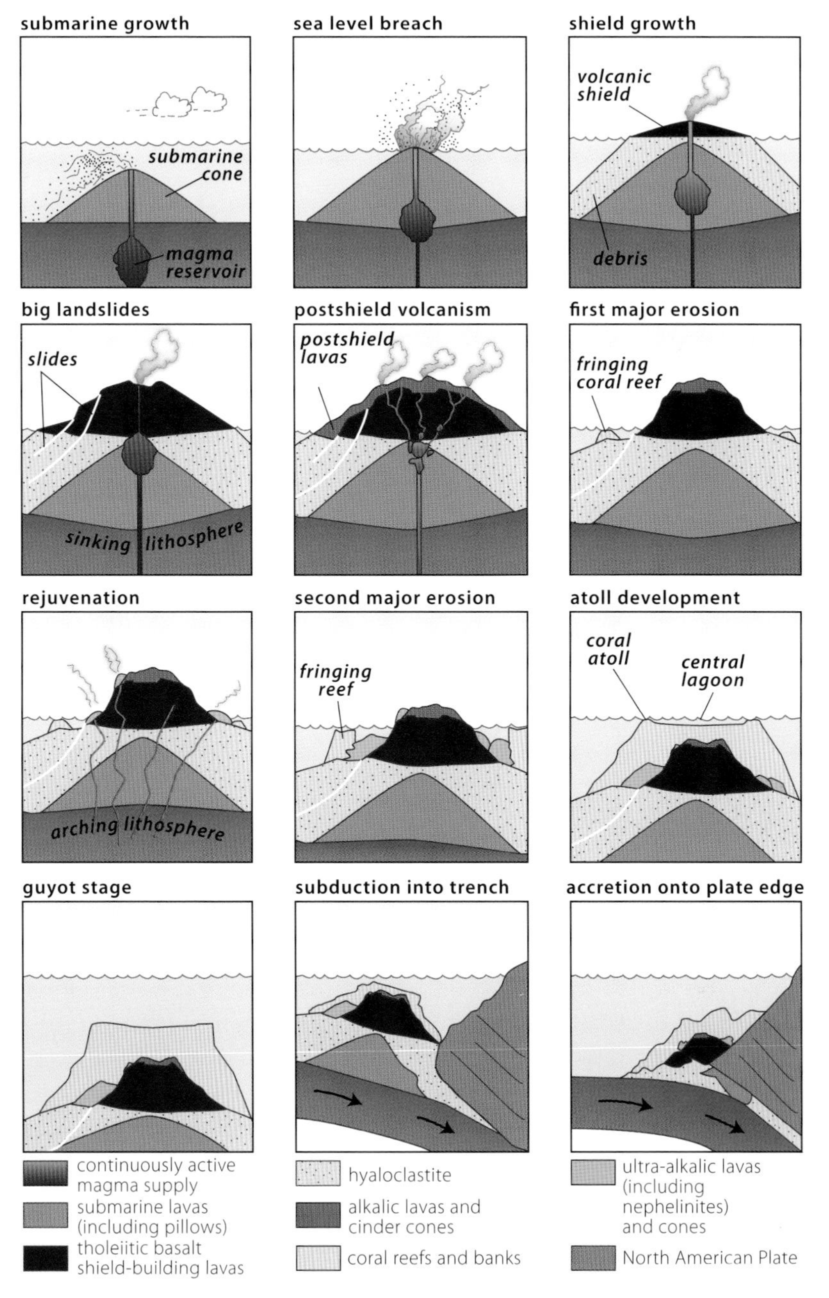

Stages in the development and decay of a typical Hawaiian island.

Youthful Submarine Stage or Preshield Stage. The first 100,000 to 200,000 years of a Hawaiian volcano's life take place entirely underwater as magma establishes a well-defined conduit to the surface. The first batches of molten rock originate at depths of as much as 40 to 60 miles, tapping the deeper margins of the approaching hot spot. The rate of eruption is slow-paced at first, perhaps only one eruption every few decades, but picks up as the drifting plate transports the new volcano into position directly overlying the hot spot. At this point, the sources of magma production are shallowest—only 25 to 35 miles deep—and the pace of eruptions accelerates. The volcano may be in nearly continuous eruption, frequently repaving its surface with fresh lava.

Until the volcano builds above sea level, forming new land, its flanks will be steep. Flows cannot travel as far owing to the chilling of lava underwater. Frequent underwater landslides also tend to occur, making the volcano's flanks even steeper and more rubbly.

As the young volcano takes shape it will acquire rift zones, the giant seams in its flanks opened by gravitational sagging of its mass toward the deepest parts of the surrounding sea. As the volcano gets bigger and begins impinging on its giant neighbors to the northwest, its rift zones grow longer and may adjust their orientations slightly as some sides of the mountain grow more stable and others weaken.

The overall chemical composition of the lavas erupted also changes. The small batches of magma derived from great depths early in the submarine stage produce alkalic basalts. Once the volcano becomes centered close to the hot spot, the magma changes to a tholeiitic composition and the pace of eruptions accelerates.

In shallow water, where water pressure is low, steam rapidly develops and eruptions transition from outpouring of submarine lava to steam-driven explosions. Piles of glassy lava shards and fragments accumulate and bury the underwater mountainside. Called hyaloclastite, or "glassy broken rock," this somewhat unstable cover eventually builds a platform above sea level that protects vents from flooding, but waves can initially wash the new land away with ease. Once water no longer has access to vents, however, lava begins to pour out and armor the hyaloclastite, and a new island emerges.

Shield Stage. Frequent, highly fluid eruptions of tholeiitic lava, in large part from lengthy rift zones, gradually develop a broad, gently sloping mountain called a shield volcano. The shield stage is the main growth spurt of a Hawaiian volcano, erupting 80 to 95 percent of its total volume as it moves over the center of the hot spot. Calderas and smaller craters collapse and then fill again as magma supply to the shallow magma reservoir waxes and wanes. Here the volcano reaches maturity.

But there is trouble brewing owing to the exuberant growth rate. The weak hyaloclastite base underlying the shield lavas, the vigor of swelling and erupting rift zones, and other factors related to gravity tugging on the flanks cause stupendous pieces of the active volcano to slump gradually or collapse all at once into the ocean. An island can lose as much as a third of its mass in a single massive landslide, leaving towering sea cliffs as much as 1 mile high along a shattered coast. Almost every Hawaiian island shows some evidence of partial collapse. The largest of these disintegrations take place when a volcano grows to maximum size.

Mega-landslides in Hawai'i come around, on average, once every 300,000 to 400,000 years—fortunately not frequent enough to be a concern.

Massive sea cliffs dominate the coast from Pololū to Waipi'o Valleys, the result of an ancient giant landslide from Kohala Volcano, Hawai'i.

Late-Shield Stage. Plate movement is carrying the volcano away from the hot spot that gave it birth. Eruptions become less frequent during the late-shield stage and alkalic lavas reappear. Alkalic magma results from a smaller amount of melting, deeper in the mantle. When magma is stored longer in smaller batches, it cools and the lavas become less fluid. Perhaps the main change that terminates the life of the shield volcano is breakdown of its central summit magma conduit.

Postshield or Old Stage. During the postshield stage, eruptions continue but at an ever-declining rate. Separate batches of magma now find their way to the surface along pathways independent of the dying central conduit. Eruptions can break out practically anywhere on the upper slopes of the volcano, though rift zone alignments may remain crudely apparent. Postshield volcanism is shown by the eruption of alkalic lavas and the building of cinder cones, burying any shield-stage calderas.

On a few postshield-stage volcanoes, trachyte can erupt in addition to basalt. Trachyte develops from large bodies of slowly crystallizing alkalic basalt magma. It is a silica-rich residual liquid. The remaining liquid trachytic magma also contains a high concentration of gases and dissolved groundwater—ingredients for powerful volcanic explosions if the delicate balance maintaining stability in the magma chamber is disturbed. A trachytic eruption typically begins with explosions of pumice and ash, followed by outflows of pasty trachyte flows much thicker than ordinary Hawaiian basalt flows. The light-gray trachyte rock contrasts starkly with the much darker-gray basalt—an indication of its higher silica chemistry.

Erosional Stage. As plate movement carries a volcano a few tens of miles northwest of the hot spot, noticeable signs of erosion of its flanks appear, especially on the wet windward (northeast) side. By the time it is positioned a hundred miles away, the volcano will quite likely have died and become deeply eroded into knife-edged ridges and deep stream valleys. Tectonic subsidence will become less significant by then, too, with broad coral reefs and limestone coastal platforms developing.

Rejuvenated or Post-Erosional Volcanism. A last volcanic hiccup can still interrupt the slow disappearance of an island back into the sea. Flexing of the lithosphere beneath the island, caused by the great weight of the younger, still growing islands to the southeast, reduces pressure on the underlying mantle, triggering small amounts of partial melting at great depth. Resulting small batches of rare, highly alkalic magma can erupt practically anywhere along the wide crest of the Hawaiian Arch and across the landscapes of deeply eroded islands, a process called volcanic rejuvenation. By this time all earlier conduits for magma eruption are clogged, so these scattered, ultra-alkalic outbursts tend to take place where eruptions have never occurred before.

Rejuvenated volcanism creates small lava shields, stubby lava flows, cinder cones, and—near the shore—low, broad ash cones. Lavas include some alkalic basalts like those erupting during postshield activity, along with others, especially nephelinites, which are much more strongly enriched in sodium.

Rejuvenation of volcanic activity on older Hawaiian Islands can bring up garnet peridotite as scattered chunks called mantle xenoliths, embedded in lava and ash deposits. Their presence shows that as Hawaiian volcanoes near extinction their melting sources become deeper, as garnets are only stable at more than 35 miles depth. Xenoliths are uncommon, but some Hawaiian flows and volcanic ash beds contain great numbers of them.

Not all Hawaiian volcanoes experience rejuvenation. It is an erratic, unpredictable phenomenon.

CALDERAS, VENTS, AND DIKES

Periodic filling and drainage of a magma chamber as a volcano continues to erupt weakens the overlying crust, causing it to fracture and sometimes collapse. The result is a deep, flat-floored depression with sheer walls, known as a caldera (from the Spanish word for cauldron). Individual calderas are typically miles across, though those found on Hawaiian volcanoes are much smaller than many continental ones. Frequent eruptions can fill the caldera to overflowing, with a new caldera replacing it later.

The cliffs rimming a caldera reveal ways that shallow magma can migrate to erupt from a chamber. These include dikes, which are simply magma-filled cracks, and sills, which are places where magma intrudes between older layers. Dikes and sills are both dense, sheet-like, intrusive rock bodies. The difference is simply one of orientation relative to surrounding rocks. Dikes tend to be vertical or steeply inclined. They cross preexisting layers. Sills are flat lying, or nearly so, and intrude along the contacts between different older layers, wedging them apart. A swarm of dikes marks the location of a rift zone.

A sill can inflate as additional magma pours in, swelling into a bulbous mass termed a laccolith. Sills and laccoliths are not as well exposed and easy to see in

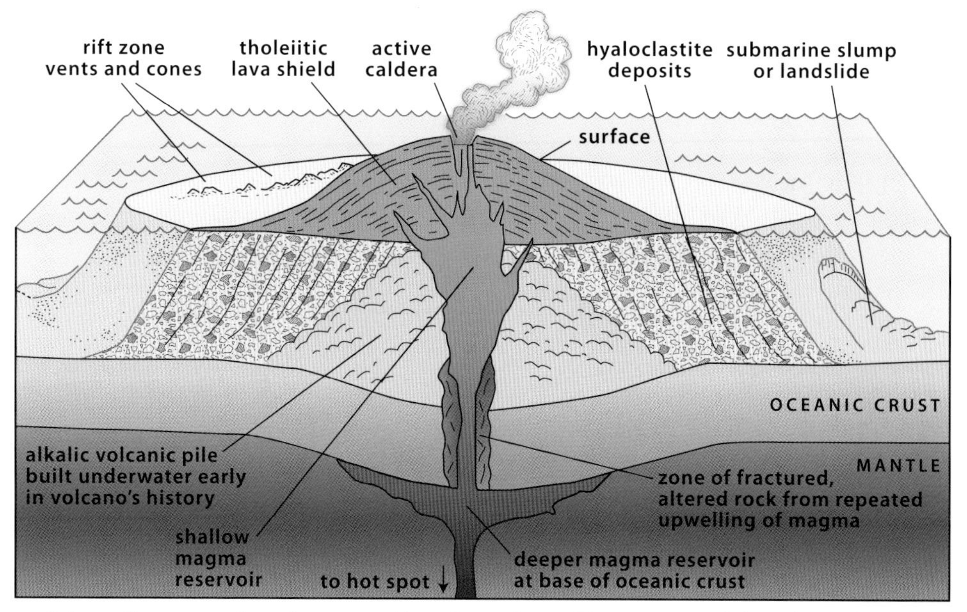

Schematic cross section of a mature Hawaiian volcano. The structure of the volcano formed above sea level contrasts sharply with the structure that develops underwater during the early stage of volcanic growth.

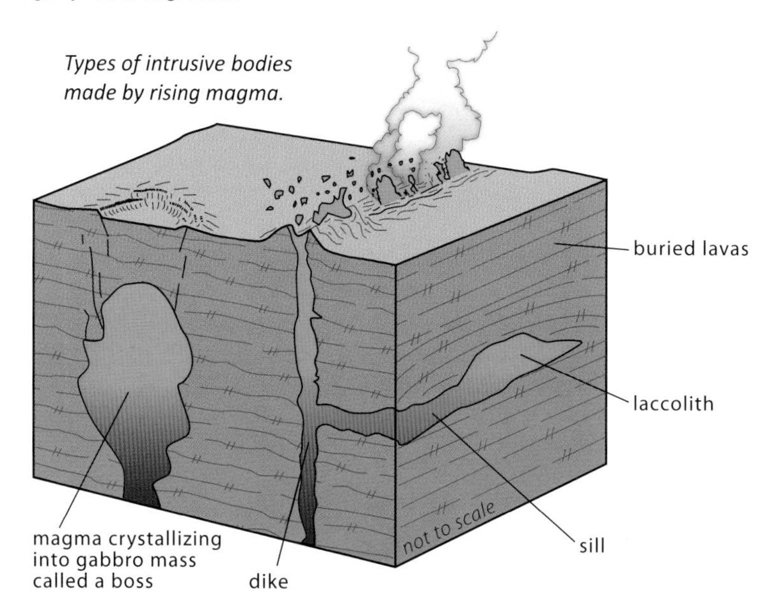

Types of intrusive bodies made by rising magma.

Hawai'i as dikes, but you can spot them in places like the walls of Kīlauea's summit caldera on the Big Island.

As a magma-filled dike approaches the surface, it causes the ground to arch and split open, forming a vent through which an eruption occurs. Most vents in Hawai'i, at least initially, take the form of long cracks or fissures that can stretch for miles. An example is the Pu'u'ō'ō eruption at Kīlauea, which began as a fissure through which lava continuously gushed—a low "curtain of fire" 4 miles long. This phase of the eruption lasted only a few months, though, before activity concentrated near the downslope end of the fissure, where it was mostly centered for the next 35 years—long enough to build a cone at one point over 800 feet tall! The underlying dike remained molten throughout this time, linking the summit magma chamber with the vent, 12 miles away.

A prehistoric laccolith exposed in the wall of Kīlauea caldera.

A fissure eruption from Kīlauea Volcano in 2018 spewing low lava fountains and lava flows. The still-molten spatter piles up and welds together to build a spatter rampart along the fissure. —Photo by S. Isgett, US Geological Survey

Vents are easy to recognize in the field, because prolonged exposure of cooling lava to escaping heat bakes these openings brick-red in many places. They will generally have spatter—blobs of still-molten lava that pile up, ooze, and weld together—on one or both sides of the fissure, which may or may not remain open as lava drains back in at the end of the eruption. Spatter ramparts and spatter cones have complex surface textures, very glassy when young. Other features, mentioned below, can also help you recognize vent features as you explore Hawai'i's young volcanic landscapes.

LAVA FLOWS

Lava is simply magma that erupts by gushing and flowing out of the ground. Most Hawaiian lavas are very fluid owing to their great temperature, as high as 2,200 degrees Fahrenheit—hotter by far than most lavas erupted from US mainland volcanoes. Fluid lava flows that spread out in all directions can build low, broad mounds around vents, called lava shields. Blobs of fountaining lava can also pile up around a single vent to form small, steep-sided spatter cones, or if accumulating for a considerable distance along a fissure, spatter ramparts.

Two basic types of lava flows erupt from basaltic volcanoes. Both have Hawaiian names: 'a'ā and pāhoehoe. 'A'ā flows have a very rough, clinkery surface. Many people equate the sound of the Hawaiian name *'a'ā* with the pain one imagines from walking barefoot across this surface, the so-called "ow-ow" lava. Although this translation seems to make sense, *'a'ā* means "to burn" or "to glow," as staring eyes. From a distance, an active 'a'ā flow at night may look like many glowing red eyes on the hillside.

The other type of lava is called pāhoehoe, characterized by smoothly rounded, rippled, or ropy looking surfaces. The word *pāhoehoe* might originally refer to the pattern of ripples seen when you sweep a canoe paddle through the water. It is easy

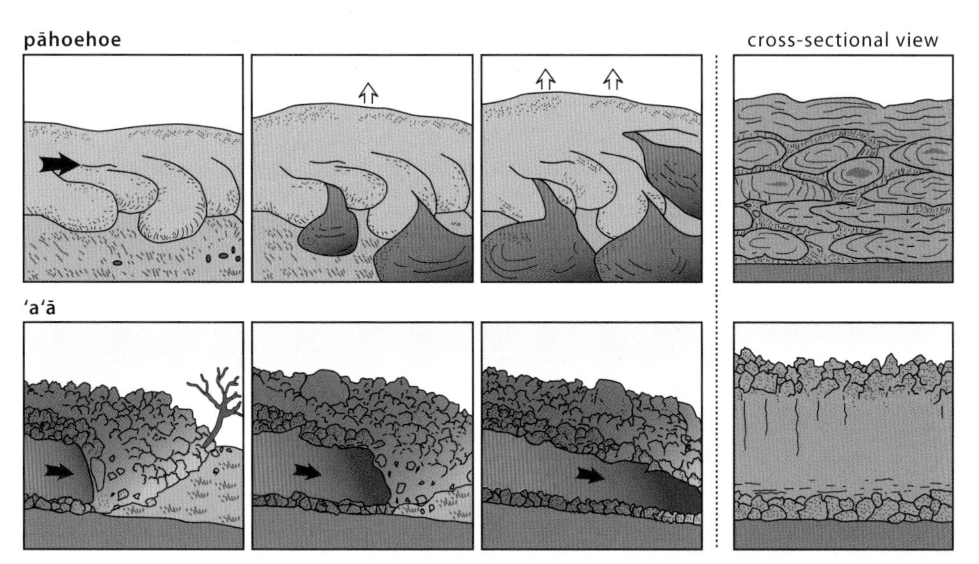

Upper frames: Flowing pāhoehoe and pāhoehoe flow in cross section (upper right frame); Lower frames: Flowing 'a'ā and an 'a'ā flow in cross section (lower right frame).

Smoother ropy pāhoehoe and rubbly ʻaʻā lava textures on the steep southern slope of Kīlauea Volcano.

to see why this association was made, looking at the wrinkly surfaces of many flows. In some places, pāhoehoe flows have a smooth, flat surface.

These two lava types can have identical compositions, but their viscosities, or stiffnesses, may differ greatly. ʻAʻā is typically more viscous than pāhoehoe, sometimes because it is cooler when it erupts or flows a long distance from the vent. But the deciding factor is shear stress—the tearing or smearing out of a body, like the spreading of a deck of cards. Lava pouring rapidly down a steep, rough slope experiences high shear stress or tearing force and tends to form ʻaʻā, whereas on flat ground the crust is not torn and pāhoehoe is favored.

Individual ʻaʻā flows exposed in roadcuts or by erosion characteristically have clinkery tops and bottoms, with a thick, dense layer of lava sandwiched in between. While the entire flow experiences shearing as it moves downslope, only the much cooler surface and sides of a typical ʻaʻā flow will be stiff enough to be torn up by the shear stress. This shredding creates the rough surface of spiny rubble, or clinkers, so characteristic of ʻaʻā.

The hotter molten flow core is largely unaffected by shearing, though gas vesicles within it are smeared out or elongated, leaving a dense rock once it cools. The massive core tends to override its own rubbly crust, pieces of which cascade down the terminus as the flow advances, much like the rolling tread of a tractor or tank. The clinkery bottom zone is in many places oxidized to a rust-brown color. ʻAʻā flows can be anywhere from 1 foot to 50 feet thick or more, depending on the steepness of the slope, and have been known to knock over trees and buildings.

In contrast to ʻaʻā, pāhoehoe flows advance by feeding tongues, or toes, of lava one after the other, each forming a crust, like a natural sock, which swells and splits as additional lava enters. The toes merge into larger lobes and new toes emerge from

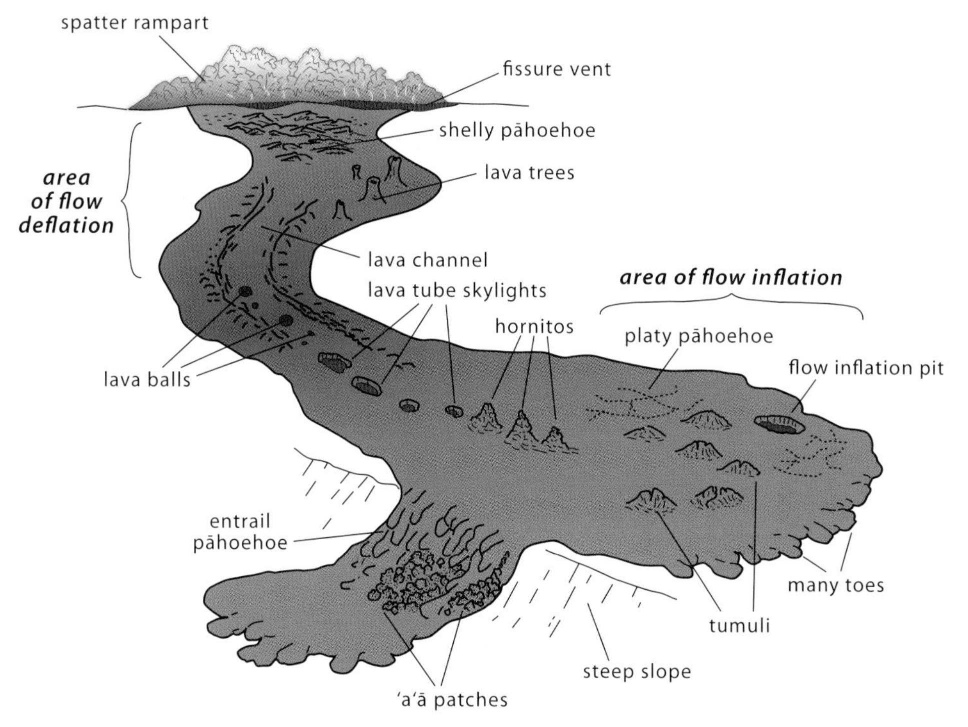

Typical cooled pāhoehoe flow features. Features found close to a vent in the upper part of a flow, where deflation prevails, contrast sharply with the inflation-related features found farther downslope.

Large ʻaʻā flow moving down a road in 2018. The glowing molten core is exposed along the advancing front while spiny, blocky clinker from fist to boulder size tumbles off and is run over by the flow. —Photo by A. Lerner, US Geological Survey

Roadcut through a Mauna Loa ʻaʻā flow showing the classic sandwich structure: a dense core with clinkery rubble above and beneath.

the ruptured edges of the lobe as it advances. Seen in a roadcut, then, a pāhoehoe flow with numerous toes resembles a great pile of stacked sandbags of many different sizes, some of which partly drained to form small hollows as the flow settled into final position. Pāhoehoe lava generally has more abundant and spherical vesicles than ʻaʻā lava. At higher volumes, the lava might pour out in larger streams. The cooling surface crust of pāhoehoe tends to remain flexible and may be crinkled and folded into ropes or other textures as it moves. The combination of bulbous toes and flatter layers exposed in some roadcuts may appear quite complex, making it difficult or impossible to know whether you are looking at one or several stacked pāhoehoe flows in cross section. Separate ʻaʻā flows, on the other hand, are much easier to distinguish once you learn to recognize their basic structures.

Once pāhoehoe lava has traveled more than 1 mile from the vent, it may begin to show inflationary behavior on flat ground, particularly if it is fed by a lava tube from upslope. The surface and base of the advancing flow solidify, but the center remains molten and begins to push the solid crust upward as more lava fills it. Cracks may develop in the thickening surface crust and can get quite large. These flows are generally referred to as inflated or tube-fed pāhoehoe. On level ground with high volumes of lava, flat-topped sheet flows may form as the crust rises in a uniform manner. Uneven terrain and localized swelling can create tumuli (singular is tumulus), uplifted mounds of crust. Inflated pāhoehoe flows can eventually grow to thicknesses of 50 feet or more, though the process may take weeks.

When a pāhoehoe flow loses much gas and heat or encounters a steep slope, it may undergo a transition to ʻaʻā. A highly jumbled flow surface develops showing a mix of clinkers and upturned, fractured slabs of wrinkly, pāhoehoe-style crust when this happens. Geologists call this intermediate lava slabby pāhoehoe.

In roadcuts, cliffs, and streambanks, you may see another feature typical of ʻaʻā or ponded pāhoehoe flows: neat palisades of vertical rock columns. Geologists have understood for more than a century that the columns express a pattern of shrinkage fractures in the rock. Some people who have watched lava flows in action report seeing those fractures form as the lava chills solid, while it is still hot. They are known as columnar joints or cooling fractures.

In general, columnar joints split open perpendicular to the cooling surfaces relative to the ground and atmosphere. It makes sense, then, that the columns stand vertically in lava that flowed across level, flat ground. If the flow fills a gorge, however, the columns will flare at depth outward toward the buried walls, even approaching a horizontal orientation. Seen from above, the pattern of shrinkage cracks looks like the nearly geometric fracturing you see in sun-cracked mud, or in crazed porcelain. The fractures outline polygons that have four to seven sides, most commonly five. Displays of columns viewed on-end like this occur where stream erosion has stripped away the clinkery surfaces of ʻaʻā flows, revealing the tops of jointed flow cores. The Hilo District on the Big Island is the best area in Hawaiʻi for viewing columnar jointing, though this feature can be seen in many other locations, too.

Inflated pāhoehoe lava engulfed and uplifted a school bus near Kalapana on Kīlauea in 1991.

Columnar jointing in thick basalt flows along the Wailuku River in Hilo. The flows filled an ancient gorge around 10,000 years ago and chilled rapidly against the walls, creating columns that angle away from the vertical. —Photo by James Anderson

Lava Lakes

At Kīlauea especially, lava can erupt constantly for years, filling the floors of craters sometimes to overflowing with roiling lakes of molten rock. These lava lakes release dense clouds of toxic volcanic gas, called vog ("volcanic fog"), that drift in dilute form for hundreds of miles downwind. If a lava lake hardens solid while confined within older crater walls, it tends to form a smooth, gently hummocky surface, like a buckled parking lot. This easy-walking surface is typically pāhoehoe. A good example is the floor of Kīlauea Iki Crater in Hawai'i Volcanoes National Park.

If the lava lake overflows, however, the thin, gas-rich spillovers of lava harden into a treacherous surface full of cavities called shelly pāhoehoe. Be wary of this type of lava around any vent—it cuts like broken glass!

An active lava lake forms a crust that is quite thin, like the skin of a soup or pudding. Currents of lava, driven by escaping gases, break the crust into plates that slide past and collide with one another, rather like miniatures of the lithospheric plates of Earth's surface at a tectonic scale. Many lava plates are swallowed or "subducted" back into the depths of the lava lake. New ones form to replace them as the molten lava loses heat to the atmosphere. Active lava lakes existed at the summit

Billowy shelly pāhoehoe from 2018 in Puna is treacherous to cross. Gas-rich, near-vent lava flows form thin crusts that separate from the liquid lava and roll into hollow shells.

Active lava lake in Halema'uma'u Crater at Kīlauea Volcano in 2016. A smooth, floating crust forms on the surface, broken into plates and moved around by the roiling lava underneath. The bright area on the far right is where gas is escaping from under the crust as it "subducts" back under, causing the lava to splash. —Photo by T. Orr, US Geological Survey

of Kīlauea for most of the nineteenth century, giving authors like Mark Twain much to write about. The most recent lava lake activity began anew there in 2021 in the "fire pit" of Halema'uma'u.

Lava in the Forest

When molten lava pours into a forest, it is chilled by the moist wood of larger trees as it flows around their trunks. If the flow level recedes, a pillar or cast of hardened lava may remain standing, encasing a hollow shaft inside where the trunk burned away. Geologists call such cylindrical monuments lava trees. If the flow level does not recede, a hole in the lava marking a former tree trunk remains behind. This is called a tree mold. The depth of the tree mold indicates the thickness of the flow surrounding it.

Hundreds of lava trees and tree molds adorn the surface of many pāhoehoe lava flows, especially close to vents in formerly forested areas. In places, the lava engulfing the trees was so fluid that it penetrated the fine seams between plates of charcoal, forming a delicate cross-stitched pattern in the walls of shafts where trunks once stood and burned away.

Lava trees occur in partly drained shelly pāhoehoe flows, usually close to vents. They may develop a cap of spatter if close enough to a spattering vent. 'A'ā flows are generally too viscous to produce lava trees and only rarely create tree molds.

Imprint within a lava tree from a 1973 lava flow on Kīlauea Volcano in Hawai'i Volcanoes National Park. The fine detail shows where fluid lava filled the cracks in the charred surface.

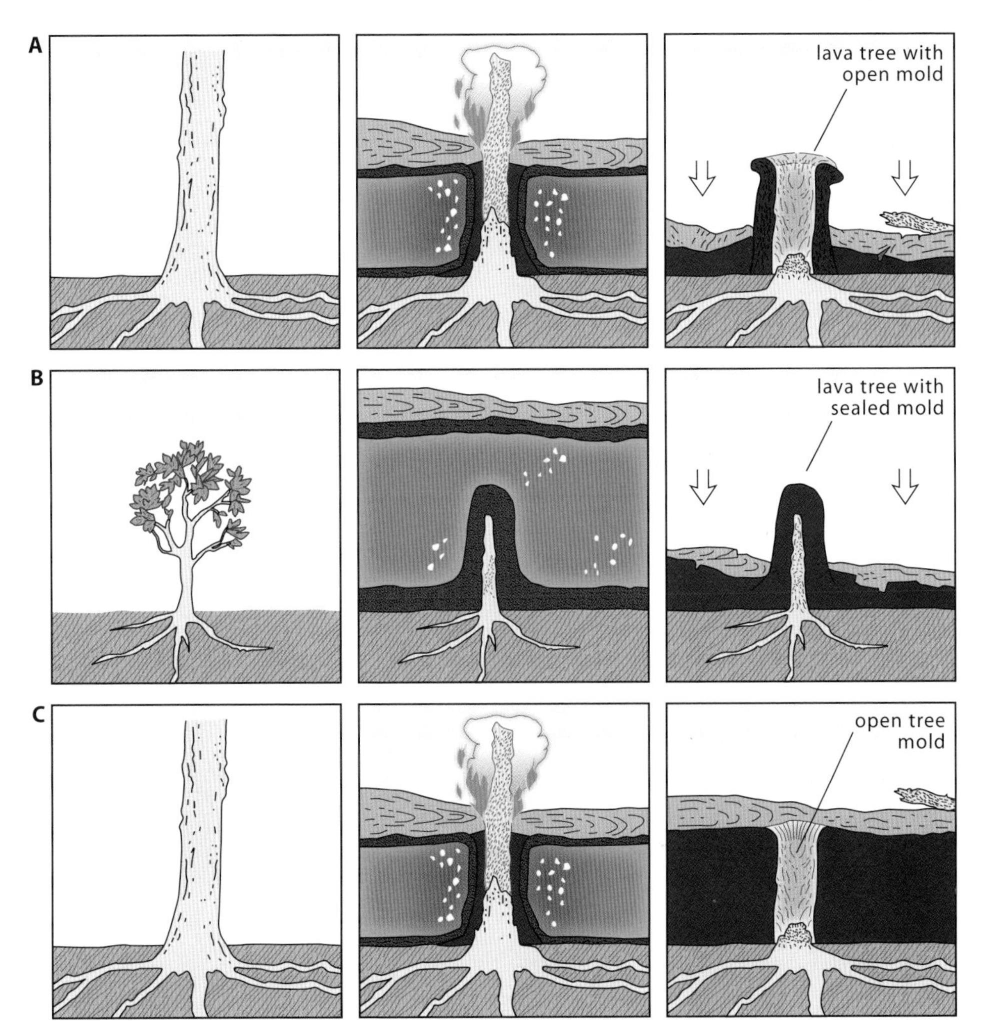

(A) Formation of a lava tree with a mold open at the top; (B) Formation of a lava tree without development of a mold open to the surface; (C) Formation of a tree mold with no enclosing lava tree.

Lava Channels and Caves

As the margins of an active lava flow cool and harden, they confine the molten lava to a narrow central channel. The centralized stream of lava may narrow and deepen further as surges of liquid lava overtop channel banks and harden into levees. During times of strong flow, big pieces can break from unstable levee walls and then roll, tumbling downstream in molten lava. The pieces grow larger as smooth coats of fluid lava chill around them, forming lava balls. You can see rounded lava balls scattered across the surfaces of many Big Island flows, especially on Kīlauea.

The channel at the center of a lava flow can also crust over as the molten rock loses heat to the air. The lava continues flowing under the thickening crust in a subterranean "pipeline" called a lava tube (some geologists refer to these conduits

Skylight in an active lava tube in 2007. The lava channel at the bottom of the photo is growing a secondary crust from contact with cooler air. The glowing walls and roof above the active lava stream are hot enough to partially melt and drip or ooze. —Photo by Ben Gaddis

Soda-straw stalactites in Emesine Cave. These hollow tubes of partial melt grow drop by drop, driven by vapor bubbles escaping when the pressure in the lava tube is reduced. They are encrusted by a metallic magnesium iron oxide formed from interaction of the melt with air in the hot tube.

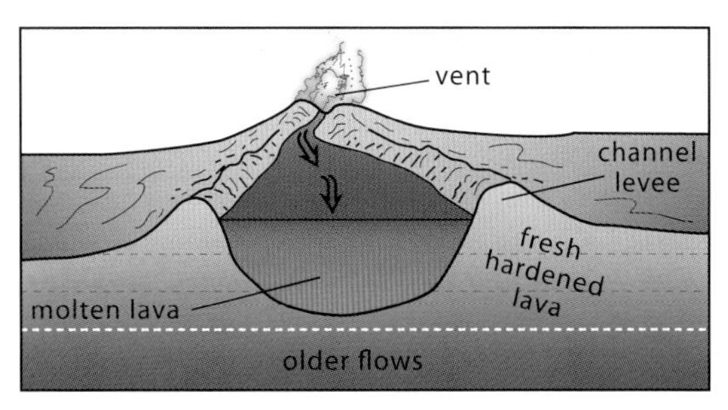

Sequence in development of a lava tube leading to cave formation (bottom).

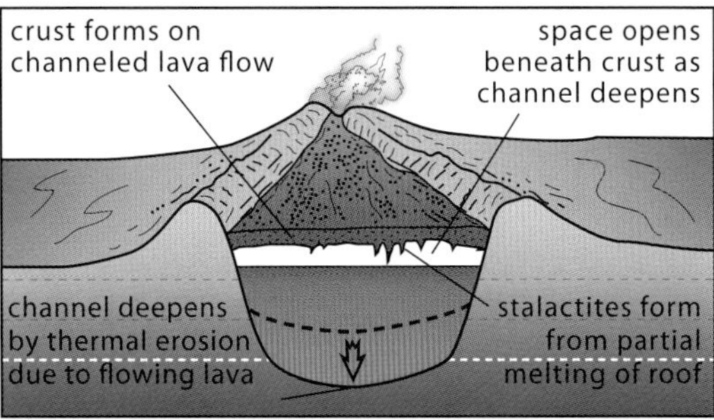

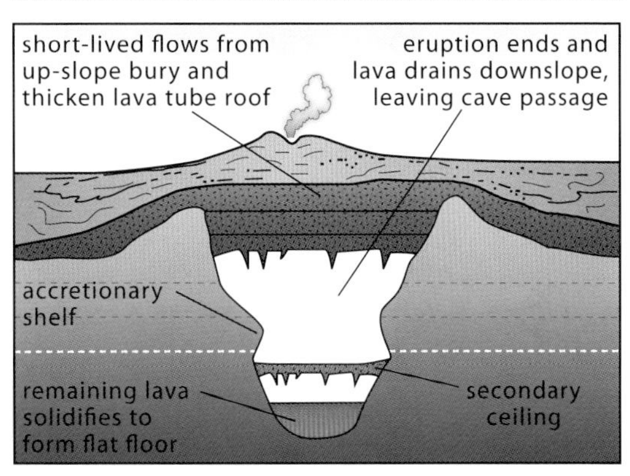

as "pyroducts"). If the flow stays active for more than a few weeks, the combined heat and physical force of the flowing lava begin to erode the base of the lava tube, causing it to deepen. The intense heat above the lava river causes the walls and roof of the tube to partially melt, creating drip and flow textures. When the supply of molten lava from the vent wanes, the subterranean channel may partly drain out, leaving behind a lava tube cave that can be large enough for people to walk through. Lava tube caves can be truly enormous. One on the Big Island stretches through multiple passages for more than 45 miles!

Cross sections of lava tube cave passages are rarely perfectly circular. Most have wide, flat, or partly collapsed ceilings, or slot-like cross sections, resembling narrow, roofed-over ravines. Many have smooth shelves lining their walls, and some have branching side passages, making them seem like labyrinths. In some cases, two (or more) cave passageways are vertically stacked, separated by a thin secondary ceiling. The many features that you can see in the floors, walls, and roofs of a lava tube cave tell a detailed story of its fiery formation.

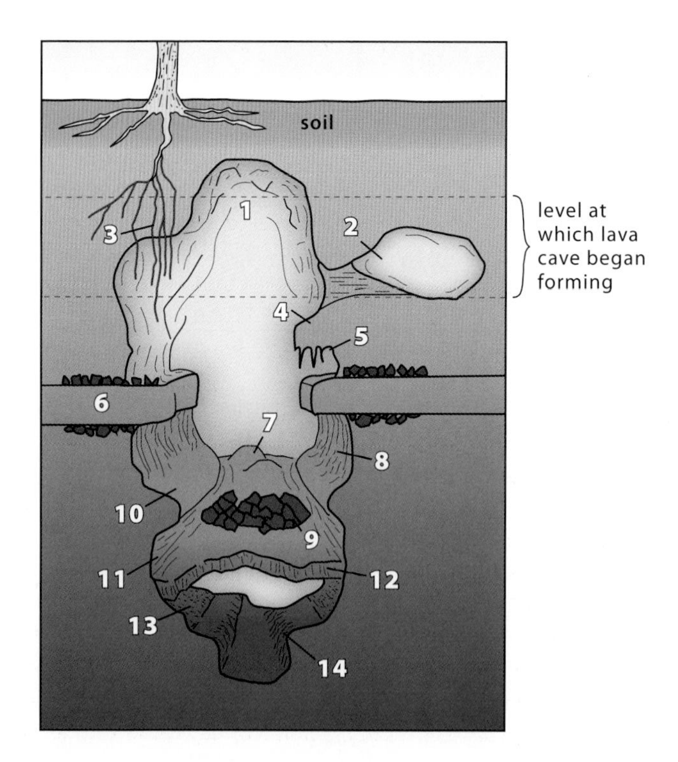

Features in a typical lava tube cave formed during eruption of a lava flow:

1) Dome formed when the ceiling cracks and breaks up.
2) Upper side passageways. The whole cave began forming at this level (above the dashed line), then deepened as lava continued flowing through it, eroding the bottom.
3) Tree roots dangling from the ceiling. Many are reddish brown.
4) Shelf made by lava draining from side passageways (2).
5) Lava stalactites.
6) Erosional shelf; in many cases this is the resistant core of an ʻaʻā flow exposed in the cave walls.
7) Lava balls embedded in a smooth cave floor.
8) Glassy ripples left where a lava stream flowed past walls that were on the verge of melting.
9) Blocks that fell from above when the dome (1) formed.
10) Smooth-sided shelf formed by accretion of lava to passage walls. The level of flowing lava in the passage held steady here for a while, allowing the shelf to form.
11) Lines left by flowing lava as the level of the lava stream dropped.
12) Secondary ceiling.
13) Glazed, sparkling walls from partial melting by passing lava stream with lots of trapped heat.
14) Resmoothing of shelf from trapped heat and partial melting.

PYROCLASTS AND CONES

Lava doesn't always simply flow out of vents when they open in Hawai'i. Pent-up steam and gases can blast magma out in explosions of hot particles called pyroclasts ("fire-broken fragments"). They pile up around a vent to build pyroclastic cones. Geologists classify pyroclasts according to their sizes. The largest chunks are called blocks. Most are simply pieces of older lava rock blasted from the walls of a vent during an explosive eruption. Blocks ranging up to the sizes of small houses are known on some volcanoes, though they rarely get bigger than table size in Hawai'i, and most are much smaller. Good examples of blocks can be seen in the deposits around the summit of Haleakalā on Maui and around the summit of Kīlauea at Uēkahuna or near Keanakāko'i.

If large, liquid blobs of lava are tossed out instead, these strike the ground as smooth-sided pieces called volcanic bombs. Some streamlined bombs taper at two ends and are called fusiform or spindle bombs because of this shape. Others hit the land surface with a profound splat and are appropriately named cow-dung bombs. They are, of course, much more fluid than fusiform bombs. Bombs range from a few inches to several feet across. Look for them on the slopes and around the rims of sparsely vegetated cones on Mauna Kea and at Haleakalā.

Gravel- to pebble-size volcanic fragments exploded from a vent are called lapilli. A special kind of lapilli composed of chunks of basalt full of vesicles is termed cinder or scoria. Lapilli pile up around vents to form cinder or scoria cones. Exposed to the air or steam while remaining very hot for many hours or days after hitting the

Volcanic blocks thrown out by explosions that smashed the fence around an overlook of Halema'uma'u Crater in 2008. —Photo by T. Orr, US Geological Survey

A volcanic bomb and cinder from a cinder cone near the Mauna Kea Visitor Information Station, Island of Hawai'i.

ground, the cinder slowly oxidizes, turning a brick-red color. Faster-cooling cinder farther from a vent remains dark gray, like ordinary basalt lava. Scattered bombs and blocks are commonly mixed with the cinder.

Cinder cones can rise a couple thousand feet and may approach a half mile in diameter. They are usually a struggle to climb, with slopes as steep as 30 to 35 degrees, the so-called angle of repose. Anything piled more steeply tends to collapse under the pull of gravity. Cinder cones are generally the products of single explosive eruptions lasting from months to a few years.

Pumice is another kind of lapilli so full of vesicles that it appears foamy. Famous for its abrasive qualities, pumice forms like froth in a stein of beer when large concentrations of gas rush from a vent, blowing the magma into spongy bits.

Towering fountains of lava from a vent produce showers of pumice-like material from their crests that may carry in the wind for miles. Individual fragments, generally golden in color, may be up to several inches across and consist of over 90 percent tiny, interlocking vesicles. This special kind of light-weight lava foam is called reticulite. If you find some and have a pocket hand lens, take a close look at these amazing rocks. But be careful and remember: you are handling fragile natural glass.

Lava fountains are also associated with Pele's hair, thin strands of fine, golden glass that resemble blond or light-brownish human hairs and may collect in mats, blown miles downwind. This spun glass—similar to fiberglass—forms as tiny droplets of lava stretch and quench while traveling through the air. The droplets may form glassy black Pele's tears, so named for their shape. Lava racing down channels or over falls may also spin into glass threads at the edges of gradually forming crusts.

Volcanic ash is fine rock dust or sandy grit also produced in explosive eruptions. These tiny fragments may be glassy if the explosions are fed by fresh magma. Even small Hawaiian eruptions can produce some ash, though generally very little if lava is simply flowing or fountaining out of the ground. Some titanic ash eruptions have taken place, however, even at Kīlauea, a "drive-in volcano" famous for its normally approachable eruptive activity.

In 1790, Kīlauea exploded with great violence, sending billowing clouds of dark ash laden with pumice as high as 6 miles above its summit. All this ashy material collapsed back to the ground or blasted across the surface in deadly hot, gas-charged clouds called pyroclastic surges. At times so much steam escaped, and so much electricity charged the atmosphere above the vent, that ash particles clumped together as they fell, forming spherical mud pellets up to marble size. These accretionary lapilli carpeted the ground for miles downwind.

Activity like this has also taken place at many locations where vents have opened up right at the shorelines of older Hawaiian Islands, as at Koko Head on O'ahu.

Pele's hair and Pele's tears, glassy material ejected from fountaining eruptive vents on Kīlauea. Tears represent the tiny droplets of falling lava, while the hairs are strands of glass spun in the air as molten droplets fall through the sky or as lava bubbles burst on an active lava lake surface.

Accretionary lapilli—concentrically layered ash balls about 2 millimeters across—from the 1790 explosive eruption of Kīlauea Volcano. This example is found along the Ka'ū Desert Trail in Hawai'i Volcanoes National Park.

There, the escaping ash carried by pyroclastic surges around vents built up broad cones with low rims and wide craters—ash cones. In addition to ash layers, ash cones also can include great swarms of blocks.

Beds of loose volcanic ash harden if they are hot enough for individual ash fragments to weld together, or as percolating groundwater cements them over a much longer span of time. A bed of hardened ash is called tuff, and the term *tuff cone* refers to older ash cones, such as Diamond Head (Lēʻahi) near Waikīkī, that have hardened and become more resistant to erosion.

Craters of Explosion, Craters of Collapse

Pyroclastic cones develop cup-shaped explosion craters as magmatic ejecta blasts out from vents. Such craters are bordered by rims composed of ash, bombs, blocks, or cinder, depending on the type of eruption. If a strong wind is blowing during the eruption, this material may pile up on the downwind side of a vent, giving its rim a high point directly downwind. In extreme cases, the rim may not even completely enclose the crater but simply forms a horseshoe-shaped collar to one side. Cinder cones usually erupt lava flows, typically ʻaʻā, as their eruptions lose gases, and the flowing lava can undermine the flank of the cone, also producing a lopsided or horseshoe-collar rim. Variations in cone shape such as this are especially visible on Mauna Kea.

On younger volcanoes such as Kīlauea, pyroclastic cones are rare. Those that occur tend to be dominated by spatter and cinder, and may form a steep-sided crater if the center collapses. Lava ponding in these craters during long-lived eruptions overflows and tends to eventually bury the cone, so its final form may look more like a lava shield.

Nohonaohae cinder cone with horseshoe-shaped crater, Mauna Kea.

Craters are still numerous on the summits and rift zones of young volcanoes, however. These lack rims and tend to have vertical walls with sloping floors composed of rubble or filled with pads of later-erupted lava, usually platy pāhoehoe. Kīlauea Iki Crater is an excellent example of a crater of collapse that forms where shallow pockets of magma drain away, the magma tapped by new vents opening

Hi'iaka Crater, a pit crater along Kīlauea's east rift zone. The slopes are steep and shed talus; the smooth floor was covered by lava from a fissure eruption in 1973. A small fissure vent is visible in the lower right.

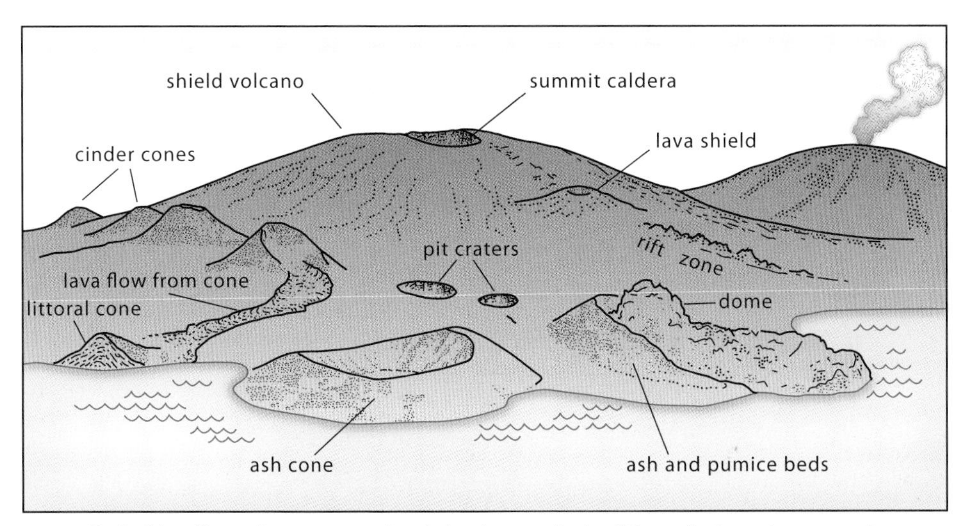

Typical landforms in a young volcanic landscape. Each of these features has examples described in the text and road guides.

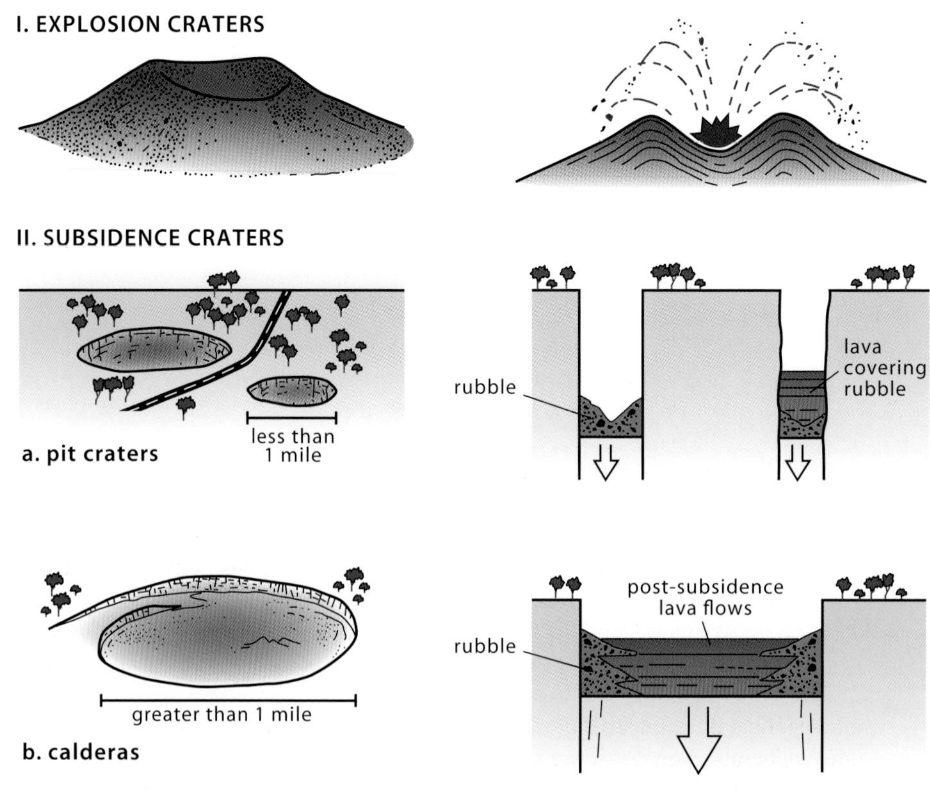

The shape of a volcanic crater depends upon the processes that form it: either explosive activity in the case of rimmed craters or collapse into rimless, steep-walled pits and basins. A few craters even show hybrid-forms, with an early explosive stage followed by terminal collapse.

elsewhere downslope. The ground, losing support, gives way and drops like the floor of an elevator to fill void space opened beneath. No eruption necessarily takes place as the crater develops, at least in the general vicinity. The collapse process can take place nearly all at once or slowly, lasting decades.

Craters of collapse measuring less than about 1 mile wide are called pit craters. Geologists distinguish bigger ones like at the top of Kīlauea and Mauna Loa as calderas, which we described earlier.

SOILS

Although landslides and wave attack do much to erode the islands, much slower weathering processes are at work constantly to transform landscapes. Rocks are chemically unstable when exposed to air and water. Little-noticed-but-ongoing chemical reactions partially dissolve the rock, leaving a granular residue behind that colonizing organisms rapidly exploit as a source of nutrients. The result over time is soil. A fully mature soil usually takes several thousand years to form. Because soils are vital to our survival as a growing medium for plants and crops, they are among our most important natural resources. Yet they are also fragile. Floods, landslides, and construction work can eliminate mature soils—centuries in the making—in just minutes.

The rate of soil formation depends strongly upon climate, and climate varies quite a bit from place to place in Hawai'i. Soil formation is much faster on the windward (northeast) sides of the islands than on drier leeward (southwest) sides. Lava flows that erupted 200 years ago in the arid North Kona District of Hawai'i Island still look almost perfectly fresh and support few plants, whereas flows from only a few decades ago in the Puna District to the east, with its heavy rainfalls, wear a coat of gray lichens shaded in low forests of shrubs and trees.

The main agent for weathering is ordinary water, which becomes slightly acidic as it incorporates nitrogen and carbon dioxide from Earth's atmosphere. Rainwater has an acidity similar to that of black coffee. Solid lava partly dissolves and is otherwise chemically altered in moist, warm climates, in some instances losing more than half of its original constituent matter. Acidic drainage waters carry away great amounts of calcium, magnesium, sodium, potassium, phosphate, and silica that was originally locked up in the lava. The remaining rock becomes weak and punky. You can scoop it out of a roadcut with a spoon. Its residual material has converted to the clay minerals so vital for soil fertility. These include pale-brown smectites and white kaolinite. Various other mineral residues may also remain, including yellow limonite and yellow-brown goethite. Once plant roots work their way in and organic matter accumulates, the transformation to soil is complete.

Soil scientists distinguish soil types, or orders, based on their contents of clays and other minerals. Almost all the soil orders found on the continents also occur in Hawai'i. Andisol is the most common soil type on younger islands. It weathers over thousands of years from beds of volcanic ash and pumice. Ranging from beige to charcoal-gray, this soil is very rich in organic matter, retains water well, and is one of the most fertile soil types in the world. Around 10 percent of the world population depends upon andisol for growing food, though Hawai'i is far from achieving the agricultural self-sufficiency that it could.

On older parts of the islands—for instance, Molokai, Kaua'i, and central O'ahu—weathering has progressed much further to create oxisols, also called laterite, containing large amounts of aluminum-rich kaolinite, generally spiked with ferric iron that imparts a rich red-brown color to the soil. Laterites lacking iron are light colored and are called bauxite, the only ore of aluminum. It is mined intensively in South America, Australia, Jamaica, and China, though no one proposes that such mining should also begin in Hawai'i. Hawaiian trachytes consist mainly of feldspar minerals and groundmass that contain a lot of aluminum and no iron, so they readily weather into bauxites on the wet sides of the islands.

All rocks are fractured, most of them in regular patterns. Water seeps preferentially into the fractures and reacts with the surrounding rock, converting it into soil. Groundwater chemical reactions corrode angular chunks of rock from two directions at edges, and three directions at corners. So, an originally angular rock mass becomes rounded as the edges and corners weather more rapidly than the flat faces. A good many Hawaiian soils enclose rounded residuals, called corestones, of the original rock. If the soil erodes, the corestones lag behind to litter the surface. They are widespread on the island of Lāna'i. Many large rocks in Hawaiian streambeds became

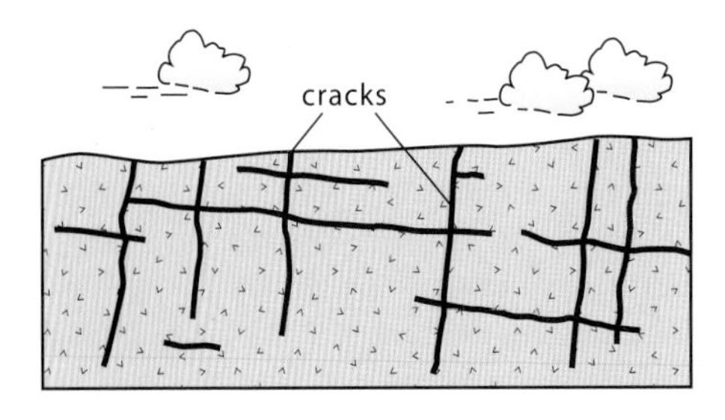

eroding soil

lag corestones

Evolution of solid rock with joints (fracture planes) into a soil bed with corestones—the unweathered residues of the original rock material. This process takes tens of thousands of years in Hawaiian lavas.

Rounded corestones exposed in an outcrop on Kohala Volcano, Hawai'i. Intense spheroidal weathering from water infiltrating along joints broke down and oxidized much of the rock, leaving only rounded remnants of the original stone.

rounded not through stream transport but in place, as residual stones. Later rainfalls, soil creep, and landslides bring these stones into the streams.

When flowing lava buries a soil, soil formation ceases. The process of soil formation begins once again at the new surface. But the old, buried soil, called a paleosol, can still be seen wherever deep erosion exposes it. Paleosols retain the brown and yellow colors of their clays and may even include fossil plant roots in the form of delicate white calcium carbonate threads. Look for paleosols at the bases of some ʻaʻā flows where erosion has exposed them. They may look very red at the contact where the heat from the lava baked the soil.

FLUTED CLIFFS AND AMPHITHEATER VALLEYS

In addition to the beautiful beaches and active volcanoes, one of the most striking aspects of Hawaiian landscapes is towering, ruggedly carved cliffs, including famous Nuʻuanu Pali on Oʻahu and the Nā Pali Coast of Kauaʻi. The wet northeastern sides of the Hawaiian Islands all feature cliffs, or *pali* in the Hawaiian language, many of which originated as giant landslide scars. They are awesomely high and steep, their surfaces fluted with immense vertical grooves that look as if a monster had raked them with a set of giant claws. In fact, fluted cliffs covered with vegetation exist on most rugged tropical islands.

The fluted pattern reflects a unique aspect of tropical weathering and erosion. On a steep cliff face, corrosive rainwater clings to shady surfaces longer than it does to sunny, windy ones, where it readily evaporates. In time, rocks in the shady areas weather and dissolve, leaving wide, deep pockets and gullies, whereas the sunny areas form ridges. The pattern smooths out into regular fluting because the smaller ridges and gullies hidden in the shadow of the larger ridges remain moist and corrode away completely, leaving only the larger ridges and gullies behind.

In addition to fluting in cliff sides, spectacularly broad amphitheaters have formed at the heads of some large valleys, with long, thin waterfalls cascading down steep

Fluted cliffs on the northeast side of the Koʻolau Range; viewed from Heʻeia Fishpond, Oʻahu.

slopes. Other tropical settings have similar amphitheater-headed valleys, but you won't see them in more temperate climates. The Hawaiian valley walls are so steep that they can barely hold soil. The ridges between valleys are narrow, with knife-edge crests. Deep sediment deposits fill the valley floors, much of it dumped there by landslides and waning floods.

The streams draining the Hawaiian Islands are generally clear. They ordinarily carry so little sediment that they seem unlikely to be entirely responsible for forming those huge valleys. They must have had some help. As with the fluted cliffs, rock dissolution plays an important role in forming the amphitheaters. Most of the rainfall and moisture helping to weather valleys in Hawai'i accumulates high on mountain slopes. When the upper end of the valley grows more rapidly than the lower part, it eventually develops into an amphitheater.

Ancient volcanic calderas also play a role in the formation of some of Hawai'i's most spectacular amphitheater valleys. The old calderas are composed of lavas altered and softened by hot gases while eruptions were still active. Streams erode such soft rock easily, hollowing out the calderas into wide erosional bowls sometimes several thousand feet deep.

With all this steep topography, waterfalls abound. Even small streams are interrupted by delightful little falls and cascades that form because some rocks erode more easily than others. Volcanic ash erodes quickly, as do buried soils and the rubble zones above and below 'a'ā flows. Dikes and the massive cores of 'a'ā flows resist erosion and become the lips of waterfalls.

Running water uses abrasive particles of sediment to carve bedrock in the same way a sandblaster uses sand. Neither water nor wind alone can carve rock. The clear streams draining the wet sides of the islands are poorly equipped to carve solid masses of rock. During floods, however, streams fill with sediment washed from surrounding slopes and can become very effective agents of erosion.

DROWNED VALLEYS AND SINKING COASTS

Streams, landslides, and associated dissolving of rocks on slopes create canyons that transform an island landscape as it ages from smooth, gentle lava slopes to the rugged topography described above. As an island sinks into the ocean, however, seawater inundates the canyon bottoms, and fresh sediment washed into the water will cover them to just above sea level. The resulting coastal valleys with broad floors framed by steep slopes were ideal land for early Polynesian settlers and farmers. You can mentally reconstruct a valley's original depth by projecting the slopes of its walls downward to where they meet below the surface. In most cases, the sediment fill turns out to be hundreds of feet deep.

In prehistoric Hawai'i, indigenous residents took advantage of some drowned valleys and embayments by building fishponds, large coastal enclosures to trap or encourage the gathering of fish. Stone breakwaters, in some instances hundreds of feet long, formed the enclosures, sealing off inlets except for narrow sluice gates to accommodate tides and permit smaller fish access back and forth with the open sea. Hawaiians perfected this technology, the most sophisticated aquaculture in Polynesia.

In due time, sinking of the island carries so much of the valley below sea level that its head can no longer supply enough fresh sediment to maintain a broad valley floor downslope. Maps of the underwater topography of the Hawaiian Islands show

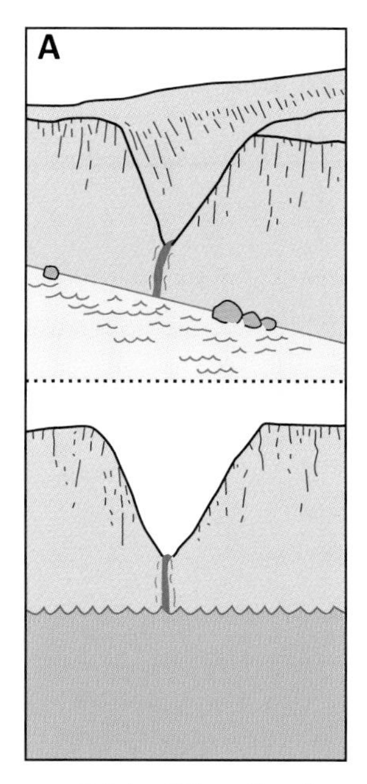

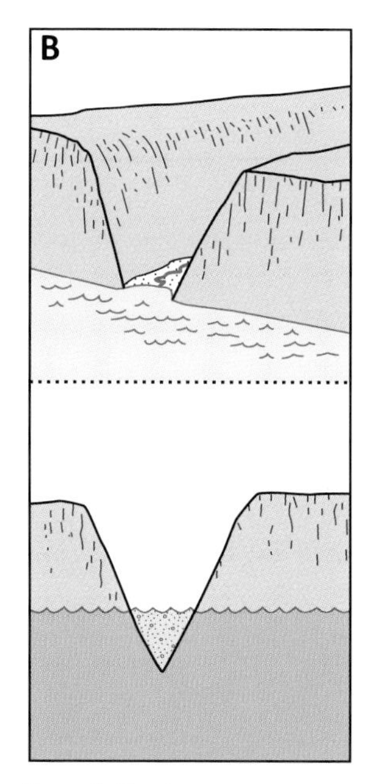

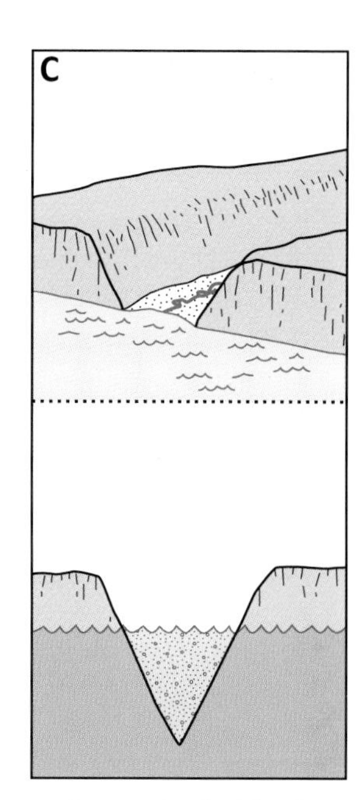

Origin of Hawaiian drowned valleys. A. Erosion creates steep stream valleys. B. As the island sinks, seawater inundates the valley. C. The valley fills with sediment to the level of the sea. The lower block in each frame is a cross-sectional view.

numerous former stream canyons that have completely drowned this way. Dead, ancient reefs off the northwest coast of the Big Island, as much as 3,000 feet below today's sea level, trace the outlines of now deeply submerged shorelines.

TSUNAMIS

Despite their attractiveness as places to live, the sinking stream valleys of Hawai'i are potentially treacherous places because of the funneling effect they have on giant sea waves called tsunamis. Most tsunamis start when sudden movement on a fault shifts a large area of the ocean floor vertically to the accompaniment of an earthquake. The movement of the seabed displaces a large volume of water and forms a wave that races across the ocean at speeds of several hundred miles per hour, depending on the depth of the water. Such waves pose no problem to ships at sea, where crews do not notice their passing. But they slow down as much as ten times and build to monstrous heights when they approach land and enter shallow water, especially in bays and inlets.

Seismic energy travels through the solid Earth much faster than waves at sea, and seismograph stations can give warnings of possible tsunamis caused by large, distant earthquakes hours before they reach Hawaiian shores. Rock transmits seismic energy about as far in a minute as a tsunami ripples across the deep ocean in an hour. Large local earthquakes, however, may generate tsunamis that reach nearby coasts quickly, with little or no warning. The southern half of the Big Island is the most seismically active part of the state, and such earthquake-generated tsunamis have drowned dozens of residents on the Big Island within historic times.

When seismograph stations issue a tsunami watch, Civil Defense personnel in Hawai‘i prepare for action. In many cases no tsunami develops, possibly because the fault movement shifted an area of the ocean floor horizontally. Most often, only vertical displacements cause tsunamis. If a wave forms near the quake epicenter, the watch is upgraded to a warning and sirens along the coasts in Hawai‘i will sound, alerting people to evacuate. You can see these sirens, bright-yellow horns or stacks of dark-green "doughnuts" on telephone poles, near many beaches.

A tsunami arrives in a rapidly falling tide followed by a sudden onshore rush of turbulent white water many feet thick. They do not curl and break like the giant waves made famous by surfers. Observers have compared incoming tsunamis to broad, flooding rivers. They are powerful enough to bend metal poles, rip up coral, topple all but the largest buildings, and wash cars and boats miles inland. A typical tsunami consists of multiple withdrawals and resurgences of water, as many as a half dozen over a period of several hours. The first incoming wave is not necessarily the most destructive. Very high and very low water levels occur at intervals of about 10 to 25 minutes. People have made the costly mistake of rushing out to catch the fish that lie exposed when an outgoing tsunami wave suddenly reveals large expanses of nearshore seabed; the water level soon returns much higher than it was before it dropped. It is equally costly to stop running for high ground after surviving the first crest because subsequent waves could be even bigger.

Historical tsunamis causing significant damage in Hawai‘i struck in 1868, 1946, 1956, 1960, and 2011. Minor tsunamis that do little or no significant damage occur on average once every decade or so, given the frequent earthquakes that take place all around the edge of the Pacific Ocean.

WIND, WAVES, AND BEACHES

Ordinary waves take their energy from the wind, then expend it in doing the work of shaping the coast. The prevailing northeasterly trade winds consistently drive the heaviest waves against the northeast coasts of the Hawaiian Islands, though storms far out at sea raise heavy swells that may come in from any direction.

People walking along a beach typically see waves that come at only a slight angle to the shore, regardless of which way the wind may be blowing or the offshore currents flowing, or how the coast twists and turns. The approaching waves conform themselves to the outline of the shore before they reach the land. When a wave enters shallow water, it slows down as it begins to drag on the seafloor. Meanwhile, any part of the wave still in deep water races ahead until it gets into shallow water, where it, too, begins to slow down. Imagine a wave approaching a coast obliquely: it will pivot like a marching band as one end slows in shallow water while the end still in deep water lunges forward.

Occasionally, waves approach the beach exactly head-on, at a right angle, with no change in the direction of their approach. They then arrange the beach sand into a row of scallops with sharp points projecting into the surf, called beach cusps. The waves sort coarser sand into the cusps and leave finer sand in the low areas between them. Beach cusps last only as long as the wind is constant. When the wind shifts, the waves quickly erase them as they again wash onto the beach at a slight angle.

Anyone who has tried to snatch a choice seashell out of the waves quickly finds out what happens when the waves approach at a slight angle, as they normally do.

You try to grab the shell out of the incoming swash, only to miss it and see the backwash carry it out into the foaming water. When the next wave brings it back almost in reach, you must walk a few feet down the beach for the next grab. If you miss several times, you may find yourself a considerable distance down the beach. The wash of every incoming wave sweeps the shell obliquely onto the beach, and then the momentum of the moving water carries it farther down the beach in the backwash.

Every particle of sand on the beach is also moving down the coast. You can think of the beach as a river of sand flowing along the shore. On some days the waves move the sand one way; on other days, the other way. Most beaches have a prevailing wind direction, creating a longshore current.

People commonly build walls, called groins, across the beach to trap the moving sand. Groins generally trap very well. A row of groins converts a smooth, narrow beach into a much larger beach with a map outline of something like the teeth on a ripsaw. However, the sand trapped on the growing beach never reaches its natural destination. If you look downshore from a set of groins, you will almost certainly find the beach there eroding because the groins are starving it of the normal supply of sand.

Although breakwaters and piers do not look like formidable barriers, they trap sand as effectively as groins and cause beach erosion farther downshore. Sand grains do not move under their own power—waves move them. Anything that interferes with the waves will affect sand movement. Anything that traps sand in one place will starve the beach somewhere farther down the line, causing beach erosion. Communities on some Hawaiian coasts import sand to maintain their beaches.

Whether a beach consists of sand, pebbles, or cobbles depends partly on the size of particles available and partly on the size of the waves moving them. In big storms, and on the windward sides of the islands, waves winnow out the small particles so that only cobbles and boulders remain on the beach. Sandy beaches are more abundant on the leeward sides, where waves are smaller. On all Hawaiian beaches, the smaller waves of calmer seasons carry sand onto the beach, burying the big rocks until large waves in the next storm uncover them again. It is a long-standing pattern: waves store the sand offshore during heavy weather, then spread it across the beach when the weather improves.

On the youngest Hawaiian shores, the beaches are made of black sand supplied where lava flows entered the ocean and shattered in steam-driven explosions. Waves and currents sweep these glassy lava bits along the shore, where they collect in coves to form the beaches. Because the sand supply is not continuously renewed, black sand beaches tend to wash away after just a few centuries.

On somewhat older Hawaiian shores, for example on East Maui, streams wash dull black basalt and red oxidized cinder to the coast. The basalt sand grains form beaches ranging from black to pale gray and even brick red. They tend to last a lot longer than eruption-formed black sand beaches, probably many thousands of years.

Even longer lived and more common are the well-loved resort beaches formed of pale coral grains. Coral reefs flourish along the oldest coasts, on islands that are no longer sinking very fast. Schools of the coral-eating fish *Zebrasoma flavescens*, or yellow tang, are common sights for snorkelers and divers. Yellow tang and other corallivores such as parrotfish excrete large quantities of fine white to beige coralline sand, which wave action quickly carries to shore. Protected from winter wave action by fringing reefs or wide, shallow channels, these become the largest beaches in Hawai‘i.

Black sand beach near the Kapoho lighthouse, formed by lava flowing into the ocean during the 2018 eruption of Kīlauea Volcano.

Coralline beach sand consists of calcium carbonate, which does not dissolve in tropical seawater but does dissolve when rainwater soaks through it. As the rainwater infiltrates and evaporates, it leaves behind a cement of calcium carbonate that binds together remaining undissolved sand grains, forming a pale solid mass called beach rock. Most coral sand beaches have beach rock foundations. When sea level drops, wind and rain quickly erode exposed beaches, leaving the beach rock behind. Patches of partly eroded beach rock are common along older Hawaiian shorelines, many marking a time of somewhat higher sea level 125,000 to 118,000 years ago.

Eroded beach rock shows many sloping layers. Each is a past beach face buried as the beach acquired more sand. The layers slope toward the ancient shore.

CORAL REEFS

Beach rock typically overlies wave-cut benches in basalt, and in some places it overlies reef rock—the remains of ancient coral reefs exposed above sea level. The kinds of coral that build reefs live only in water consistently above 65 degrees Fahrenheit. Many corals live in association with algae that require sunlight, so the water must also be clear and shallow. Other kinds of algae and many kinds of animals living on the reefs also contribute to their development.

Corals are animals related to anemones and sea jellies. They live partly by snatching microscopic animals from the passing water and partly on the largesse of the algae that live in their tissues. The algae are photosynthetic plants that take in carbon dioxide, use the carbon in building their tissues, and release free oxygen. The corals use the oxygen for their own metabolism and consume some of the algae.

The algae, in turn, use the carbon dioxide the corals produce and benefit from the sheltering reef they provide.

Individual coral animals, or polyps, look like minute anemones, with tiny tentacles that wave in the water. Some tentacles are stingers. The polyps live in colonies in which individuals are simply clones of one another, sprouting like buds from a continuous membrane. They take dissolved calcium from seawater, combine it with some of the carbon dioxide they produce metabolically, and deposit it as calcite, the basic mineral matter of the reef. If you look closely at a piece of reef coral, on occasion washed up on a beach, you can see the little dimples in which the individual polyps nestled. Think of a reef as a biological "apartment house" framed by the secreted calcium carbonate coral.

Coral polyps spread widely by shedding enormous numbers of eggs into the passing seawater. The eggs develop into larvae, which drift in the water until the time comes for them to develop into polyps. Then they settle on whatever hard base they may encounter, where they attempt to start a new colony. The chances that any individual larva will survive all the hazards of the ocean are almost vanishingly small. This means that it takes a long time for a reef to start on new volcanic islands. Even after 5 million years, Kaua'i has few large reefs.

Established reefs face many hazards. Storm waves and tsunamis may break them up or spread sand or mud across the coral, suffocating it. Agricultural chemicals washing into the ocean can kill coral, and so can sewage, which nourishes a smothering bloom of algae. In addition to coral-eating fish with their rasping beaks, worms, sponges, sea stars, sea urchins, and boring clams all bore holes in coral reefs.

On a longer time scale, reefs also face the challenge of changing sea level that accompanies the coming and going of ice ages. Sea level slowly drops as glaciers grow during an ice age, leaving the old reefs high and dry, forcing the reef zone offshore to the new coast. Ice ages tend to end suddenly, however, and sea level rises rapidly as the glaciers melt. The meltwater quickly submerges reefs beyond the depths at which corals and their associated algae can live. As islands sink, extinguished reefs slowly drop into even deeper water. Fossil reefs, including some that grew in previous ice ages, girdle the submerged slopes of most Hawaiian Islands. Around a dozen flank the northwestern coast of the Big Island alone, ranging in age from 400,000 to 15,000 years.

Even more challenging to the future health of Hawaiian reefs is a general warming and acidification of the ocean underway worldwide. In recent years Hawaiian sea temperatures as much as 3 to 6 degrees Fahrenheit warmer than historically average have killed coral and associated algae. El Niño conditions have become especially stressful. What remains is dead, broken, bleached reef rock on the seabed, though some species are more tolerant of higher ocean temperatures than others. Marine heat waves struck Hawai'i in 2015 and 2019, killing as much as 30 percent of the corals in the islands. Fortunately, much have recovered since then, but the long-term outlook is not favorable because the climate continues to warm.

Corals are highly competitive animals, so different species dominate in different localities in the reef. They come in shades of white, pink, yellow, brown, blue, purple, and black, but all fade to the white of bare calcite after the polyps die. Corals also come in many shapes: some massive, others rounded or branched like antlers. Sea fans look like flattened trees and grow in a rainbow of colors, whereas lettuce

Coral and fish at Richardson Ocean Park in Hilo on the Big Island.

corals look like crinkled leaves. Mushroom corals are shaped like the cap of a broad mushroom several inches across, with many thin ridges radiating spoke-like from a central stem. Corals growing in shallow water tend to be massive and rounded; those growing in deeper water, below the damaging reach of waves, are more delicate and branching. The ecological importance of coral reefs can hardly be overestimated. Although present on only about 1 percent of the Earth's surface, they provide a nursery and habitat for approximately a quarter of all known marine fish species.

The sunlit coloration of the sea around reefs varies from a stunning turquoise where the water is shallow and floored with coralline sand, to deep cobalt blue farther out where the water is much deeper. Upwelling of seawater is not as strong in Hawaiian waters as it is along many continental margins, and the ocean lacks suspended nutrients necessary for supporting the plankton that turns the sea green along many continental shorelines. Despite the abundance of life you can see while snorkeling on reefs in Hawai'i, the ocean here is for the most part the marine equivalent of a desert. This, too, helps explain the striking cobalt blue of the open ocean.

DUNES

The sediment supply present in beaches and reef lagoons is truly enormous. Some people associate sand dunes with deserts, but sand dunes are closely associated with beaches in all climates, wet or dry. Waves wash sand onto the upper beach at high tide or during heavy storms. When the upper beach dries in the sun, the sea breeze blows sand off it and into coastal dunes behind the beach. The dunes blow inland until they are beyond the reach of the ocean's strong salt spray, where plants can grow and stabilize them.

It is a marvel that sand dunes exist at all. Why does the wind sweep the sand into neat piles instead of scattering it across the countryside? Imagine what would happen if the wind were blowing sand across an asphalt parking lot where you had laid a small blanket. The sand would catch on the blanket because it is soft; the grains bounce onto it, but not off. As a pile of sand accumulated on the blanket, it would catch more sand for the same reason that the blanket did—because it is soft. For a sand dune, softness is the essence of existence.

Changing sea levels complicate the life of sand dunes. When sea level drops, as in an ice age, coral reefs are exposed to the air and die. Wind sweeps across them, blowing large volumes of sand inland to build big dune fields. Then, when sea level rises again, the ocean floods the old reefs, greatly reducing the supply of sand. Plants then cover and stabilize the dunes.

Just as beach rock forms when infiltrating rainwater cements sand grains with calcite, calcareous sand dunes also turn into rock. Sand dunes typically are much thicker than beach deposits; calcite cement generally penetrates only a little way into a dune, leaving the center soft. The outer rind of cemented sand blocks further penetration of rainwater.

Where erosion opens the interiors of old dunes, you may see many thin layers of hardened sand intersecting one another in an intricate pattern. Each layer is a former surface of a dune. Shifting wind causes dunes to slope one way, then another, accounting for the differently angled layers. Geologists call the overall pattern cross bedding.

Sand dune field at Mo'omomi Preserve on Molokai that probably began forming toward the close of the last ice age. The roots of beach naupaka and other plants stabilize migrating dunes.

Fossilized calcareous sand dunes now exposed as jagged limestone at Mo'omomi Beach, Molokai, show characteristic cross bedding that developed when the dunes migrated downwind as they grew. These sands originated from coral reefs that were exposed high and dry to strong wind erosion when sea level was lower toward the close of the last ice age.

Hawai'i's dune fields are especially interesting for paleontologists. They have preserved the skeletons and remains of many birds, including extinct species of birds that once thrived in the islands. This includes the now extinct flightless duck, giant geese, Hawaiian sea eagle, and a peculiar type of owl with very long, stilt-like legs. What survival strategy compelled this adaptation? Few places in the Hawaiian Islands preserve evidence of past life as well as the stabilized dune fields, especially those on Molokai and Kaua'i.

Kaua'i

Kaua'i is the oldest, most deeply eroded, and most heavily vegetated of the major Hawaiian Islands. Many people consider it the loveliest. The top of a single enormous shield volcano, Mt. Wai'ale'ale, reaches 5,148 feet. Kaua'i is roughly 33 miles long, 25 miles across, and has an area of 555 square miles. Despite appearing to be geologically simple, the rocks and terrain of Kaua'i tell a complex and unusual story.

Wai'ale'ale Volcano began to build above sea level sometime between 6 and 5 million years ago. Eventually, it grew into a gigantic shield of tholeiitic basalt, like Mauna Loa on the Big Island. Thin flows of pāhoehoe and 'a'ā lava, up to 15 feet thick, piled up to shape the volcano's gently sloping flanks. Eruptive fissures radiated from the summit toward all points of the compass, many bundled in two rift zones trending west and northwest.

Pioneer Hawaiian geologists Harold Stearns and Gordon Macdonald erroneously concluded that Wai'ale'ale developed an uncommonly large summit caldera, which they named Olokele caldera, about 13 miles across. Such large calderas are known in continental areas, but none this size have been found elsewhere in the Hawaiian Islands. They supported their idea by pointing out the massive sequence of horizontal lava flows about 2,600 feet thick on the Olokele Plateau at the center of the island. Such a thick stack of flows must surely have ponded in some sort of basin, which Stearns and Macdonald believed could only have been a giant caldera. They interpreted a similar stack of flows in Hā'upu Ridge south of Līhu'e as the remains of a smaller, satellitic caldera.

The idea that the Olokele Plateau and Hā'upu Ridge are caldera remnants poses problems, however, because both are resistant highlands. Other extinct calderas in Hawaiian landscapes are preserved as steep-walled amphitheater valleys with wide, flat floors, their interiors of altered, fractured flows susceptible to rapid erosion—much faster than surrounding mountainsides.

More recent sonar surveys detailing the bathymetry of the ocean floor surrounding the island provide geologists with information showing a much different history for Kaua'i. Several fields of rubble on the ocean floor appear to be massive pieces that broke off the eastern part of the island and avalanched into the ocean. The island stood about 1,000 feet higher above sea level when this sliding occurred and so was much larger than it is now. One avalanche mass fell to the north, the other went south. Debris from them swept up the slope of the Hawaiian Arch as far as 60 miles from the island. They must have been moving very fast underwater and surely generated large tsunamis.

The scarps from these catastrophic avalanches split the summit of Wai'ale'ale Volcano a short distance east of Waimea Canyon along a line roughly parallel to the canyon. What remained of the eastern part of the island dropped, and a strip of the volcano about 3 miles wide and 15 miles long collapsed along parallel faults to become the Makaweli graben, a downdropped fault block that extends from the northwestern part of the island to the southern shore.

Another tremendous avalanche later broke a large chunk off the north flank of Wai'ale'ale Volcano and carried it into the ocean. A towering sea cliff remained and has since eroded into the world-famous Nā Pali Coast. Landforms today suggest that slumping, sliding, or both may have impacted the western and southwestern sides of the island to a lesser extent. Once reaching maturity, Wai'ale'ale literally fell apart on all sides.

Wai'ale'ale attempted to repair itself. Eruptions continued after the slides, building a new volcanic shield in the dropped eastern region of the island. Lava streamed down the new shield to fill the Makaweli graben, and it ponded against the towering slide scarps that faced east in the Olokele area, building the thick stack of massive flows in the Olokele Plateau. Lava also ponded against irregular slide and slump blocks near Līhu'e, forming the great stack of old flows in Hā'upu Ridge.

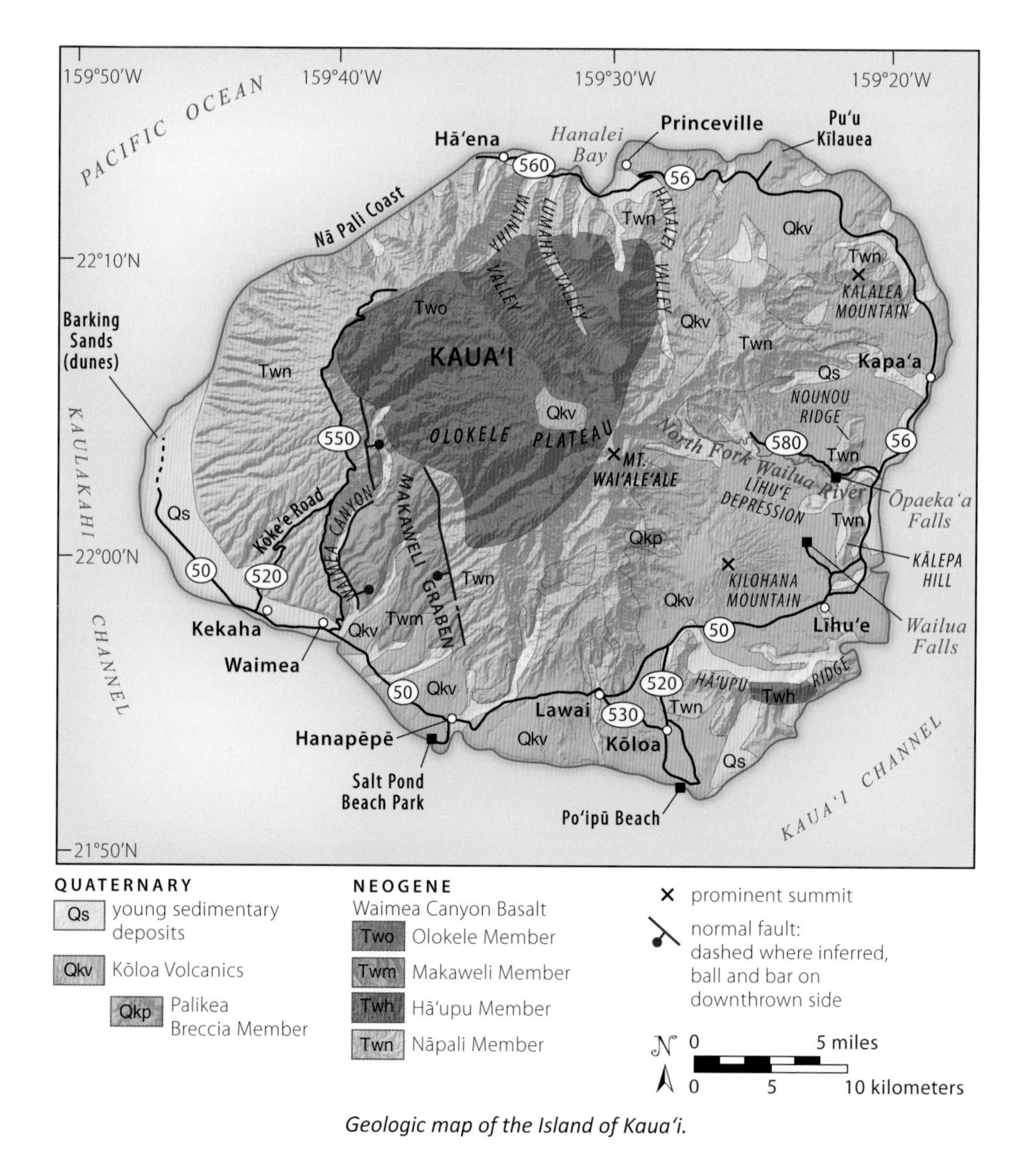

Geologic map of the Island of Kaua'i.

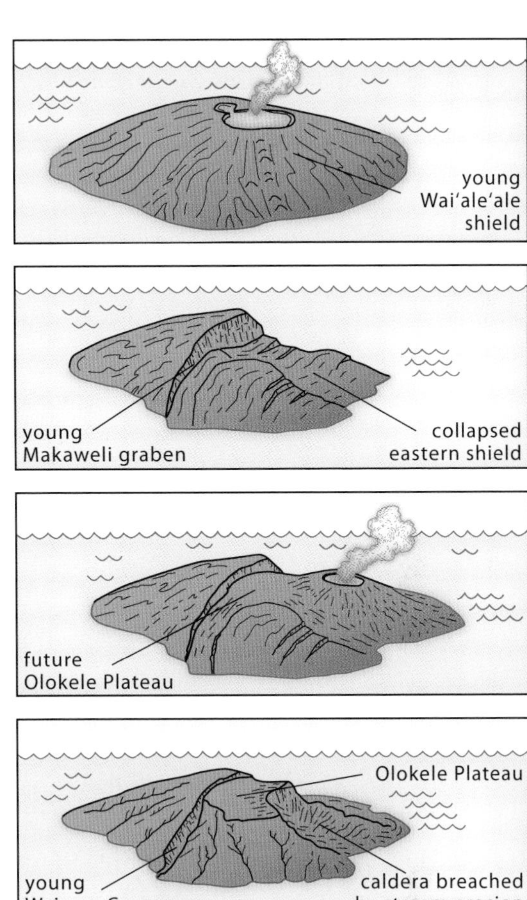

Evolution of Kaua'i

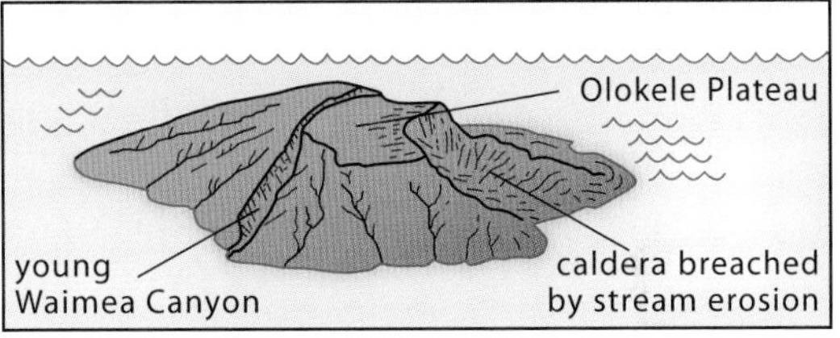

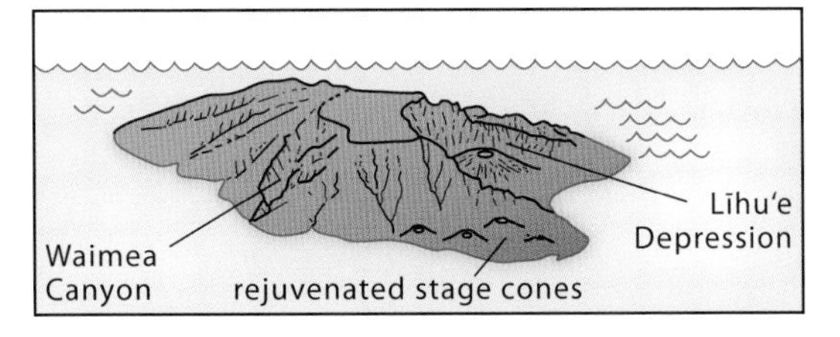

The new shield developed a caldera, called the Līhuʻe caldera, which ultimately eroded into the Līhuʻe Depression, a broad basin centered 5 miles northwest of Līhuʻe. Gravity surveys have revealed that the Earth's gravitational field is slightly stronger than normal above the floor of the Līhuʻe Depression, indicating that a large mass of dense rock probably exists beneath the floor. Many geologists think it is a mass of gabbro, the crystallized magma chamber that Waiʻaleʻale developed after the big avalanches.

Geologists call Waiʻaleʻale shield lavas the Waimea Canyon Basalt. One part is the Nāpali Member. The thick lavas of the Olokele Plateau are the Olokele

Member. Similar thick lavas probably flowing from the plateau filled the Makaweli graben and are called the Makaweli Member. The lava flows of Hā'upu Ridge are also considered a distinct member of the Waimea Canyon Basalt.

Rapid shield building and rebuilding continued at Wai'ale'ale until 4.0 million years ago. It probably took less than 1 million years for the volcano to build the island to its maximum extent. As activity waned, the lavas became more alkalic and the eruptions more explosive. This later activity was underway by 3.9 million years ago, adding little mass to the island. Meanwhile, tectonic subsidence and erosion began substantially to reduce the island's size.

Let's imagine those first stages of erosion. Gigantic slices of the volcano continued to avalanche away, extending the shoreline seaward in some places, shifting it inland in others. Waves eroded the shorelines into sea cliffs, except where coral reefs protected the coast. Streams dissected the originally gentle volcanic slopes into a landscape of jagged ridges and deep valleys—a landscape of exotic tropical beauty.

The sloping flanks of Wai'ale'ale eroded far more rapidly than the ponded lavas in the Olokele Plateau and Hā'upu Ridge. The flank flows were thin, inclined seaward, and full of weak layers of permeable rubble. Streams easily undercut them along the rubble layers, triggering landslides that followed the tilt of the flows. Only a few strip-like remnants of the original flanks of the shield volcano survive.

The thick Olokele and Hā'upu basalt flows stand in bold erosional relief today, even though the lava originally ponded in low areas. The network of deep stream valleys dissecting the Kaua'i highlands still does not completely drain the Olokele Plateau. Some of the island's heavy rainfall accumulates in high swamplands instead of running off. The swamps support a lush montane forest that shelters some of the rarest bird and plant species in the islands.

VOLCANIC REJUVENATION

Kaua'i would be smaller than it is and even more rugged were it not for a period of rejuvenated volcanism that took place between 3.65 and 0.15 million years ago. These youngest eruptions were generally small and widely scattered. Geologists call the rocks produced by them the Kōloa Volcanics.

About forty of these late volcanic vents have been identified on Kaua'i. The oldest are in the west and northwest parts of the island; the younger ones, in the east and southeast. The largest is Kilohana, the gently sloping mountain immediately west of Līhu'e. A small shield volcano 6 miles across, it has a summit crater about 250 feet deep. Most of the vents from the rejuvenated stage line up along north–south trends, with fewer trending northeast to southwest.

The Kōloa Volcanics, mainly basalt, poured into many valleys, completely filling some of them and diverting their streams to erode new channels elsewhere. Much of the landscape in eastern Kaua'i expresses a long interplay of such streams and volcanic eruptions. Lava erupting next to the shore added new land to Kaua'i, including much of the island's coastal flatlands.

The alkalic basalts from the rejuvenated volcanic stage include some exotic rocks greatly enriched in sodium, such as nephelinite, melilite basalt, and basanite. These lavas also include xenoliths, most of which are bits from the mantle beneath Kaua'i, a part of the Earth we would know little about without them. Geologists love these "strange rocks."

LATE EROSION OF KAUA'I

In the past half million years or so, stream erosion, taking advantage of the waning pace of eruptions, has dissected much of the Kōloa volcanic landscape, creating new waterways and valleys. Small rejuvenated eruptions have continued to take place during this span of time, most recently just a few thousand years ago—a blink of a geologic eye—but for the most part erosion has returned to dominate Kaua'i's changing landscape.

An example of recent erosion is Hanalei Valley, near Princeville. Streams have cut through a section of Kōloa lavas more than 2,000 feet thick here. The valley today is as much as 4 miles wide and 3,000 feet deep and has nearly vertical walls stretching from the coast 12 miles inland to the summit of Mt. Wai'ale'ale. The neighboring valleys to the west, Lumaha'i and Wainiha, are equally spectacular, as is Waimea Canyon on the opposite side of Olokele Plateau. Together they testify to the power of weathering and erosion in a wet, tropical climate.

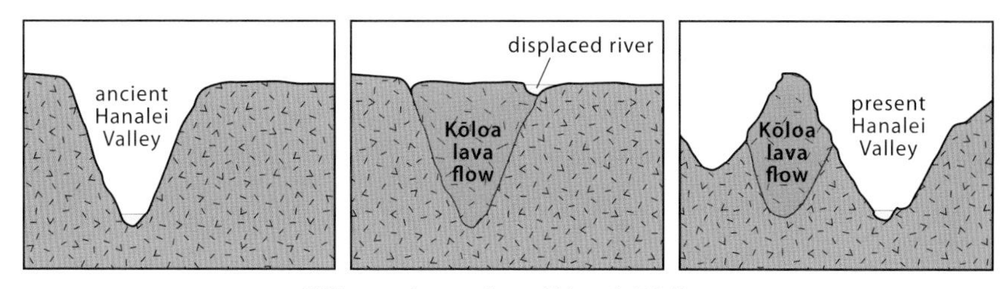

Filling and recutting of Hanalei Valley.

About 6 to 5 million years ago, when Kaua'i first emerged from the ocean, its steamy volcanic summit was perched directly over the Hawaiian hot spot, where the Big Island is now. Five million years from now, the ocean may reclaim Kaua'i as the Pacific Plate beneath cools and subsides during the journey to the Kuril-Kamchatka Trench. By then, Kaua'i will probably be a mere coral atoll awash at high tide. Enjoy Kaua'i while it lasts!

HI 56 AND HI 560 (KŪHIŌ HIGHWAY)
LĪHU'E—PRINCEVILLE—HĀ'ENA

38 miles

See maps on pages 48 and 54.

Līhu'e lies on a small coastal plain built of lava flowing mostly from the Kilohana lava shield, the largest of the rejuvenated volcanoes of the Kōloa Volcanics. You can easily see the shield, which rises more than 1,000 feet, broadly profiled against the steeper cliffs nearby as you look west from Līhu'e. Kilohana occupies the southwest part of the Līhu'e Depression. A jagged crest of older Waimea Canyon Basalt, Hā'upu Ridge, forms the southern edge of the depression, trending east–west 1 to 7 miles southwest of the city. The highest point on the ridge, Hōkulei Peak, is 1,650 feet tall.

The origin of Hā'upu Ridge is something of a mystery. Its flows should be dipping toward the sea, away from the ancient summit of the Wai'ale'ale Volcano, as they

WAILUA FALLS

About 3 miles north of downtown Līhu'e, at the start of HI 56, the winding HI 583 leads to Wailua Falls, along the South Fork of the Wailua River. The falls plummet 170 feet over a thick ledge of hard lava erupted from a nearby Kōloa vent. The lava entered the South Fork, developing a very thick intracanyon flow narrowly confined by the gorge walls. The flow roughly composes the rocky wall immediately beneath and surrounding the falls. Irregular cooling fractures below the lip of the falls give the stone a jagged appearance in places. This could be a result of quick cooling of the flow when it encountered water or dense steam, and indeed some geologists have described other evidence that this flow entered a deep pool or marsh at the bottom of the canyon during the eruption. Unfortunately, present vegetation is too thick to reveal this. Near the banks of the plunge pool directly beneath the overlook, you can see layers of older, slightly weathered pāhoehoe, probably of Nā Pali vintage, that formed the original walls.

Like other easily visited waterfalls on Kaua'i, Wailua Falls is spectacular, but it is by no means the largest. Individual falls draining Mt. Wai'ale'ale's deeply eroded Weeping Wall to the west approach 3,000 feet in height, among the tallest in the world! But they can be reached only by arduous wilderness hiking.

do everywhere else around the island. But here the flows are thick and nearly horizontal. Possibly they piled up against a large landslide scarp to the south, formed during the partial collapse of Wai'ale'ale late during the shield stage of volcanism. Because of their thickness and level orientation, they would have resisted erosion as the surrounding landscape wore down over several million years, creating today's ridge.

The other sides of the Līhu'e Depression are less well defined. The fluted ridges on the flank of Wai'ale'ale loom 5 to 10 miles to the west, and to the north are Kālepa Hill and Nounou Ridge, lower than Hā'upu. They possibly mark the eastern rim of the deeply eroded Līhu'e caldera. There, the older flows dip seaward, indicating that they accumulated just outside the caldera. The floor of the basin itself is a tropical, rolling lowland, home to many of Kaua'i's residents and critical water reservoirs.

The port of Līhu'e is at Nāwiliwili Bay, which originated during past ice ages when sea level stood much lower. Hulē'ia Stream cut a small valley into the emergent landscape. Later, sea level rose and flooded the valley, creating the modern bay. The combinations of port, flatland, and abundant water supply facilitated the city's development.

Līhu'e to Kapa'a

The Kūhiō Highway, HI 56, hugs the shore between Līhu'e and Kapa'a, providing access to many beaches. The beige beach sand consists mainly of fragments eroded from offshore reefs. Coconut palms and other tropical vegetation thrive in the warm, moist climate.

Just south of milepost 6, the highway crosses the Wailua River. You can turn onto HI 580 for a 3-mile drive west to view 150-foot-tall 'Ōpaeka'a Falls north of the bridge. The Kaua'i Historical Society and Bishop Museum have restored the centuries-old temple of Holoholoku Heiau near the highway intersection. HI 580 passes through the notch that the North Fork of the Wailua River eroded across Kālepa Mountain–Nounou Ridge. Kōloa lavas poured through the notch, so it must have formed before rejuvenated volcanic activity began. The river is now eroding through these younger flows.

ʻŌpaekaʻa Falls tumbles across basalt of the Kōloa Volcanics. Just below the falls you may be able to spot vertical dikes of basalt cutting through horizontal flows of Kōloa lava. The dikes fill fissures, at least a few of which fed lava into the uppermost flows. Rocks on the slopes of the adjoining ridges form bold, jaggedly eroded cliffs in places. These are all the older Nāpali Member. Some of the many dikes cutting across these older lavas along Kālepa Hill–Nounou Ridge dip toward the center of the Līhuʻe Depression. Geologists Stearns and Macdonald interpreted them as typical cone sheets, curving dikes shaped like segments of the cones that generally enclose calderas. Exposures of cone sheets are rare. They are found only in deeply eroded volcanic regions such as Scotland, where they were first recognized. They are almost certainly present around the rims of some of the larger calderas in the Galapagos Islands. If these are indeed cone sheets, they are the only ones identified so far in the Hawaiian Islands.

Between the Wailua River and the town of Kapaʻa, HI 56 crosses old beach deposits. If you look inland, you can see the northern continuation of Kālepa-Nounou Ridge. The sharp silhouette along the skyline outlines the profile of the Sleeping Giant, whose head lies to the south, feet to the north. The generally gradual descent of the main ridgeline toward the coast preserves one of the few remnants of the original slope of Waiʻaleʻale's shield. Almost everywhere else on the island, the slope has been lost to erosion.

Kapaʻa to Princeville

About 1 mile north of the little resort town of Kapaʻa, watch for a scenic lookout a half mile from milepost 9 on HI 56. Immediately north and just below the lookout is a wave-cut bench of basalt, a relic from when sea level stood higher than now. This high stand in sea level, about 35 feet, likely occurred during the Eemian interglacial epoch, between 125,000 and 118,000 years ago. The bench formed at the energetic base of the waves that rose to crash and rebound against the shore. Imagine what this shoreline might look like if sea level stood 35 feet higher today, a future possibility given climate change. Elsewhere along this shore is a much lower shoreline bench, just a few feet above sea level, that was likely cut during a less high sea level stand only 4,000 to 3,000 years ago.

Look for benches like this elsewhere as you travel along the eastern and southern coasts of Kauaʻi. These widespread coastal features indicate that the island has maintained a stable level relative to the underlying mantle for many thousands of years. This contrasts greatly with the continuous tectonic sinking of the younger, heavier Hawaiian Islands hundreds of miles to the southeast.

Between mileposts 13 and 14, the highway passes through the small community of Anahola. Look inland from the vicinity of milepost 14 to see a jagged ridge, 2,000-foot-tall Kalalea Mountain, another eroded remnant of the original shield volcano.

Near milepost 22 you may be able to spot the lighthouse at Kīlauea Point. The low ridge to the right, Puʻu Kīlauea, formed during a Kōloa eruption more than 150,000 years ago, though it appears younger. It is an important seabird sanctuary today, a nesting ground for the red-footed booby. About 0.4 mile west of milepost 23, at the

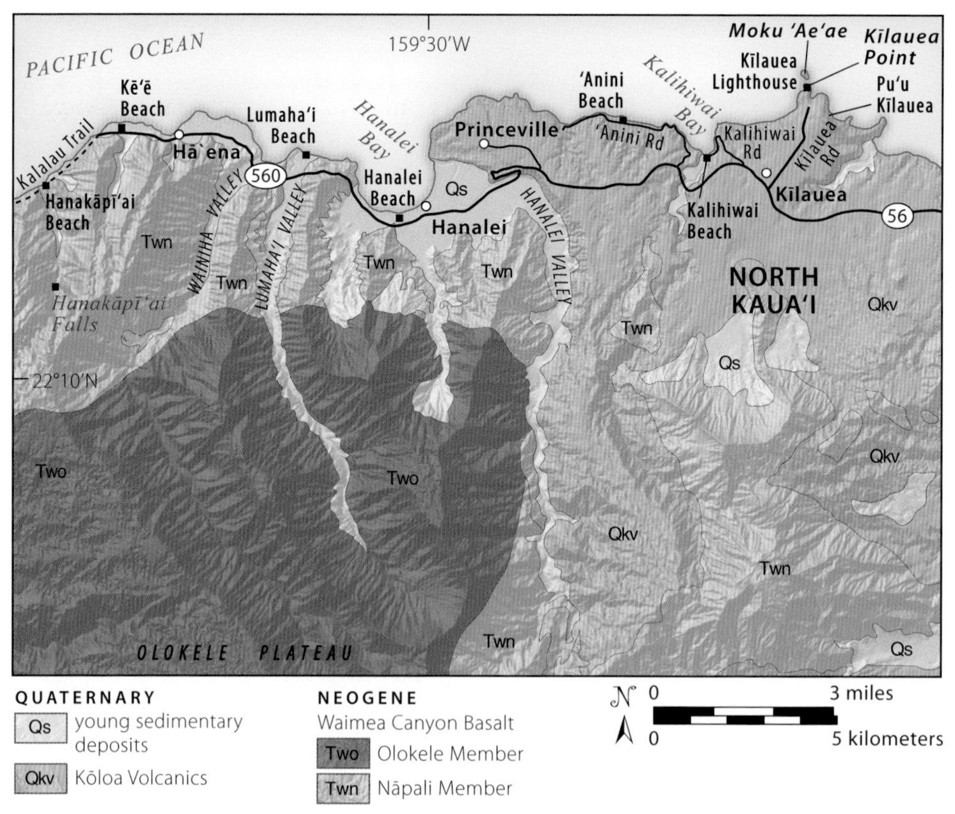

Geologic map of the northern part of Kaua'i between Kīlauea Point and Hā'ena.

Kīlauea Point is composed of pyroclastic layers erupted from the Pu'u Kīlauea tuff cone. The wave-cut bench lining the shoreline below the parking area is the same telltale Eemian interglacial bench that hugs the shore near Kapa'a.

community of Kīlauea, a side road leads across 2 miles of coastal tableland to Kīlauea Point and the edge of the sanctuary, skirting the western flank of Pu'u Kīlauea.

The eruption of Pu'u Kīlauea began when rising magma most likely met groundwater at a shallow depth and interacted with seawater in an open vent, flashing it into steam in a series of violent explosions. The cone is a disorderly mixture of volcanic ash, fragments of white limestone from the reef, and blocks of older Kōloa lava ripped out of the vent by the steam explosions. The blocks include chunks of rare melilite basalt and nephelinite, unique in Hawai'i to rejuvenated-stage eruptions. The ejecta accumulating around the vent during the final phase of the eruption possibly cut off the ocean from the magma, bringing an end to the steam blasts. The remaining molten rock simply fountained out of the ground, spreading a coat of spatter across much of the cone. The most accessible view of this volcanic potpourri is from the overlook at Kīlauea Point.

Kīlauea Point, capped by light-gray steam-blasted pyroclastic layers, and the adjoining small islet of Moku 'Ae'ae form the far western end of the 1.5-mile-wide Pu'u Kīlauea tuff cone. The ocean occupies most of the crater, the remaining rim of which lies out of sight east of the point.

HI 56 meets Kalihiwai Road about 1 mile west of the town of Kīlauea. The village of Kalihiwai is 1.1 miles from the turn. The Aleutian tsunami of 1957 heavily damaged Kalihiwai and destroyed an important bridge, separating the east and west ends of Kalihiwai Road. You will have a good view of the coral reef extending offshore from the beach cliff as you approach Kalihiwai Bay. The line of breakers marks the offshore edge of the reef.

A vertical basalt dike about 15 feet wide is visible in the cliffs at the east end of Kalihiwai Beach. The fissure it fills was probably the plumbing that fed molten lava into one of the Kōloa basalt flows. The dike shows prominent horizontal columns,

Dike (arrow) intruding eastern wall of bay at Kalihiwai Beach.

Xenoliths of dunite (olivine-rich rock) in nephelinite beach cobbles at Kalihiwai Beach, Kaua'i north shore.

which are basically like the vertical columns in lava flows. Both form at right angles to cooling surfaces as molten rock shrinks and hardens.

The black boulders concentrated toward the eastern end of the beach look like ordinary basalt, but they are nephelinite, a rare rock type so rich in sodium that it contains abundant crystals of nepheline rather than ordinary plagioclase feldspar. The boulders eroded from a nearby Kōloa lava flow. Some of the boulders contain fragments of peridotite, xenoliths 1 to 2 inches across, which appear as angular pieces even darker than the enclosing basalt. A determined search will also turn up a few xenoliths of dunite, a yellowish-green rock consisting almost entirely of olivine. Olivine weathers readily as the iron reacts with atmospheric oxygen, which explains why some of the xenoliths have the rusty color of iron oxide. All the xenoliths are fragments of the mantle, direct evidence of the nephelinite's magma origins.

Between mileposts 25 and 26, ʻAnini Road provides shoreline access from HI 56. Many connoisseurs consider ʻAnini Beach Park the finest snorkeling locale on Kauaʻi. The long, exposed offshore reef protects the shallow inshore water from high surf except during rough winter storms. The sandbar at the mouth of a stream near the western end of the beach is a good place to look for seashells.

Kōloa lavas built the broad, flat peninsula at Princeville, the site of an old Russian fort and today the site of a modern resort and golfing mecca.

Hanalei Valley to Hāʻena

The highway west of the Princeville peninsula follows the most deeply eroded flank of Waiʻaleʻale Volcano along the breathtaking north shore of Kauaʻi, crossing rocks of the Nāpali Member and related sediments the rest of the way. Watch 0.2 mile west of Princeville Center for a scenic lookout where the road begins to descend into Hanalei Valley. The view inland up the length of the valley reveals a lush tropical paradise that, unfortunately for residents, is flood prone. Fields of kalo (taro), a Hawaiian dietary staple, cover the floodplain of the Hanalei River below. The Hanalei National Wildlife Refuge occupies the bottomland extending 1.5 miles upriver.

The low, rugged, wide ridge along the east side of Hanalei Valley was once the northern end of a canyon hundreds of feet deep. Eventually, thick flows of Kōloa lava filled it to the brim. The new lava was more resistant, so the canyon's stream shifted and eroded along the west side in thinner, weak flows of the Nāpali Member, creating the modern Hanalei Valley. The eastern ridge is a mold of the ancient canyon cast in basalt.

The highway curves through a big switchback, "the Loop," as it enters Hanalei Valley. The old whaling port of Hanalei is at the mouth of Hanalei Valley, between mileposts 2 and 3. You can charter a boat here to explore the spectacular Nā Pali Coast stretching 15 miles westward. Hanalei Beach is the wide strip of calcareous sand west of the pier. Swimming here can be rough.

Watch between mileposts 4 and 5 for the pulloffs to Waikoko and Lumahaʻi Beaches. The trail to Waikoko Beach, the first one encountered, begins at a prominent bend in the road. The scattered outcrops of basalt punctuating the beach contain depressions filled with sand extremely rich in olivine. Waves swishing sand around in the holes have washed away the lighter mineral grains, leaving dense concentrations of the heavier green olivine crystals like miniature placer deposits.

A little farther along, the highway skirts the shoreline at Lumahaʻi Beach, made famous in the movie *South Pacific*, near the mouth of the Lumahaʻi River. While a good place to stroll the sand and explore more outcrops of basalt, this is not a safe place to swim owing to waves, currents, and backwash.

Hanalei Valley was displaced to its present position when an older valley was filled with lava (the low ridge on the eastern [left] side, in the background).

Just west of milepost 5 look for big roadcuts in a stack of thin, irregularly layered flows weathering into reddish-brown soil. Some, possibly all, of these flows were ʻaʻā that poured down a steep slope. Two basalt dikes, each about 2 feet across, cut steeply through the flows. The fissures they filled were probably magma plumbing for lava flows lost to erosion.

Farther west, the highway curves across Wainiha Valley. On a clear day you can see the rim of the Olokele Plateau on the skyline at the head of the valley.

Just west of milepost 8, the road passes through the community of Hāʻena, which was devastated in the great Aleutian tsunami of 1946. The highest wave crested here at 45 feet. Nearby, cars are often parked around a short, inconspicuous road that leads to Tunnels Beach. Patches of coral reef almost reach the sandy shore, marine life is abundant, the water calm. The outer of the two parallel reefs catches most of the surf.

About 1 mile west of Hāʻena the road reaches Hāʻena Beach Park. Snorkeling is usually rewarding on the reefs here, where convict tangs and many other fish congregate to graze the coral. Look across the street to see Maniniholo Dry Cave. Maniniholo has a wide entrance but not much depth. The sandy floor is rarely wet, hence the alternative name for the cave. Waves eroded Maniniholo from flows of blocky ʻaʻā and ropy pāhoehoe basalt before the beach grew large enough to protect the sea cliff. Possibly much of this erosion took place during the Eemian high stand in sea level, more than 100,000 years ago.

Dikes (the protruding vertical ridges) slice through thin lava flows in a roadcut near milepost 5, west of Hanalei. The flows, several million years old, are weathering into soils reddened with oxidized iron.

The public highway ends at a parking lot with a shuttle stop and a boardwalk leading to Kēʻē Beach and the start of the Kalalau Trail along the Nā Pali Coast, another 0.3 mile to the west. This parking lot is as far as you can drive along the north shore. The public trail from the parking area passes through an active loʻi or taro garden along a boardwalk to Kēʻē Beach. If you walk along the road, now closed to vehicles, you will pass the Waikapalaʻe Cave. The floor of this cave extends below the water table and is therefore called a wet cave. Waves eroded

Waikapala'e Cave was eroded by waves when the sea was higher. Like other sea caves along Hawaiian shores, it is a short passage with a wide mouth.

Kē'ē Beach viewed from Kalalau Trail. The offshore fringing reef and a band of resistant beach rock paving the middle part of the beach are visible.

NĀ PALI COAST

Whether seen on foot, from roadside lookouts, the ocean, or the air, the uninhabited northwest coast of Kaua'i is one of the most dramatic meetings of land and sea in the world. Geologists believe the steep, plunging cliffs were established several million years ago near the end of alkalic eruptions from Wai'ale'ale. A large piece of the original north side of Kaua'i broke off in a giant avalanche, leaving a cliff as much as 2,700 feet high. The thick lava flows that cap the Olokele Plateau effectively resist erosion far better than the Nāpali flows, supporting the steep slopes.

Hanakāpī'ai Beach showing wave-cut bench and multiple shield flows in cross section. The oldest known volcanic rocks in Hawai'i's main islands come from outcrops such as these along the Nā Pali Coast. They are nearly 6 million years old.

Nā Pali Coast from near the start of the Kalalau Trail. The coast stretches for approximately 15 miles, and the trail provides access to about half this shoreline.

The Kalalau Trail starts at the end of the road, winding for about 11 miles along the coast. A satisfying 2-mile (one-way) day hike to Hanakāpī'ai Beach provides spectacular views down onto Kē'ē Beach, approximately 0.5 mile from the trailhead, and along the Nā Pali Coast. The modern trail was originally built in the late 1800s, replacing a similar path linking early Hawaiian shoreline settlements. Engineers paved beginning portions with cobblestones in the 1930s. Outcrops of 'a'ā and pāhoehoe flows can be easily seen along the trail. Near the east end of Hanakāpī'ai Beach are two distinctive sea caves eroded into the cliffs and a cemented outcrop of colluvium, soil and rock debris washed from upslope. For more of an adventure, hike to Hanakāpī'ai Falls, approximately 1.5 miles upstream from the beach.

Waikapala'e along horizontal lava flows, so it is surprising that the mouth of the cave is round rather than elliptical.

The trail brings you to Kē'ē Beach. A reef with a conspicuously flat top lies a short distance offshore, protecting the small curve of a beach. The old reef is partly exposed along the shore here as an area of beach rock. Dune erosion along the beach has exposed the roots of a number of large trees that have been unable to stabilize the shifting sand.

HI 50 (KAUMUALI'I HIGHWAY)
Līhu'e—Hanapēpē—Barking Sands

36 miles

The rugged Hā'upu Ridge forms the skyline south of HI 50 in the area west of Līhu'e. Rocks in the ridge are a thick stack of horizontal lava flows that may have ponded against some sort of fault scarp, possibly the headwall of an avalanche that has long since eroded away. The excellent resistance of the flows to erosion explains the height of the ridge. To the northwest along the same stretch of road, you will see the low profile of Kilohana, a shield volcano 1,133 feet high, the largest of the Kōloa volcanoes.

Less than a half mile west of Līhu'e, HI 50 meets HI 58, which leads 2 miles to Nāwiliwili Bay, the main port on Kaua'i. All the bedrock in the Nāwiliwili Bay area is part of the younger Kōloa volcanic series. In contrast to the older lavas in Hā'upu Ridge, these flows dip southward, away from the Līhu'e Depression from which they probably erupted.

At the western end of Hā'upu Ridge, HI 50 crosses the low pass of Knudsen Gap, a broad remnant of a deep valley eroded through lava flows of Nāpali Member basalt. Younger flows erupted from neighboring Kilohana and other Kōloa volcanoes partially filled it, resulting in a level floor.

Numerous Kōloa vents dot the landscape on this part of Kaua'i. These vents buried the older Nāpali lavas with younger flows and pyroclastic debris. You can see a cluster of small Kōloa cones, the Pōhākea Hills, to the north. Look for them about 0.3 mile west of milepost 12, at the junction with HI 540, Halewili Road (west of Lawai). Several other north-to-south alignments of Kōloa lava shields and cinder cones also occur in this area, but they are hard to see from HI 50.

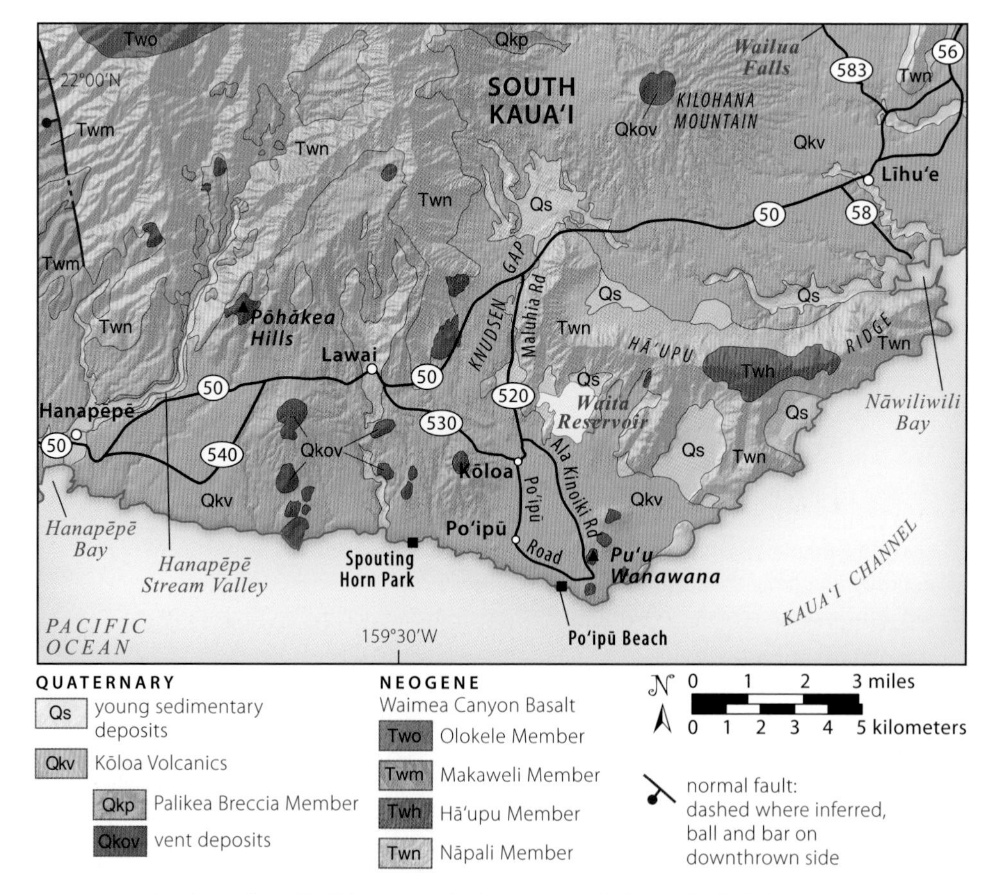

Geology along HI 50 between Līhu'e and Hanapēpē and the Po'ipū area. *For HI 50 west of Hanapēpē, see map on page 68.*

THE PO'IPŪ AREA

One of the most popular resort areas in Hawai'i is centered around Po'ipū. A few beautiful pocket beaches hide along a stretch of mostly rocky shoreline near the heritage sugar mill town of Kōloa. In the maze of roads, condo complexes, ranchland, and forestland of this district, not much roadside geology is visible, but we present some highlights here.

Evidence of the island's most recent volcanism, which produced the Kōloa series of cones, lava shields, and lava flows, can be seen in the southeast corner of Kaua'i. Many rejuvenated-stage vents around Po'ipū are oriented in north–south alignments, perhaps guided by east–west stretching of the island into the neighboring Kaua'i Channel. These formed mostly during the past few hundred thousand years. Just northeast of the intersection of Po'ipū and Ala Kinoiki Roads, right next to Po'ipū Beach Park, is the small cinder cone, Pu'u Wanawana. Its rim was breached as the vent sent lava downslope. Heading upslope on Ala Kinoiki Road, look east (right) to see Pu'uhi, a much larger, reddened cinder cone. The lava shield on the south side of Pu'uhi is among the youngest vents on Kaua'i. Unfortunately, it's not easily visible from the road.

The Kōloa rejuvenated volcanic vent Pu'u Wanawana, one of many near Po'ipū.

Spouting Horn, a blowhole in an ancient wave-cut bench at Po'ipū. Note the brown calcareous algal patch all around the mouth of the blowhole.

West from Po'ipū along the shore is Spouting Horn, one of the finest examples of a tidal blowhole in Hawai'i. The blowhole is in a nearly level wave-cut lava bench related to similar seaside benches elsewhere around the shore of eastern Kaua'i, probably formed during a high stand in sea level 4,000 to 3,000 years ago. Brown calcareous algae live around the mouth of the blowhole, where frequent eruptions of seawater sustain this microclimate, equivalent to the intertidal zone closer to water's edge.

A return to HI 50 via Maluhia (Tree Tunnel) Road (HI 520) passes another small Kōloa cone a short distance west of a prominent bend. The abundance of such cones suggests that most Kōloa eruptions were brief and explosive, each lasting only a few weeks or months. In contrast, the large Kilohana shield visible inland from Līhu'e must have taken years, if not decades, to build.

Hanapēpē Valley

East of Hanapēpē, HI 50 skirts the lower reaches of rugged Hanapēpē Valley. The lookout north of the highway, west of milepost 14, offers a spectacular view upstream into a deep canyon. The thin flows in the higher valley walls are Nāpali basalt that erupted on the flank of Wai'ale'ale. After Hanapēpē Stream eroded the valley into these shield lavas, thick flows of Kōloa basalt partly filled it. During long interludes between eruptions, Hanapēpē Stream eroded the flows, only to have its work undone when new eruptions again poured molten basalt into the valley during the period of rejuvenated volcanism. One large flow with conspicuous vertical columns stands out on the eastern side of the canyon.

The floor of Hanapēpē Valley becomes wide and flat coastward (south) of the overlook area. The stream eroded the valley when Kaua'i stood higher than it does today. As the island sank, the mouth of the valley drowned and became Hanapēpē Bay. Sediment deposited in the bay built the low plain where Port Allen stands. HI 50 meets HI 543 (Lele Road) at the western end of Hanapēpē. This road leads south for about 1 mile to some ancient salt ponds next to the old airstrip at Pu'olo Point (see map on page 68). The ponds were built to extract salt from seawater and are still used for this purpose by native Hawaiians. Clay from the red laterite soil seals the ponds to prevent seawater from draining before salt can be collected. This clay gives Kaua'i salt its distinctive red tint.

About a half mile west is Salt Pond Beach Park, a sandy beach sheltered behind small coral reefs. Coral grows right up to the shore here.

Coastal Flatlands around Waimea and Kekaha

West of Salt Pond Beach, HI 50 crosses more coastal flatland, much of it used for experimental agriculture. At Lā'au'ōkala Point, just before reaching the town of Waimea, is Fort Elizabeth State Historical Park, a wayside preserving the remains of a fortified outpost occupied by the royal Hawaiian army from 1817 to 1864.

The town of Waimea rims Waimea Bay, where Captain Cook made landfall in 1778. After crossing the bridge into the town, look for Menehune Road, the third right turn. This street leads about 2 miles north to Kīkī a Ola (the Menehune Ditch), a moat under the end of a suspension bridge walkway. Early Hawaiian immigrants built this ditch to divert river water to nearby kalo (taro) fields centuries ago.

A layer of pillow lava that flowed out of nearby Waimea Canyon is visible in the cliff above Kīkī a Ola, right where the suspension bridge crosses the street. Pillow lava forms where pāhoehoe erupts underwater. Fresh outbreaks of lava swell in cushiony buds from the surface of the flow. These quickly develop skins due to the sudden quenching in cold water. The skins resist further growth. Quenching may also shatter the quickly hardening lava into tiny bits of glass, called hyaloclastite ("glassy broken rock"), which enclose the pillows. Individual pillows commonly show cooling fractures radial to their centers.

At Kīkī a Ola individual pillows range from 1 foot to 10 feet in diameter. Hyaloclastite—now largely weathered into mudstone—is especially apparent near the top of the cliff. Here an overlying flow, probably from the same eruption, forms a capping layer. Short columns formed as lava chilled against the lower hyaloclastite and pillows. The uppermost part of the flow remained exposed to the air after the rest of the flow filled what was probably a coastal marsh.

Pillow lavas at Kīkī a Ola near Waimea town, formed when a lava flow following the drainage entered a shallow pool or waterlogged marsh.

Past Waimea, HI 50 continues to cross dry coastal plain, a strip of old seafloor with hardened sand dunes on it. The plain likely took shape during the Eemian interglacial epoch little more than 100,000 years ago, when the global climate was warmer and less land ice existed worldwide than today. At that time waves crossed what is now dry plain and lapped against the low sea cliff at the foot of the slope north of the highway.

The beige soil on the plain resulted from weathered reef sand and silt of the ancient seabed. Some sediment settled underwater. The rest blew in from the beach exposed during the latest ice age, when sea level was much lower than it is today. That ice age ended roughly 11,000 years ago. Low ridges near the coastal edge of the plain are old beach ridges and sand dunes that formed at that time, when the shoreline was located as much as 1 mile offshore from the modern coast.

Kekaha Beach Park, along the waterfront at the western edge of Kekaha, has a beautiful beige sand beach, with grains of coral and calcified algae washed in from offshore reefs. It extends for several miles along the southwest coast of the island. The surf is gentle, but the currents here are sometimes dangerous.

Look southwest from Kekaha to see the profile of Ni'ihau, the westernmost of the main Hawaiian Islands. Inland to the northeast is the gently sloping leeward flank of Wai'ale'ale. Streams have not dissected this dry side of the island as much as on the wet side, so long strips of the original surface of ancient Wai'ale'ale shield volcano survive on the drainage divides.

Barking Sands and Polihale Beach

HI 50 continues west of Kekaha to Barking Sands Missile Range and Airfield, which occupies 7 or 8 miles of the coastal zone. The road turns inland toward the ancient sea cliff shortly past the military base. Look for a left turn near the end of the highway. This well-traveled dirt road is recommended for four-wheel drive vehicles only, though passable with care in dry weather for ordinary passenger vehicles. In 4.5 miles, you'll find Polihale State Park, a spectacular, gentle beach backed by tracts of dunes held in place by vegetation. It's a northward extension of the Barking Sands Beach.

Polihale Beach with gently dipping Wai'ale'ale shield lavas in the background to the north. This is the westernmost beach in Hawai'i easily accessible to the public.

If they are properly damp, the calcareous sands making up the Barking Sands really will "bark" as you walk across them. On Polihale Beach the sand is remarkably soft, though on sunny days dangerously hot on bare feet! A walk to the far end of the beach near the picnic area leads to spectacular sea cliffs of Nāpali Member basalt. Ocean waves are still eroding this part of the cliff. Lava flows exposed in the cliffs slope gently seaward, parallel to the preserved remnants of the original flank of the volcanic shield rising above the cliffs. Though investigations of the seafloor offshore are not conclusive, it seems likely that these cliffs formed as a giant avalanche tore away part of the shield. That is certainly the case along the Nā Pali Coast, which is just a continuation of the steep, cliffy shoreline you see here farther northeast. Wave activity and stream erosion have modified this old scarp during the millions of years since.

HI 550 (WAIMEA CANYON DRIVE)
WAIMEA—PU'U O KILA OVERLOOK

25.3 miles

HI 550 begins at an intersection near the western side of the town of Waimea, just west of milepost 23 on HI 50. The road climbs steeply up an ancient sea cliff bordering the coastal plain before crossing onto a gentler slope, a remnant of the original flank of Wai'ale'ale Volcano. A large pulloff about 0.6 mile past milepost 3 provides a good view in clear weather, north up Waimea Canyon. The many lava flows seen in the canyon walls at this point are the Nāpali Member of the Waimea Canyon Basalt. They dip parallel to the flank of the old volcano.

Between mileposts 4 and 5, begin looking for exposures of red lateritic soil, the weathering product of Nāpali basalt, in beds that thicken upslope. You may spot decayed pieces of the basalt here and there embedded in the soil. At milepost 6, corestones of lava in the laterite appear on the left side of the road, the products of extreme spheroidal weathering. The highway intersects Kōke'e Road (HI 520) at mile 6.3, which is described in the next section. You can take this route as an optional return to HI 50. For now, continue directly ahead.

Lateritic soil along the Waimea Canyon Road, viewed as you head upslope toward the Olokele Plateau.

Waimea Canyon Overlook

One of the most spectacular viewpoints in Hawai'i is Waimea Canyon Overlook. Its parking area is at mile 10.5. The overlook lies on the rim of the westernmost branch of Waimea Canyon, a 14-mile-long incision into the side of Wai'ale'ale Volcano, in places as much as a half-mile deep. The canyon walls are intensely multicolored, eroded into numerous pinnacles and spires, laced with waterfalls, and often shrouded in mists that come and go. Enjoy the view!

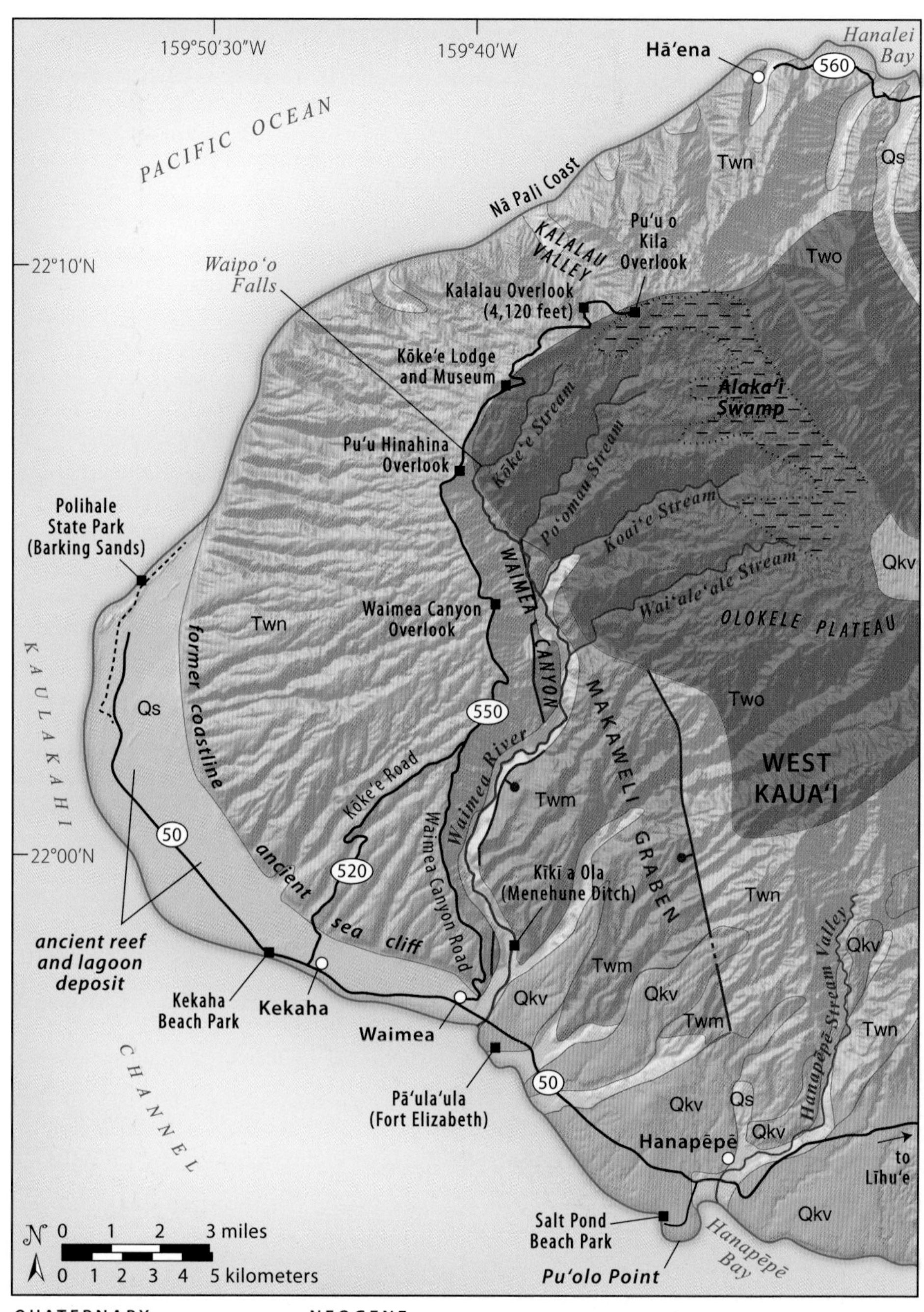

Geologic map of western Kaua'i.

Big stream canyons ordinarily branch from smaller valleys upstream, with tributaries entering from both sides. Not so in Waimea Canyon. Here the tributary valleys enter only from the east. Half of Waimea Canyon's tributaries seem to be missing. A topographic map of Kaua'i helps solve the mystery: Waimea Canyon cuts across the regional stream pattern. The strange tributary system developed because rain falling east of the canyon drains toward it, whereas almost all rain falling to the west drains away, down the western side of the island. Waimea Canyon developed this orientation because it follows the western edge of Makaweli graben, the trough that dropped between faults and then was filled with lava when Wai'ale'ale was still an active shield volcano. The Waimea River originally flowed to the sea along the graben. Lava from later eruptions displaced the river a few miles to the west, where it carved the present canyon. The ancient fault scarps, perhaps related to a mega-landslide in the eastern flank of Wai'ale'ale during its volcanic prime, guided the unusual erosive pattern shaping the modern landscape.

The upper end of Waimea Canyon, left from the overlook, roughly follows the southern edge of the Olokele Plateau. The 800-foot-tall Waipo'o Falls drops over ledges at the edge of the plateau. The high, level horizon seen across Waimea Canyon preserves a bit of the original lay of the land as it was before streams dissected Wai'ale'ale.

From Waimea Canyon Overlook you can see three prominent volcanic formations. Most of the western wall of Waimea Canyon, including the rock beneath where you stand, is composed of lava flows and hardened rubble of Nāpali Member basalt. The lower wall of Waimea Canyon, directly across from the overlook, is eroded flows in the Makaweli Member, which flowed down the floor of the graben after it formed. And the upper cliffs with numerous waterfalls across the canyon are eroded in Olokele lavas.

Looking north into Waimea Canyon. Flows of older Wai'ale'ale shield basalt dip seaward to the left, whereas younger Olokele lavas form the flat-lying flows making up the Olokele Plateau on the horizon.

Looking down Waimea Canyon from Pu'u Hinahina. Numerous nearly vertical dikes streak the distant canyon wall to the right and cross the canyon floor below.

Pu'u Hinahina Overlook

Pu'u Hinahina Overlook provides another sweeping viewpoint along Waimea Canyon Drive. Two short paths here lead to vistas. Given clear conditions, the Ni'ihau path allows you to see Ni'ihau Island, 30 miles distant across the ocean to the west. The path to the left leads to Pu'u Hinahina Overlook itself, with a long look down Waimea Canyon. To the right of the overlook, Nāpali basalt forms the towering western wall, which roughly follows the edge of the Makaweli graben. Imposing cliffs of the eastern wall in the upper reaches of the canyon are composed of Olokele lavas, which form the hard cap of the Olokele Plateau. Look for numerous dikes in the landscape below. A prominent one slices across the landscape next to the stream directly beneath the overlook. Many others, oriented radially to the ancient summit of Wai'ale'ale, track up the lower western canyon wall a little farther away. The upper canyon layers may represent eruptions of Nāpali basalt fed from the plumbing provided by these same dikes.

The Olokele Plateau

North of Pu'u Hinahina Overlook, HI 550 climbs through upland rain forest with mixed native and imported trees—gum, fir, and coastal California redwood—onto the Olokele Plateau. It passes a government geophysical laboratory used to track orbiting satellites and spacecraft, several campgrounds, and Kōke'e Lodge, before finally ending at two overlooks, Kalalau and Pu'u o Kila, at the northwestern edge of the plateau. From these overlooks you can peer down from the spectacular rim of Kalalau Valley to the Nā Pali Coast, 4,000 feet below. Cliffs flanking the valley

The heavily weathered, wet, forested tableland of the Olokele Plateau.

Kalalau Valley viewed from Kalalau Overlook. Note the steeply furrowed walls so characteristic of tropical weathering and erosion.

reveal thin flows of Nāpali basalt that poured down the flank of Wai'ale'ale when it was still a growing shield volcano. In a few places you can see narrow dikes slicing across the great stack of flows, part of the old volcano's internal magma plumbing system.

The end of the road at Pu'u o Kila Overlook is the trailhead for excursions into the Alaka'i Swamp, the soggy tableland just west of the Wai'ale'ale summit. Receiving as much as 500 inches of rain a year, this dwarf cloud forest is one of the wettest places on Earth. Consider yourself lucky if the sun shines on you here.

NI'IHAU ISLAND VIEWED FROM KŌKE'E ROAD

From Kōke'e State Park and upper Waimea Canyon you can return to the coast via Kōke'e Road (HI 552), which directly links HI 550 with the town of Kekaha. This winding, paved route descends seaward-sloping Nā Pali lavas and switchbacks across a steep pali, an ancient sea cliff now landlocked at the foot of the slope. It levels off along the mile-wide coastal plain next to Kekaha. The route provides sweeping vistas of the neighboring island of Ni'ihau 17 miles across Kaulakahi Channel. Ni'ihau has a permanent population of only 130, all indigenous Hawaiians. Ancestors of the present owners of the island purchased it from the king of Hawai'i in 1864 for $10,000. Ni'ihau's 1,000-foot-tall axial ridge, a titanic landslide scarp, rises like a wall out of the ocean. To the right just off the island's northern point is the top of largely submerged Lehua tuff cone, a Kōloa-age vent and seabird sanctuary. If the air is clear,

Six-million-year-old Ni'ihau Island, with smaller Lehua tuff cone (arrow) rising across Kaulakahi Channel southwest of the town of Waimea. Lehua is either the last-gasp product of now dead Ni'ihau Volcano or an independent rejuvenated volcano of unknown age.

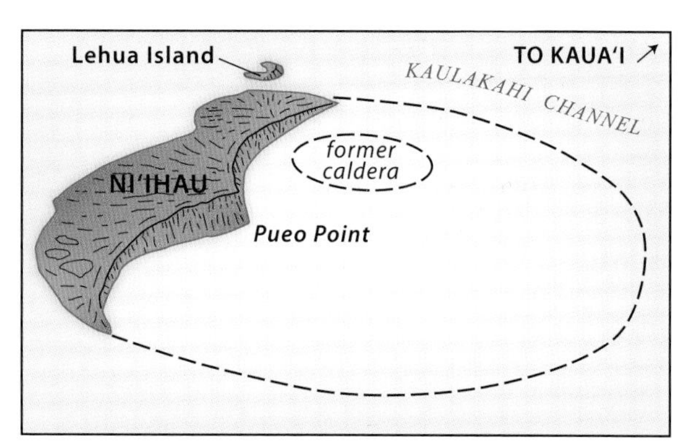

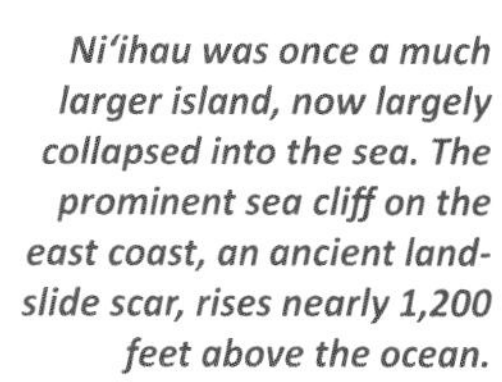
Ni'ihau was once a much larger island, now largely collapsed into the sea. The prominent sea cliff on the east coast, an ancient landslide scar, rises nearly 1,200 feet above the ocean.

you may also see a rocky mass farther out to sea to the left of Ni'ihau, almost 60 miles away. This 550-foot-tall desolate pinnacle is Ka'ula, the most remote of the main Hawaiian Islands. It lies at the summit of an almost entirely submerged shield volcano. Ka'ula was one of the first of the Hawaiian Islands sighted by Captain Cook. As you approach the lowland, Ka'ula disappears over the horizon owing to curvature of the Earth.

The descent on Kōke'e Road reveals remarkable changes in the local landscape. Red lateritic soils many feet thick from deeply weathered Nāpali flows are widespread along the upper part of the road. Little solid rock exists in this worn terrain. These soils become much less apparent as you near the coast. Dark flows, only slightly weathered, crop out in gully walls instead. The vegetative cover is much less dense than upslope. This transition reflects slower weathering rates under drier conditions in the low country.

O'AHU

O'ahu, the political and economic heart of Hawai'i, packs a large population into an area only 600 square miles—44 miles long and 30 miles wide at the maximum. Two rugged mountain ranges, the Wai'anae and Ko'olau, form the western and eastern regions of the island, respectively. Between these deeply eroded shield volcanoes lies a broad alluvial plain, the Leilehua Plateau, which stretches south to meet the sea at Pearl Harbor and north to the famous surfing beaches of the North Shore. Because the Ko'olau Range is younger, lava flows and related sediments from the east lap up practically to the foot of the Wai'anae Mountains in the west. Combined runoff from both ranges feed Waikele Stream, which enters the West Loch of Pearl Harbor. Early Hawaiians gathered at Pearl Harbor and the neighboring coastal plains for important religious festivals and other events—hence O'ahu's nickname, the Gathering Place.

Reddish soils exposed in many places on the Leilehua Plateau are laterites: extremely leached and oxidized earth that takes hundreds of thousands of years to form. Laterites are aluminum-rich soils that lack many nutrients; they have mostly dissolved, making them too infertile for many types of agriculture. Nevertheless, sugarcane and pineapple plantations flourished here until global economic conditions forced their abandonment. Since then, the plateau has become increasingly suburban and crowded with traffic.

WAI'ANAE VOLCANO

O'ahu began as a small, vigorously active volcanic islet that first rose above sea level nearly 4 million years ago. During the next million years, the young island grew rapidly as lava flows built the huge shield of Wai'anae Volcano. Wai'anae Range is the eroded remnant, which includes Mount Ka'ala, the highest peak on O'ahu at 4,025 feet. The conspicuous flat summit of Ka'ala is the high point on the skyline west of Honolulu. This mountaintop tableland, a nearly inaccessible cloud bog, is one of the few remaining original surfaces of Wai'anae, the oldest terrain on the island.

Like other young Hawaiian volcanoes, the Wai'anae shield had an enormous crater at the summit, Lualualei caldera. Two narrow, active rift zones extended northwest and southeast from the caldera. Swarms of dikes across them make up the resistant backbone of the Wai'anae Range. A dispersed zone of radiating dikes also shows evidence of flank eruptions northeast of the caldera.

The overwhelming bulk of Wai'anae Volcano is composed of thin flows of pāhoehoe and 'a'ā lava, all tholeiitic basalt. The flows dip outward from the caldera at angles ranging from less than 5 degrees along the rift zones to almost 15 degrees in the other flanks. Geologists call them the Lualualei Member of the Wai'anae Volcanics.

As Wai'anae Volcano neared completion of its massive shield, eruptions became less frequent in the rift zones and were mainly restricted to the summit. At first, thick tholeiitic basalt flows ponded in the caldera. Later, the flows became more alkalic and viscous. Geologists designate these younger, caldera-related lavas the Kamaile'unu Member of the Wai'anae Volcanics.

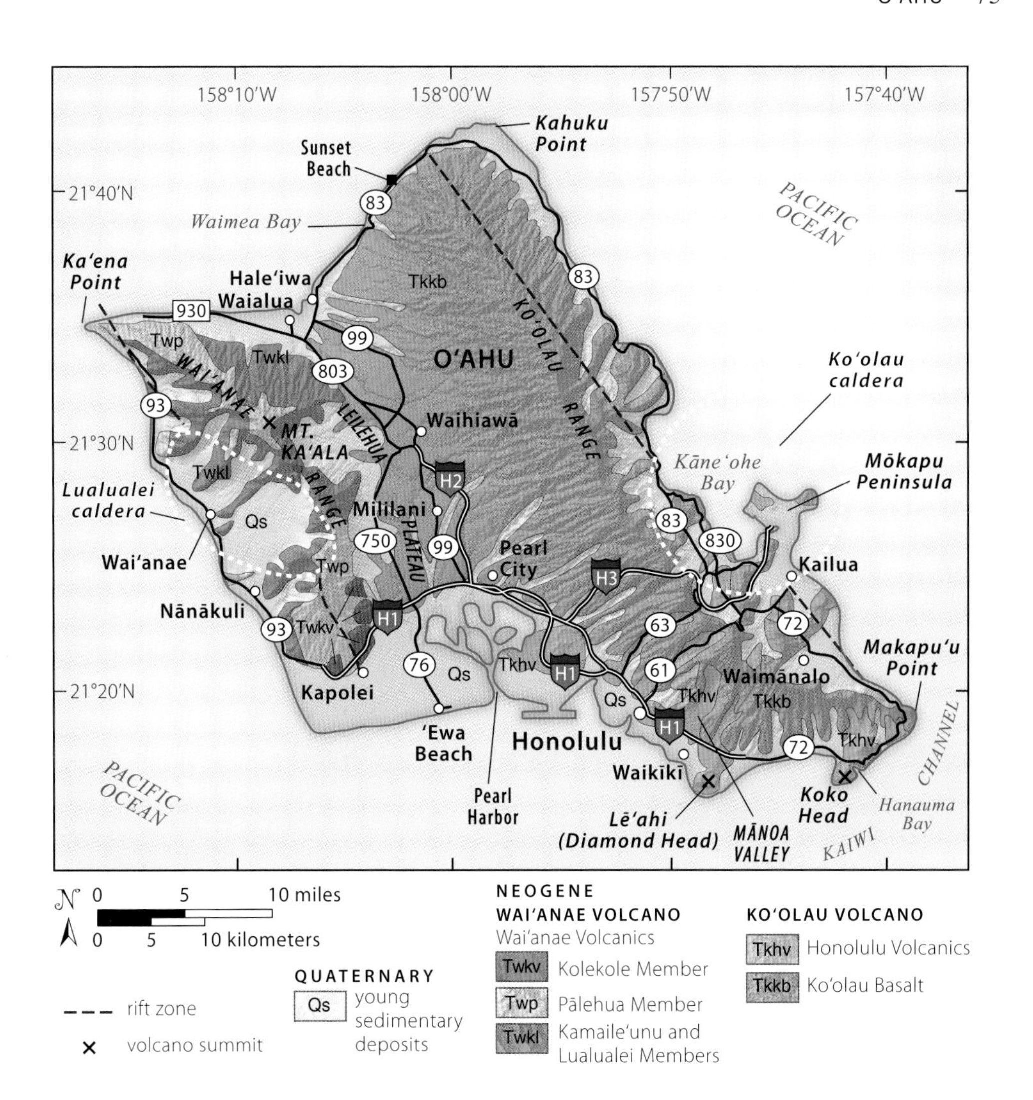

Geologic map of the Island of O'ahu.

By the time Lualualei caldera was filled, Wai'anae was erupting mainly alkalic basalt flows, although some vents produced cinder, ash, and other pyroclastic material. Throughout the period of caldera filling, the outer flanks of Wai'anae eroded, so the alkalic lava flows buried a landscape laced with young stream channels and small valleys. In the end, the cap of alkalic lavas and ash, a thin veneer called the Pālehua Member, covered most of Wai'anae Volcano.

While Wai'anae was still growing, its western submarine slope began slowly slumping into the Hawaiian Deep, much as the southern flank of Kīlauea on the Big Island is slumping into the Pacific today. Later, most of the shield volcano's western flank suddenly detached and slid violently, spreading chunks of volcano underwater as far as 50 miles from the island.

The north flank of Wai'anae was also unstable. The Ka'ena debris avalanche moved nearly 70 miles across the ocean bottom and formed an escarpment parallel to the North Shore between Ka'ena Point and Waialua. It is difficult to imagine how slide debris could move so far across the seafloor unless it was moving extremely fast. This avalanche doubtlessly generated an enormous tsunami, too ancient for its debris to be preserved on other Hawaiian coastlines.

Percolating gases, steam, and hot water altered the lava flows inside Lualualei caldera. The weakened rock weathered and eroded rapidly. A great stack of caldera-filling lava flows thousands of feet thick eventually turned into a low, broad flatland, Lualualei Valley, which extends inland from the town of Wai'anae. Erosion also stripped most of the alkalic lavas off the northern flanks of the volcano and exposed dike swarms deep within the rift zones. Thick deposits of stream gravel, conglomerate, and other alluvium accumulated in the growing valleys and lowlands.

A last gasp of volcanism at the southern end of the Wai'anae Range built six small cinder and spatter cones, which poured flows of alkalic basalt down the southeastern flank of the aging shield, interrupting the erosional dissection of the volcano. Radiometric age dates show that these basaltic rocks, the Kolekole Volcanics, erupted between 2.9 and 2.75 million years ago.

Strictly speaking, Kolekole volcanic activity was not rejuvenated volcanism because little time intervened between the eruptions of Pālehua and Kolekole lavas. Both groups of basalt probably erupted from the same magma source, but a brief hiatus separated the two phases of activity, at least locally, perhaps due to the huge Wai'anae avalanche. If so, then Wai'anae never had a spell of rejuvenated activity, unlike most other Hawaiian volcanoes.

KO'OLAU VOLCANO AND THE GREAT NU'UANU SLIDE

Ko'olau Volcano is younger and larger than Wai'anae, and covers about two-thirds of O'ahu. It grew above sea level by 3 million years ago, about when Wai'anae Volcano was becoming extinct. The first stage of growth for Ko'olau was the typical rapid eruption of shield-building flows, mostly pāhoehoe. A caldera nearly 8 miles long and at least 4 miles wide developed in what is today the Kailua-Kāne'ohe area. Two vigorously active rift zones trended northwest and southeast of the new caldera, roughly parallel to the rift zones on neighboring Wai'anae.

Ko'olau and Wai'anae Volcanoes probably started out as separate islands. The broad saddle of the Leilehua Plateau that connects them today consists mainly of alluvial deposits shed from the Wai'anae Range and lava flows that came from longer-lived Ko'olau Volcano. As volcanism waned, Ko'olau watersheds began shedding large amounts of additional alluvium onto the saddle.

The Leilehua Plateau and Wai'anae Range braced the western flank of the Ko'olau shield, whereas the volcano's unsupported eastern flank sloped steeply into the deep ocean. During the stage of most rapid volcanic growth, the eastern flank swelled whenever magma rose into the volcano, then deflated as it erupted. It was simply a matter of time before the strained mountainside would collapse, causing the giant Nu'uanu debris avalanche.

The Nu'uanu avalanche, one of the largest known landslides on Earth, occurred mostly underwater. It sped across the seafloor, carrying chunks of Ko'olau Volcano as far as 120 miles to the northeast. The slide laid down a bed of rubble 20 miles wide. One block is 18 miles long and 1 mile thick, as large as many of the ancient seamounts on the surrounding ocean floor. The catastrophic collapse must have had tremendous momentum to run all the way across the Hawaiian Deep and 90 miles up the gentle slope of the Hawaiian Arch. Like the earlier Wai'anae avalanches, it certainly caused a horrendous tsunami that likely had significant impact all around the Pacific Rim. Mathematical models of possible wave heights run as high as 650 feet for local waters. While a spooky thought, it is good to remind yourself of how very rare and unlikely such events are.

You can see one legacy of the Nu'uanu debris avalanche in the way the eastern shore of O'ahu almost follows the Ko'olau rift zones and crosses the eastern fringe of its caldera. Had the Nu'uanu avalanche not detached a large piece of the volcano, the eastern shore of O'ahu would lie several miles seaward of its present position and also have a seaward-curving map outline.

About 1.8 million years ago, probably not long after the Nu'uanu avalanche occurred, Ko'olau entered its last gasp of activity. It erupted almost no alkalic basalts after shield-building volcanism ended. During the next 800,000 years, the volcano eroded and sank thousands of feet as it drifted away from the Hawaiian hot spot.

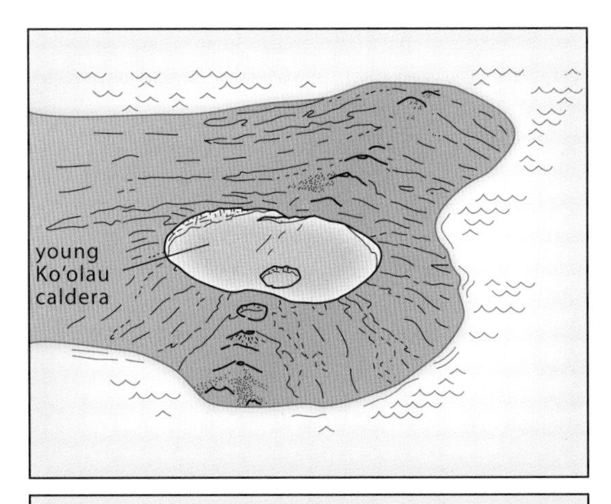

Collapse and erosion of Ko'olau Volcano. While once located near the very summit of the volcano, Ko'olau caldera is now a deeply eroded, partly submerged, and heavily populated lowland. The mountainside to the east of the caldera vanished into the ocean in a gigantic submarine landslide, scattering pieces of the broken volcano across the seabed as far as 120 miles to the northeast.

Erosion and simultaneous sedimentation created broad valleys out of what remained of the caldera floor, exposing countless dikes in the rift zones.

A vigorous episode of rejuvenated volcanism then interrupted the steady decline of Ko'olau Volcano and produced the Honolulu Volcanics. The first eruptions began 1,000,000 to 850,000 years ago on Mōkapu Peninsula, at the northern margin of Ko'olau caldera. Since then, the southern part of the range has had about forty more eruptions. The largest group of Honolulu volcanoes trends south from the Mōkapu Peninsula across the range crest to the coastal plain at Honolulu. It includes Punchbowl Crater, Pu'u Kākea (Sugarloaf), Pu'u 'Ōhi'a (Tantalus Peak), Pu'u 'Ualaka'a (Round Top), and Diamond Head (Lē'ahi). A lesser cluster of vents at Salt Lake Crater, Makalapa, and Āliamanu cones rises from the coastal plain between Honolulu and Pearl Harbor. The youngest set of pyroclastic cones erupted perhaps as recently as 40,000 years ago in the Koko Rift, a chain that stretches 5 miles between Kawaihoa (Portlock) and Makapu'u Points at the southeastern end of the island.

The Honolulu Volcanics feature some exotic varieties of alkalic basalt unusually rich in sodium, such as nephelinite and basanite. Some lava flows contain xenoliths of lherzolite and pyroxenite, mantle rocks carried to the surface as fragments. They provide geologists with a rare and important view of Earth's interior beneath O'ahu.

In *Volcanoes in the Sea*, Gordon Macdonald, Agatin Abbott, and Frank Peterson discuss the rejuvenated volcanism of Ko'olau: "The intervals of time between successive eruptions seem to have been as long as, or even longer than, the length of time from the last eruption to the present" (pp. 93–94). If they are correct, perhaps we can expect future rejuvenated eruptions on the island.

O'AHU SHORES

The geologic history of O'ahu is more than the rise and decline of its two volcanoes. Beautiful coral reefs and coral sand lagoons in the warm tropical water have also played a role. O'ahu is no longer sinking into the ocean, perhaps because the mantle below has finally adjusted to the load. Or, as some geologists believe, the ocean floor may be flexing to compensate for the weight of the younger islands growing to the southeast, thus offsetting the island's sinking and even causing minor uplift. Whatever the reason, O'ahu has the most stable coasts in the Hawaiian Islands.

Even though O'ahu is not sinking, the shorelines show evidence of recent fluctuations in sea level such as drowned stream valleys, dead reefs, lagoons, marine terraces elevated above sea level, ancient beach rubble, and beach rock. Most of that evidence is related to the coming and going of ice ages. Sea level drops as much as several hundred feet when glaciers grow on the continents, then rises as all that water drains back into the oceans when they melt. Coral deposits now exposed inland at different elevations represent different times of higher sea level relative to modern shorelines. Radiometric (carbon-14) dating of these deposits show this pattern.

Unlike beaches on younger Hawaiian Islands, O'ahu beaches are entirely composed of fine, calcareous sands derived from offshore reefs. Reefs fringe a greater proportion of total shoreline around O'ahu than any other major Hawaiian island, though Kaua'i also has a substantial network of fringing reefs.

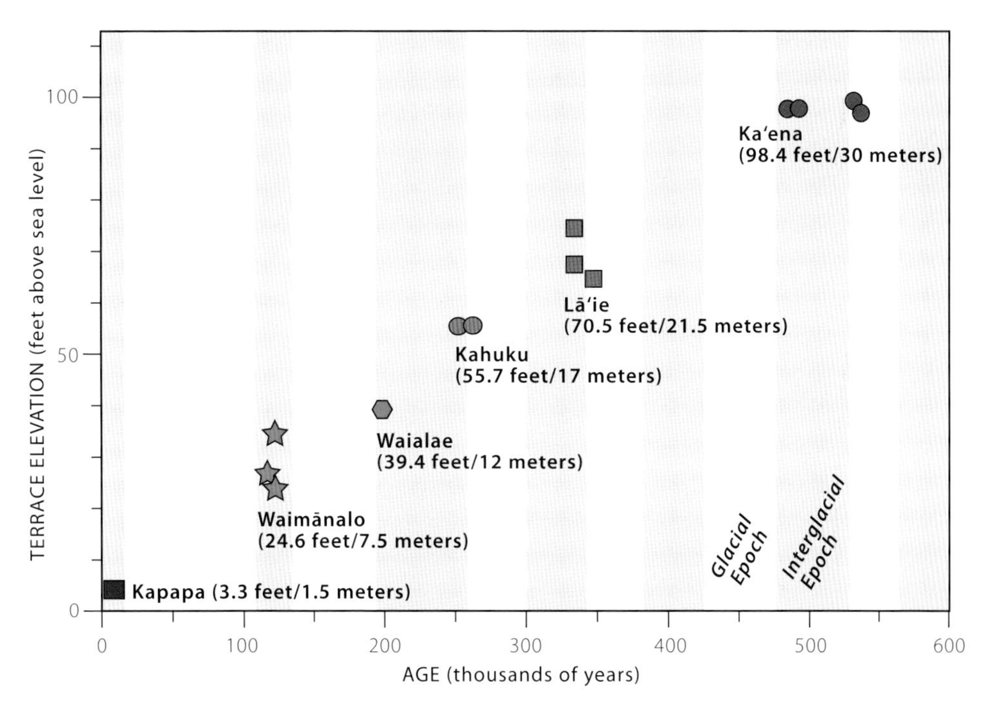

Ages and elevations of ancient marine terraces along the coasts of O'ahu. Wave action at the coast initially created each terrace as a marine bench, generally during a period of high-standing sea level between ice ages ("interglacial times"). Tectonic uplift then elevated each terrace, with this effect greatest for the oldest terraces. —Modified from McMurtry and others, 2010

GEOLOGY AROUND HONOLULU

Honolulu sprawls across a wide reef platform, built by coral and other marine organisms during the previous (Eemian) interglacial high stand in sea level, 125,000 to 118,000 years ago. Sediments eroded from interior valleys and pyroclastic deposits from younger rejuvenated cones near the coast have since largely buried this platform. Caverns riddle the limestone beneath the city, but they are largely flooded with shallow groundwater and filled with sands and muds. Channelized streams and canals divert and control drainage, initially to irrigate farmland on the Honolulu plain a century ago, but now primarily to prevent flooding in areas packed with housing and skyscrapers.

The deep valley floors cutting the southeast flank of Ko'olau Volcano have also largely filled with sediment, young lava flows, and other volcanic debris, creating many broad floors. An example is Mānoa Valley, site of the University of Hawai'i campus east of downtown. Mānoa Valley would be very narrow and as much as 300 feet deeper without this inpouring of eroded and erupted material. Favorable topography has allowed much of the city's residential areas to spread far up the larger valleys. Residential development also has climbed high up adjacent ridgelines where their crests are not too sharp and steep.

Water supply for the city primarily comes from wells drilled *sideways* into the old body of Ko'olau Volcano, which contains abundant groundwater trapped between

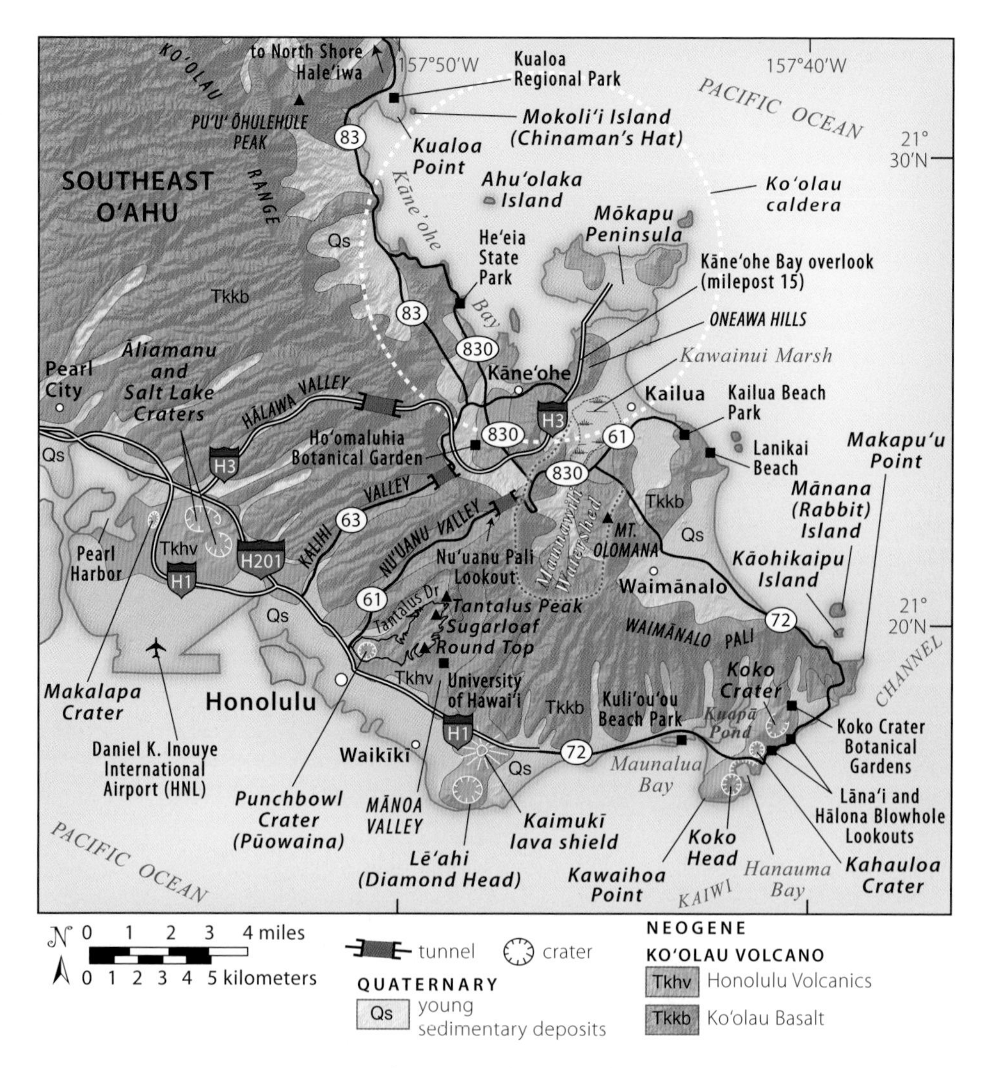

Points of interest in the Honolulu area.

impermeable dikes. Much of this water simply drains to consumers downslope, a very efficient means of water delivery.

Given the complex street grid of Honolulu (many roadsides, not much exposed geology), we simply present points of interest here that can be approached from many directions. Refer to a good street map or online resources to find your way.

Round Top Overlook

Ten-mile-long Tantalus Drive loops across the eroded flank of the Ko'olau Range above central Honolulu. Driving upslope from Punahou School, alma mater of former President Barack Obama, the drive ascends the flank of Round Top (Pu'u 'Ualaka'a), a Honolulu Volcanics cone. The first highway overlook here offers a sweeping view of Mānoa Valley, Waikīkī, and Diamond Head tuff cone. Diamond

Head is the easternmost of a chain of tuff cones, tens of thousands of years old, scattered across the Honolulu coastal plain. Between Lēʻahi and the foot of the Koʻolau Range is the Kaimukī District, a gently sloping rise that is a lava shield, another vent of uncertain age of the Honolulu Volcanics. The main campus of the University of Hawaiʻi occupies the mouth of Mānoa Valley directly below the overlook.

A roadcut across the street from the overlook reveals well-sorted, gravelly deposits of lava fragments ejected during the eruption of Round Top. This material probably

Layers of volcanic ash and gravel (lapilli) on the flank of Round Top, at the overlook above Mānoa Valley. Differences in grain size represent variable strengths of explosions, with finer ash falling during weaker explosions, or in the intervals between frequent blasts. Camera lens cap for scale (right center view).

View from Round Top Overlook along Tantalus Drive, looking southeast across the University of Hawaiʻi Mānoa campus to Kaimukī lava shield (arrow) and Diamond Head tuff cone (brown ridge at upper right). The lava shield, tuff cone, and the flat floor of the valley beneath the university campus are all deposits of Honolulu Volcanics, the youngest stage of volcanism on Oʻahu.

fell directly out of the blast clouds rising from the active vent as Round Top grew. It is crudely layered, showing successive episodes of stony pyroclastic fallout.

Lē'ahi (Diamond Head)

Diamond Head is certifiably free of diamonds. It acquired its name from calcite crystals, which nineteenth-century British sailors mistook for diamonds, creating a brief sensation. Hot water and steam circulating through the ash beds in the crater rim

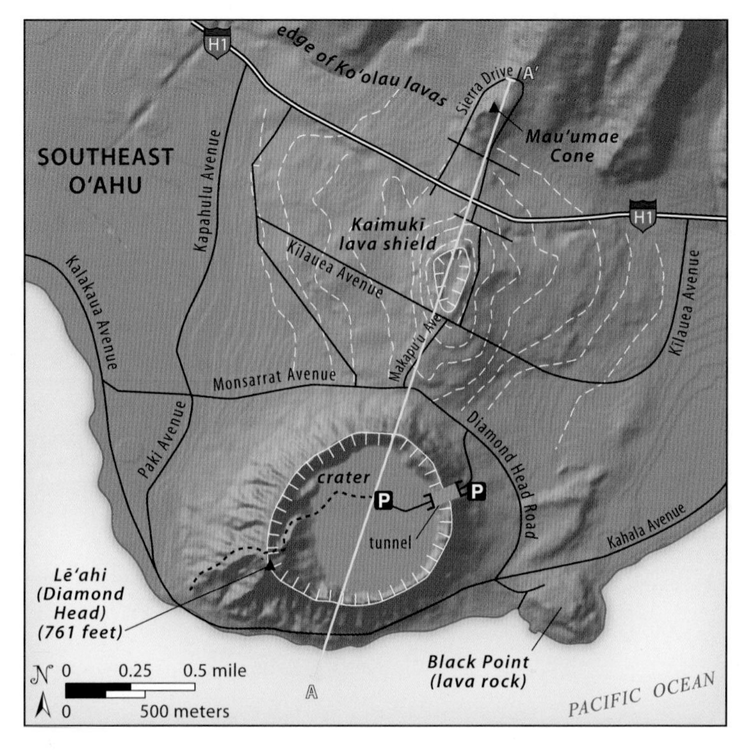

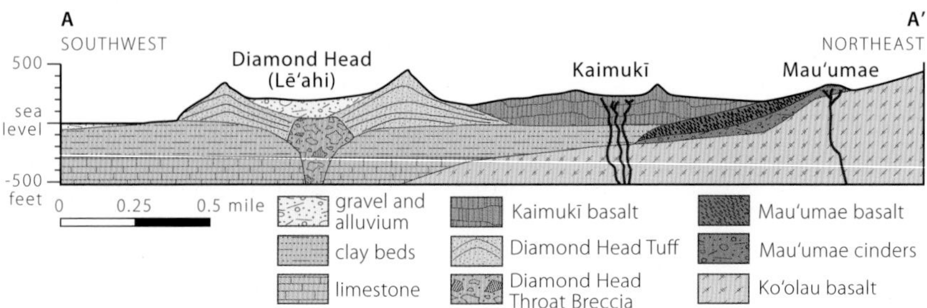

Map of Honolulu Volcanics vents in the eastern part of Honolulu, including Diamond Head, arguably the world's most famous tuff cone. The white dashed lines give contours at 40-foot intervals above sea level, showing the gentle slopes of Kaimukī shield, now a densely packed residential neighborhood. Line A–A' corresponds to the south-to-north cross section through three vents. Mau'umae is the oldest, Lē'ahi (Diamond Head) next oldest, and Kaimukī lava shield the youngest. Underlying limestone and clay beds illustrate that O'ahu had been sinking and eroding long before this rejuvenated volcanism took place. —Cross section based on section by Wentworth, 1951

dissolved the calcite from fragments of reef limestone incorporated in the volcanic debris, then precipitated it to form the crystals.

Diamond Head reaches 760 feet above sea level. It is a classic tuff cone that grew as rising magma encountered water at shallow depth. The series of steam explosions that followed ejected volcanic ash and cinders, large blocks of older volcanic rocks, and chunks of reef limestone. The volcanic debris reacted with hot water to make palagonite clay, which colors much of the volcanic cone in shades of reddish brown. You can reach a street that leads inside the crater from the northeast side of Monsarrat Avenue–Diamond Head Road. From a parking area inside the crater, a trail leads to the Amelia Earhart Monument high on the western flank of Diamond Head. At 560 feet above sea level, it provides excellent views of this famous volcanic vent that erupted so briefly. Take a flashlight for the long, dark tunnel on the trail.

Punchbowl Crater

Punchbowl Crater (Pūowaina; the Hill of Sacrifice) is arguably the most perfectly formed tuff cone in Hawai'i, though its deposits are covered with soil, vegetation, and houses. Unlike other tuff cones in the area, Punchbowl did not erupt at the shoreline or low coastal ground, but upslope instead. The water table nonetheless must have been shallow in this area for powerful steam-blast explosions to take place. The National Memorial Cemetery of the Pacific occupies its crater. The lookout within the cemetery is perched on the southern rim, looking out across the coastal plain of ancient reef and lagoonal deposits now crowded with high-rises. Punchbowl has been dated to about 400,000 years old.

Salt Lake Crater and Related Tuff Cones

A set of low, rejuvenated-stage steam-driven eruptive vents cluster at the western edge of Honolulu, a short distance inland from the international airport, including Makalapa, Āliamanu, and Salt Lake Craters. Salt Lake Crater (Āliapa'okai) is a half-mile-wide basin with low rims of tuff, occupied today by ponds, waterways, and a golf course. Salt Lake District Park borders the crater to the northwest, next to the older and larger Āliamanu Crater. Geologists do not know when the Salt Lake vent erupted. Lava suitable for dating is missing, and the tuff is altered and weathered. Geologist Harold Stearns attempted to correlate the Salt Lake tuff with fossils from drill holes in Pearl Harbor. He concluded that the tuff cones in this part of Honolulu probably formed during the latest ice age, but before about 30,000 years ago. Recent potassium-argon dating of Makalapa and Āliamanu produced ages of 470,000 and 250,000 years ago, respectively. This discrepancy leaves a wide spectrum of possibilities for future workers to resolve.

Āliamanu is a set of overlapping explosion craters nearly 1 mile in diameter. It is the most conspicuous of the eruptive centers in the Salt Lake group, with rims rising hundreds of feet above the surrounding lowland. A military reservation restricts visitor access, but a drive to the northern end of Ala Liliko'i Street takes you to Salt Lake District Park. Look for layers of tuff exposed at the foot of the slope at the far end of the playing field.

Numerous blocks are embedded in the Āliamanu tuff, including some widely scattered bright-green peridotite xenoliths. Some of these contain red garnets, rare for Hawai'i. These minerals can only have formed at depths of at least 30 miles. Please

Peridotite mantle xenolith embedded in Āliamanu tuff at Salt Lake Crater Park. The overall greenish coloration comes from an abundance of olivine crystals in the xenolith, though these are much weathered. Less-abundant, darker-green crystals are pyroxenes. Occasionally tiny red garnets may also be found in these xenoliths.

do not collect samples; they are on reservation property and should be preserved for others to see.

Like Salt Lake Crater, Makalapa is now a marshy basin with a low western rim hardly noticeable driving by on the H-1 freeway. Pearl Harbor Memorial visitor center is nearby.

Pearl Harbor

West of Salt Lake Crater, the H-1 freeway skirts the northern shores of Pearl Harbor. This famous anchorage is the composite estuary of several streams that joined at the harbor. When sea level was low during the most recent ice age, old reefs, lagoons, and other sedimentary deposits were exposed in a broad plain through which streams cut new valleys. As sea level rose at the end of the last ice age, seawater flooded much of the plain and the lower areas in the stream valleys and formed Pearl Harbor. Parts of the old divides between streams still stand above sea level, making up the peninsulas

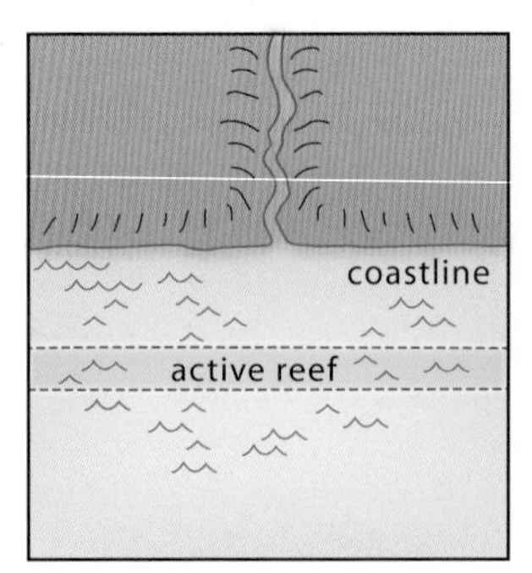

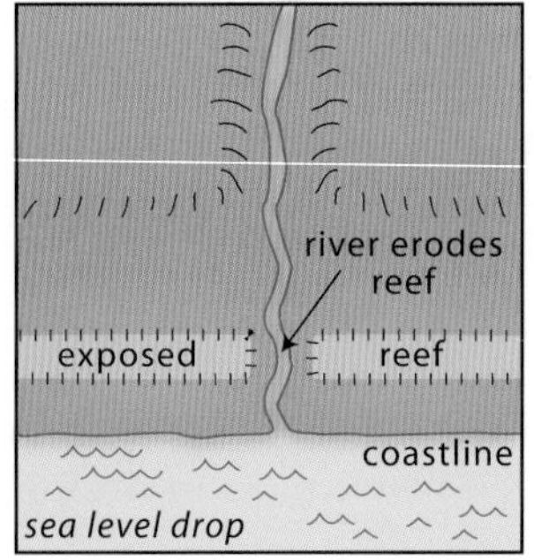

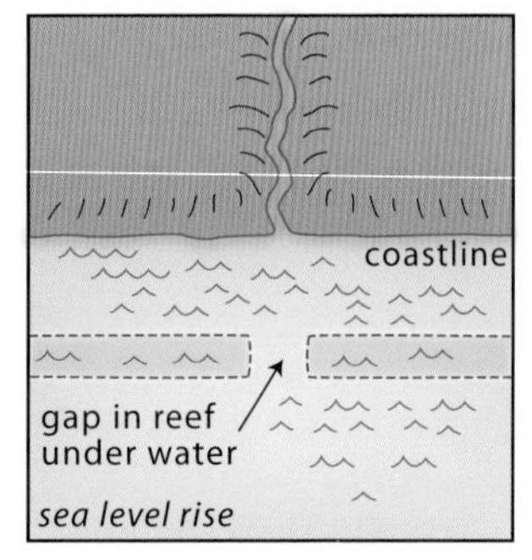

Navigational passages through coral reefs are critical for island harbors. These often form where streams have cut channels through exposed reefs during times of lower sea level. When sea level rises once more, the channels flood, providing safe passageways for modern ships into sheltered waters.

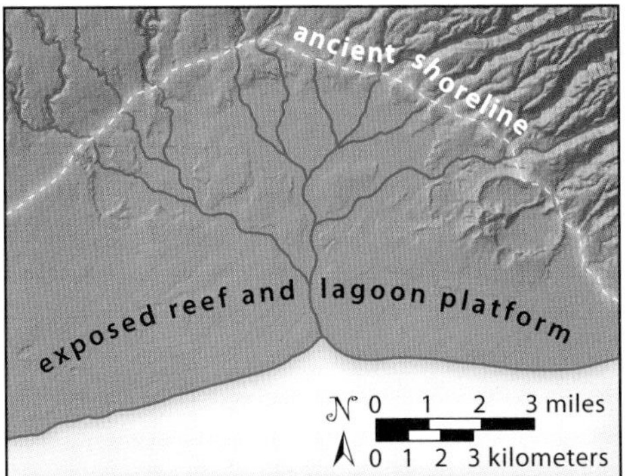

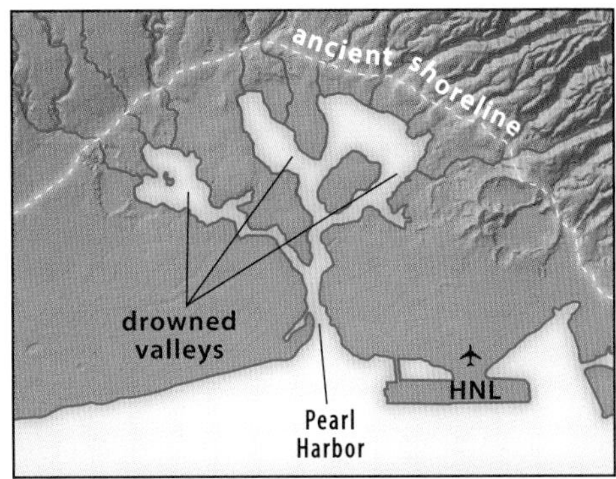

Pearl Harbor is the product of changing sea levels. A drop in sea level during the last ice age (115,000 to 11,000 years ago) exposed the ancient Eemian seabed throughout the Pearl Harbor area. Streams draining the Leilehua Plateau and Ko'olau Range cut channels across this bed, which later flooded to form Pearl Harbor's lochs and islands when sea level rose again at the end of the ice age.

and islets that separate the branches of the harbor. The narrow entrance to Pearl Harbor is the flooded mouth of the main stream, which ran to the coast when sea level was low, cutting through the fringing reef to erode the passage used by ships today.

Three Highways across the Ko'olau Range

The H-3 highway, HI 61, and HI 63 all cross the Ko'olau Range, linking Pearl Harbor and the busy heart of Honolulu to the heavily populated area around Kāne'ohe Bay and neighboring Kailua: O'ahu's Windward Coast. Each highway ascends a separate valley as it heads northeast toward the steep, narrow range crest.

The valleys have wide, gently sloping floors most of the way. Over eons, sediment has poured into them from surrounding eroding mountainsides faster than their streams can transport it away. In places, vents of the rejuvenated stage have also erupted pyroclastic material and lava flows that have made valley floors much shallower. Where the highways meet the range crest, tunnels have been constructed to avoid steep headwalls and pali slopes.

Let's consider each of these three routes in turn, beginning with the westernmost.

H-3
Pearl Harbor—Mōkapu
15 miles

The H-3 highway leads north from Pearl Harbor, following the level floor of Hālawa Valley. Unlike its neighbors to the east, the valley has not been impacted by geologically recent Honolulu volcanism. The H-3 Tetsuo Harano Tunnel through the Ko'olau Range crest to Ha'ikū Valley on the east side is nearly 1 mile long and emerges at the base of steep cliffs. This pali, the former edge of the volcano's caldera, encircles the depression that hosts Kāne'ohe Bay.

Tucked at the foot of the pali and easily accessible from H-3 is Ho'omaluhia Botanical Garden. This county park has numerous overlooks and short trails that provide close views of the fluted face of the pali that in some places towers 2,500 to 3,000 feet above the adjoining lowland. During heavy rainfalls, each of the giant drainage channels scouring the face of the great cliff has booming cascades and waterfalls. All that water can cause flooding down below, which is one reason the botanical garden exists—as a flood control measure. A lake in the lower part of the garden is a catchment basin helping to protect the town of Kāne'ohe downslope

The H-3 freeway follows the Oneawa Hills north to Mōkapu Peninsula. About a half mile short of reaching the peninsula (north of exit 14), on a west-facing slope, a lookout provides a beautiful view of Kāne'ohe Bay and the northern rim of the caldera basin. Lighter blue to buff patches of water are shallow coral reefs, with deeper blue marking channels and depths in between them. Low, flat islands offshore are the elevated terraces of dead reefs from past high stands in sea level. A swarm of northwest-trending dikes resisted erosion to create Pu'u'ōhulehule, a 2,265-foot-tall triangular mountain looming over Kāne'ohe Bay to the northwest. The dikes continue right across the floor of the bay into the Oneawa Hills, parallel to the shoreline. Soils and vegetation make the dikes hard to view locally, however. The offshore position of the dike swarm indicates that Ko'olau's former shield summit and magma chamber were also located in the area of Kāne'ohe Bay.

Dike in roadcut along H-3 freeway in the Oneawa Hills near Kāne'ohe Bay overlook. Changing vegetative cover, soil and rockfalls, and other natural processes cover or expose dikes frequently in this area. Keep your eyes peeled if you wish to see one.

HI 63 (LIKELIKE HIGHWAY)
MIDTOWN HONOLULU—KĀNE'OHE BAY
7 miles

HI 63, the Likelike Highway, named for a beloved Hawaiian princess, ascends Kalihi Valley eastbound, crossing the Koʻolau Range crest through 2,800-foot-long John H. Wilson Tunnel. Construction of the tunnel through weak, fractured rock was difficult, including a collapse that killed several workers and delayed opening of the highway by a few years. This is the shortest route linking downtown Honolulu with the Windward Coast, and though the geology is obscured by dense vegetation, some vistas of the cliffy, intricately furrowed landscape of the Koʻolau Range are outstanding. A small, rejuvenated-stage volcano, Kamanaiki, perches at a ridge crest flanking the upslope end of Kalihi Valley, not visible from the road. Active around 600,000 years ago, it erupted an alkalic basalt flow that paves the valley floor most of the way. Kamanaiki is one of the northernmost vents of the 7-mile-long Tantalus rift zone, a corridor of Honolulu Volcanics cones that includes Mt. Tantalus to the east.

As HI 63 approaches Kāneʻohe it skirts Hoʻomaluhia Botanical Garden, described in the H-3 road guide above.

HI 61 (PALI HIGHWAY)
HONOLULU—KAILUA
11 miles

HI 61, also called the Pali Highway, ascends Nuʻuanu Valley, which contains a nephelinite lava flow most of its length. Nephelinite is a rare variety of basalt so rich in sodium that it features crystals of nepheline instead of plagioclase feldspar. Makuku cinder cone, one of the vents that erupted the nephelinite, lies about 5 miles upslope from the junction of the Pali Highway with the H-1 freeway, though vegetation obscures it. It is left (west) of the highway where it passes Nuʻuanu Reservoir, near the southern end of Nuʻuanu Pali Drive. The grade of the floor of Nuʻuanu Valley roughly matches the original slope of the Koʻolau shield volcano, thanks in large part to the Makuku eruption.

Near the end of his campaign to conquer the Hawaiian Islands, Kamehameha the Great forced the army of a local chief up Nuʻuanu Valley along the present route of HI 61. When they reached the head of the valley, they were trapped between Kamehameha's warriors and the Pali, a formidable cliff. They were unable to rally, and when Kamehameha drove three hundred of them off the cliff, the Oʻahu forces surrendered.

The Nuʻuanu Pali Lookout, along HI 61 at the sharp crest of the Koʻolau Range, is a good place to view the pali. The Nuʻuanu Pali Lookout overlooks the entire amphitheater-shaped Koʻolau caldera. This huge, fluted cliff was left standing as less resistant rocks inside the caldera eroded. The low ridges in the caldera are stream divides that become much higher as they cross onto the more resistant rocks outside the caldera. Intense drainage from heavy rainfalls and high rates of tropical weathering account for the spectacular grooved appearance of the amphitheater walls. Not seen from here is the dense network of northwest-southeast-trending dikes that cross the caldera, linked to the volcano's two rift zones. Later rejuvenated volcanism built the cones that form Mōkapu Peninsula enclosing the eastern end of Kāne‘ohe Bay.

The head scarp of the titanic Nu‘uanu slide that took away the eastern half of the Ko‘olau Volcano now lies submerged several miles offshore. Sedimentation and fresh volcanic rocks from rejuvenated activity have partly buried it.

On the east side of the Pali Tunnels, HI 61 enters the Maunawili watershed. Maunawili Stream and its tributaries drain a 10-square-mile watershed immediately south of the pali-lined caldera of Ko‘olau Volcano. Maunawili Falls and its popular plunge pool may be reached by hiking from the Pali Highway or the community of Maunawili, although access suffers from abusive visitor traffic and is sometimes closed.

Floods racing out of Maunawili Valley through its narrow opening at Olomana pour into Kawainui Marsh downslope and in the past have threatened the beachside town of Kailua on the other side. Construction of a flood control levee and channelized drainage canals now protect the town. You can stroll along a well-maintained public path, the Marsh Trail, following the crest of the levee between Kailua Road and Kaha Street in Kailua, a birdwatcher's delight and a great place to look inland at the spectacular watershed and neighboring cliffs.

Kawainui Marsh used to be an extension of the sea, a larger version of Kailua Bay, before a sand spit that grew soon after the end of the last ice age isolated it from the ocean. Sediment deposited by flooding largely filled it to create today's swampy landscape. Kailua is built upon the spit and related calcareous dunes that postdate the ice age.

The US Army Corps' flood-control berm and drainage channel at the Kawainui Marsh in Kailua.

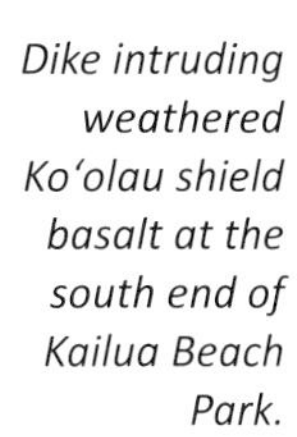

Dike intruding weathered Ko‘olau shield basalt at the south end of Kailua Beach Park.

Kailua Beach Park lies at the eastern corner of Kailua town, before reaching Lanikai. A wide beach with great swimming options, the geologically interesting part is the rocky eastern end, where you can inspect up close the 4,000- to 3,000-year-old elevated reef terrace and find several well displayed dikes intruding the underlying Ko‘olau lavas. The basalt lava shows spheroidal weathering in places. Be wary of tides, waves, and slippery surfaces, but the stroll out along this shoreline is well worth the effort.

HI 72 (KALANIANA‘OLE HIGHWAY)
Honolulu—Koko Rift—Waimānalo

10 miles

See map on page 80.

The latest outburst of eruptions on O‘ahu took place no more than a few tens of thousands of years ago, only yesterday in geologic terms, and formed a 5-mile-long belt of young cones, craters, and flows along the southeastern shore of the island. Local geologists call this belt the Koko Rift. While exact dates of individual eruptions are hard to determine, there is ample reason to believe that future eruptions are possible in this area.

Why did so many eruptions take place here? The magma originated in the mantle. We know this because volcanic rocks here contain scattered peridotite xenoliths. As the magma approached the surface it exploited a zone of tension, or potential splitting in the crust, possibly localized where gravity strongly pulls the southeast corner of the Ko'olau Range toward the nearby depths of Kaiwi Channel. This proved to be the easiest pathway for the magma to reach the surface.

The youngest tuff cones in Honolulu, somewhat older than the Koko Rift, also line up, perhaps predetermined by stretching of O'ahu's crust into the deep sea southwest of the island. Eruptive fissures opened where buoyant batches of magma wedged their way upward into these zones of weakness.

Because of heavy highway traffic it is usually better to explore the Koko Rift traveling from west to east, making turnoffs to the right (ocean side) of the highway safer and easier. With this in mind, approach Koko Rift on HI 72, driving east from Honolulu.

The highway hugs the coast to Hawai'i Kai. There you can turn off at Kuli'ou'ou Beach Park to look across Maunalua Bay at Koko Head, the southwesternmost Koko Rift cone. The cone rises about 650 feet above the bay, jutting seaward at Kawaihoa (Portlock) Point. The steam-blast explosions that built the cone took place in shallow water or in a coastal flat with a shallow water table. A modern analog is the eruption of Surtsey in the Atlantic Ocean near Iceland from 1963 to 1967.

Look for patches of green olivine sand on Kuli'ou'ou Beach. The fine crystals weathered and eroded from the flanks of Koko Head and perhaps neighboring cones. Shallow longshore currents brought them here to mix with other sediments. The water is often cloudy from suspended mud and silt offshore, so swimming is not great.

Koko Head tuff cone viewed from Kuli'ou'ou Beach Park in Hawai'i Kai. The erosional runnels streaking the flank of the cone formed from rain runoff and have taken thousands of years to develop. Multiple overlapping explosion craters truncate the cone's top.

From Kuli'ou'ou Beach Park, HI 72 crosses the mouth of an estuary, called Kuapā Pond, that formed where growing Koko Rift cones diverted stream channels into Maunalua Bay. The combined drainages cut a deep valley near shore that flooded as sea level rose at the end of the last ice age—a drowned valley.

The highway crosses the Koko Rift between Hanauma Bay and Kahauloa Crater. At the divide is the entrance to Hanauma Bay State Nature Preserve. The bay is the flooded crater of one of the tuff cones that coalesce with nearby Koko Head. Overlapping of cones suggests that they all formed during the same large eruption, perhaps taking place over just a few years. If so, the last big spasm of explosions probably took place in the area of the bay, not only because the deepest crater exists here, but also because it seems to cut the rims of the smaller, shallower craters nearby. Alternatively, the eruption that formed Hanauma Bay could have taken place sometime after Koko Head formed. Recent radiometric dating of the Koko Rift volcanic rocks produced ages ranging from 100,000 to 60,000 years old, with Hanauma at the younger end.

Past Hanauma Bay, HI 72 skirts south of Kahauloa Crater, which is now largely filled with younger deposits and used as a rifle range. Roadcuts upslope reveal the finely multilayered ash deposits making up the cones. The best place to view these safely is Lāna'i Lookout, where the highway passes the seaward flank of Kahauloa cone.

The hardened ash layers (tuff) adjacent to this lookout are magnificent examples of tuff cone deposits, some of the finest in the world. Some layers show gentle cross bedding, indicating that the ash accumulated not only by falling directly out of explosion clouds, but also by surging in powerful gusts of very hot steam and gas down

Eroding seaside tuff layers on the flanks of Kahauloa and Koko Crater cones in the Koko Rift. Viewed from Lāna'i Lookout.

Cross bedding from pyroclastic surge activity during deposition of the tuff layers below Lāna'i Lookout. Close-up shows layers of accretionary lapilli (pisolites) in tuff at Lāna'i Lookout. Camera lens cap for scale.

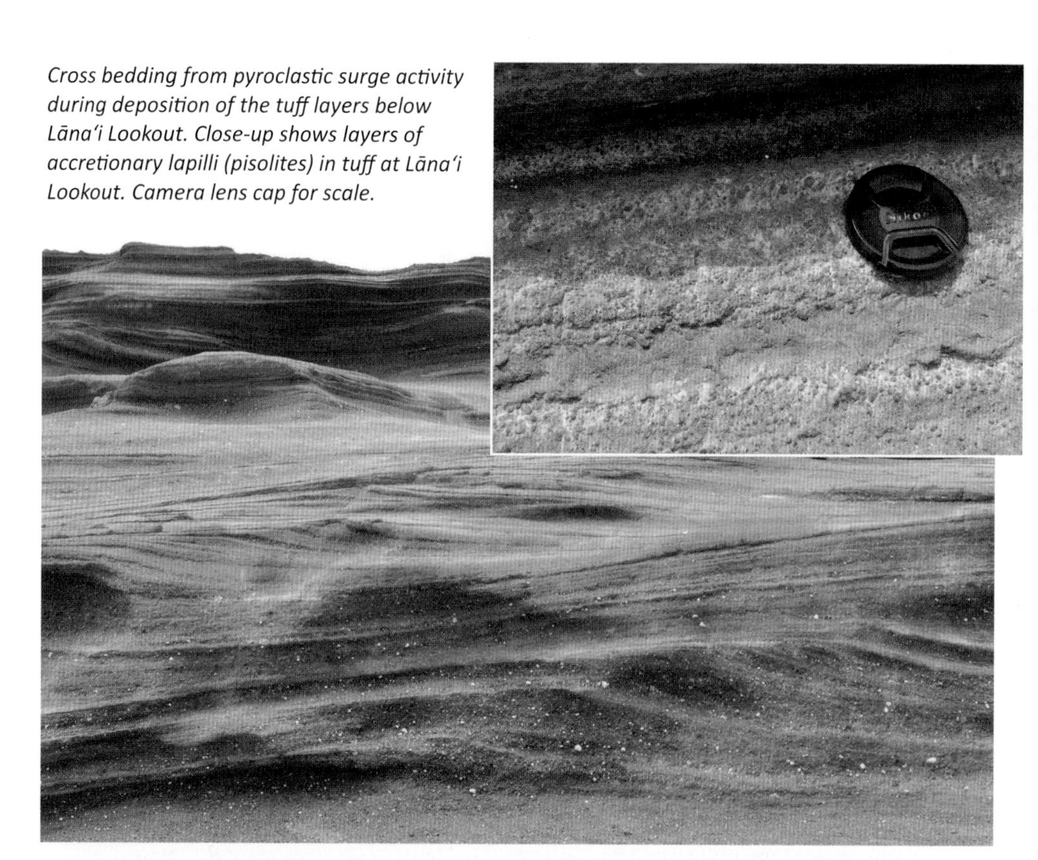

Sag in tuff layer left when a small fragment of basalt impacted the soft surface of the ash during the eruption of nearby Kahauloa Crater.

the flank of the growing cone. Such pyroclastic surges are especially dangerous phenomena during volcanic eruptions. They can be strong enough to blow down buildings. Bits of white coral reef fragments and dark, older basalt may also be seen mixed in. Look for sags where falling fragments depressed softer layers underneath. You may see layers containing accretionary lapilli—fine mudballs formed by ash dropping through electrostatically charged steam clouds.

At the shoreline below Lāna'i Lookout is a wide, nearly level, wave-eroded terrace or platform cut into the tuff. The platform is elevated too high above water level to have been eroded by today's wave action. Likely it formed during the somewhat higher stable stand in sea level from 4,000 to 3,000 years ago.

The next highway lookout is at the Hālona Blowhole, where waves washing into eroded cavities in the tuff trap and pressurize air, causing it to spray water like a geyser through narrow openings in the platform surface. Blowholes are common features on wave-cut platforms throughout the Hawaiian Islands.

The cone rising above the Blowhole Lookout is Koko Crater, at 1,200 feet the tallest of the Koko Rift cones. Pu'u Mai is the name for the summit. Two cones coalesced to form Koko Crater. The southwestern crater is as large as Hanauma Bay, though well above sea level. During the eruption, a strong wind blew ash and associated debris toward the southwest, piling it up to build the highest rims in that direction. Koko Crater formed earlier than the craters to the southwest; Kahauloa Crater cuts the southwestern flank of the cone, so Koko Crater must have formed before it, possibly by a few tens of thousands of years.

Lava beneath the golf course and next to HI 72 east of the Kealahou Street turnoff erupted from a Koko Rift cinder cone upslope more than 100,000 years ago. We know Koko Crater already existed at the time because its northeastern flank diverted the flow.

HI 72 turns inland and ascends a beheaded, now dry valley, Kealakipapa, to reach O'ahu's Windward Coast at Makapu'u Point. A paved trail leads from the highway to an overlook on the headland that provides spectacular views southwestward along the Koko Rift, eastward across the Kaiwi Channel to Molokai, and northwest along the coast to Kailua-Kāne'ohe. The flat shelf of lava underlying Makapu'u Beach Park, immediately west of the lookout, is a remnant of the Kaupō Flow, which erupted around 100,000 years ago. Makapu'u Beach is a popular body surfing locality, but the shore break can be treacherous. Swimming conditions are best toward the western end of the beach. You may also see blowholes active along the shore nearby.

Offshore to the north are two additional young cones, Kāohikaipu Island, and farther out, Mānana (Rabbit) Island. Kāohikaipu is a remnant of a cinder cone showing little interaction with seawater during its formation, despite being largely submerged today. Mānana originated as an ash cone like other large Koko Rift cones. No radiometric ages have been determined for these islets, but Mānana is compositionally distinctive from other Koko Rift vents, including xenoliths of nephelinite with rare sodalite crystals. It probably developed from an earlier batch of magma unrelated to the other Koko Rift eruptions.

Northwest of Makapu'u Point, the highway skirts the furrowed Waimānalo Pali that increases in elevation as you drive toward the former summit of Ko'olau

KOKO CRATER BOTANICAL GARDEN

Koko Crater is capped with a hardened bed of dark ash, more resistant to erosion than the softer, lighter-colored layers underneath. This layer, several feet thick, may be a product of case hardening, a process in which the shallow infiltration of water during weathering precipitates dissolved minerals such as calcite and silica that form cement when the water later evaporates. The precipitating substances bind together otherwise loose grains, forming a hard "case" that protects the softer layers underneath. At Koko Crater, erosion has formed channels that are slowly eating their way upslope from the base of the cone, and as a result, the capping layer is undercut and fragmented in many places. The only intact part clings to upper slopes around the summit.

You can see good examples of this differential erosion by continuing on HI 72 to Kealahou Street and following it to the Koko Crater Botanical Garden. The loop trail through the garden circles the 200-acre interior of the crater. Look upslope for the resistant rim of tuff above. Some pieces have tumbled downslope almost to trail level. Tuff beds exposed along trail sides include fragments of white reef rock and older basalt, some forming sags in the layering. The microclimate of Koko Crater, quite different from most parts of O'ahu, favors exotic plant species from regions as far away as the Mexican desert and the African veldt. It is possible for you to imagine that you are thousands of miles from Hawai'i here.

The inner wall of Koko Crater, showing its prominent erosion-resistant cap layer at the rim. Viewed from the Koko Crater Botanical Garden loop trail.

Bits of white calcareous reef rock embedded in ash layers along the Koko Crater Botanical Garden loop trail. Pen for scale.

Makapu'u Beach and the Waimānalo Pali, looking north from Makapu'u Point. The small peninsula at the north end of the beach is a rejuvenated Koko Rift flow, no more than 100,000 years old.

Volcano. This cliff face was initiated by the giant Nu'uanu landslide probably around 2 million years ago. The pali has since eroded inland from an original location several miles offshore. Flood and landslide debris and other sediments form the plain at Waimānalo.

HI 83 (KAMEHAMEHA HIGHWAY)
Kāne'ohe Bay—Hale'iwa
40 miles

As elsewhere on O'ahu, heavy traffic along HI 83 makes it difficult to pay close attention to roadside geology, except for a few selected points of interest. HI 83 follows the steep Windward Coast, parallel to Ko'olau's northwest rift zone, which lies hidden in forested, wild mountain terrain 2 to 4 miles inland. Where the highway turns west past Kahuku it crosses the rift zone, but heavy vegetation and weathering still make it impossible to see roadside dikes.

Mountainsides crowd in close to the highway on the Windward Coast, in part due to the intense erosion caused by heavy rainfalls, and in part due to the modification of Ko'olau Volcano by the stupendous Nu'uanu submarine slide 2 million years ago. Streams tumble over numerous waterfalls, though no substantial cascades can be seen from the road. (At the time of this writing, 1,500-foot-tall Sacred Falls, a spectacular example, is closed to the public. Landslides and rockfalls make interior road maintenance a challenge). Windward beaches also tend to be raw and wild with ferocious rip currents, sharp reef rock, and, as in the case of Pounders Beach near Lā'ie, dangerous shore breaks.

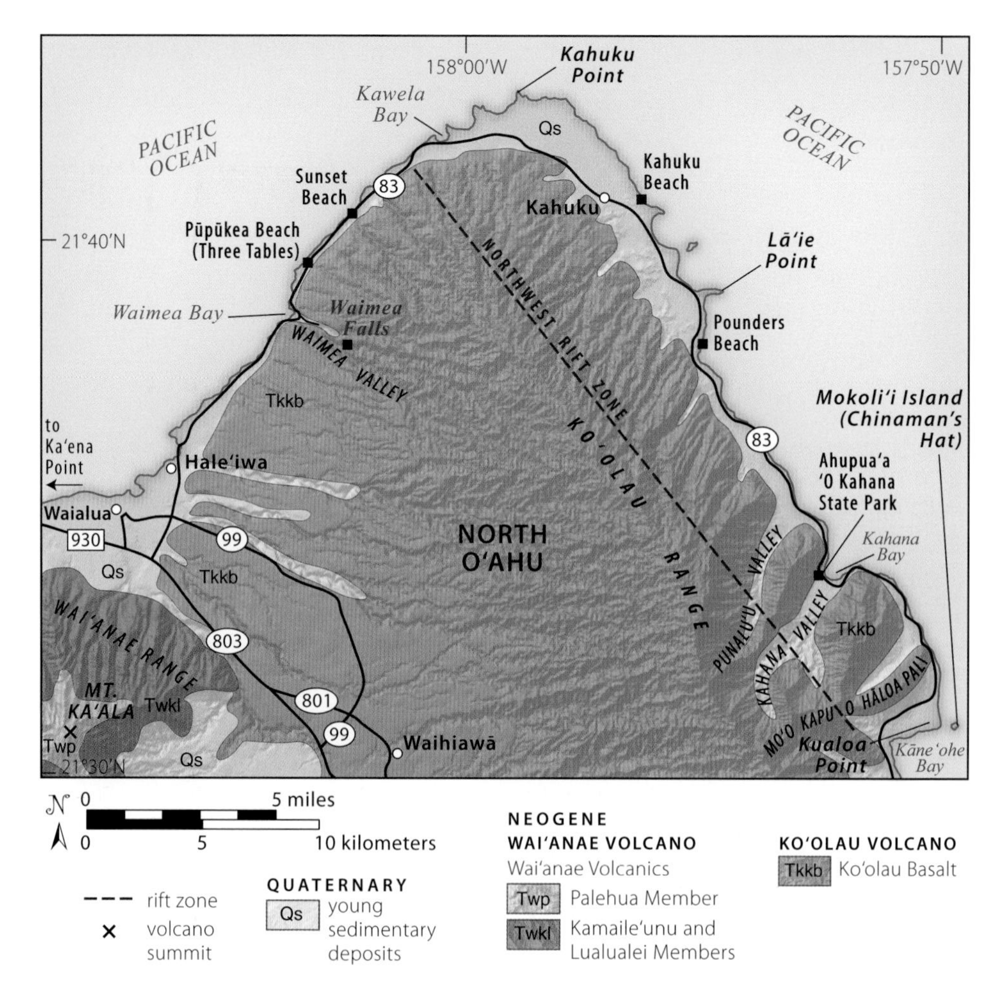

Geology along HI 83 between Kāne'ohe Bay and Hale'iwa.

HI 83, north of Kāne'ohe, leaves Kāne'ohe Bay and the eroded Ko'olau caldera at Kualoa (Āpua) Point. Formidable Mo'o Kapu o Hāloa pali rises as much as 1,900 feet above Kualoa Regional Park, a convenient place to park and enjoy the point and surroundings. Cliffs here as elsewhere show countless thin basalt flows in cross section erupted during growth of Ko'olau's shield.

Mokoli'i Island (Chinaman's Hat) is a large islet just a short distance offshore from Kualoa Point. Several factors possibly played a role in creating this unusual feature. Mokoli'i may be an erosional remnant of a resistant headland that once extended offshore. Waves tend to refract toward headlands, embracing them from both sides and concentrating their erosional energy, so they attack a headland more vigorously than a straight segment of coast. A resistant sequence of flows may survive, rising out of the sea as a sea stack long after rocks closer to shore have been pulverized and removed. Alternatively, stream erosion when sea level was much lower during the last ice age may have worn away the lavas surrounding Mokoli'i, which then

The walls of Mo'o Kapu o Hāloa pali that rise above Kualoa Regional Park at the north end of Kāne'ohe Bay, are composed of countless Ko'olau shield lavas that accumulated during the volcano's fastest stage of growth.

HE'EIA STATE PARK

Heading north from Kāne'ohe, the coastal highway, HI 830, passes He'eia State Park. Lookouts in the park present sweeping vistas of Kāne'ohe Bay. Ahu'olaka Island, directly offshore, is another example of an ancient reef terrace marking a high stand in sea level several thousand years ago. Closer at hand, magnificent He'eia Fishpond to the east of the point seals the bay at the mouth of He'eia Stream. The bay began forming during the last ice age, when sea level was hundreds of feet lower and He'eia Stream cut a deeper valley, now flooded at its lower end by return of the ocean between 18,000 and 10,000 years ago.

Giant He'eia Fishpond, built across the mouth of a valley drowned by rising sea level at the end of the last ice age, with caldera-related pali and Oneawa Hills in the background.

became an islet when sea level rose again. Maybe both factors played a role in this instance. In any case, when erosion and sea-level rise first separated Mokoli'i from the mainland, eroded sand collected on the side of the islet closest the shore. A sandy peninsula, called a *tombolo*, probably connected the islet with the mainland. With time, even the sand in deeper water washed away. The flat area of Kualoa Park may be a remnant of that tombolo.

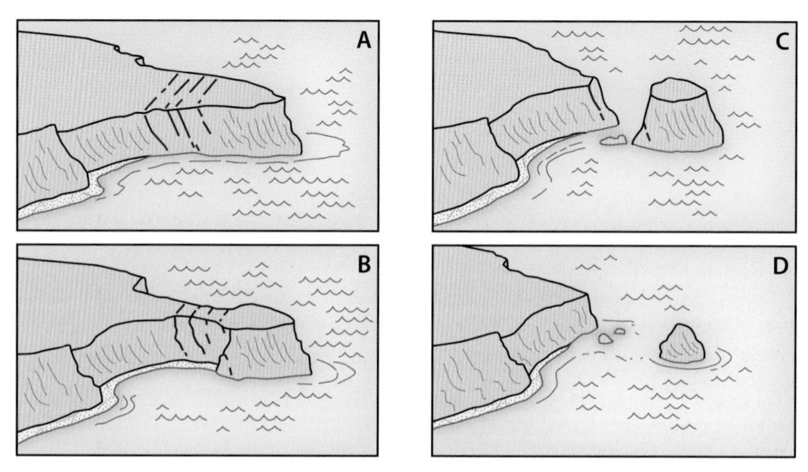

Origin of sea stacks resulting from wave erosion.

Kahana Valley (Ahupua'a 'O Kahana State Park)

Kahana Valley and neighboring Punalu'u Valley to the north are the largest valleys on the Windward Coast north of Kāne'ohe Bay. Kahana Valley, marked by a sign to Ahupua'a 'O Kahana State Park, drains into Kahana Bay, a submerged extension of the valley that was eroded when O'ahu stood higher. Kahana is the largest drowned valley on the eastern shore. You may drive inland a short distance to access a hiking trail system crossing the upper part of the valley—a good way to appreciate its amphitheater-shaped divides. Substantial bedrock outcrops are few and far between, however.

Kahana Valley is also an *ahupua'a*, a traditional Hawaiian land division that predates European contact. Ahupua'as were sections of islands from coast to mountains that would optimally contain all the resources needed to support a permanent resident population, including freshwater, timber, fertile soils, and a shoreline with a sustainable fishery. Many of Hawai'i's current place names are the names of ahupua'as inherited from centuries ago. Tradition-minded Hawaiian families occupy Kahana Valley to the present day.

Early efforts to secure a water supply for Honolulu focused on Kahana Valley. One drainage tunnel driven nearly 2,000 feet into the mountainside between 1929 and 1931 cut through 120 dikes in the northwest rift zone. The objective was to tap groundwater impounded behind the impermeable dikes, which act as underground natural dams. Tunnels such as this, in rift zones on Ko'olau, have become a major source of freshwater for the city of Honolulu.

Lā'ie Point

Lā'ie Point, at the end of the Laniloa Peninsula, may be reached by a major residential street in the center of Lā'ie town. Here you can see old, cemented sand dunes that once were heaped behind the beach and have eroded into coastal cliffs standing 20 to 30 feet above sea level. Just off the point, waves have eroded a low island out of the same material. The great tsunami of April 1, 1946, which devastated Hawaiian coasts, eroded a picturesque natural bridge on the island. Strong waves are constantly undercutting the edge of the limestone, and blocks of the cliff occasionally topple into the surf. The thin, steeply dipping layers faintly visible in the cliffs were deposited on the downwind sides of the dunes, where sand accumulated in a cornice and slipped down the face.

A natural bridge formed when a large tsunami in 1946 eroded a hole in offshore raised limestone at Lā'ie Point. The limestone is the hardened remains of ancient calcareous sand dunes active along this shoreline, probably during the latest ice age over 11,000 years ago.

O'ahu's North Shore

The stretch of coastline between Kahuku and Hale'iwa—O'ahu's North Shore—features some of the most famous surfing spots in the world, including Sunset Beach and Waimea Bay. It is not unusual to see winter waves breaking above 20 feet here, driven by storms as far away as the Gulf of Alaska. Expect traffic jams when it is high season for surfers and beach lovers!

Kawela Bay, at the eastern edge of the small town of Kawela, is one of the few places along the North Shore that is genuinely safe for swimming. The beach is several thousand feet long, with only a few rocky outcrops. Sand comes from reefs at both sides of the entrance to the bay. Even during heavy weather, the inshore swell is slight. The water is usually clear except in winter, when Kawela Stream empties sediment into the bay.

Ke Nui Road, a frontage road, begins in the town of Sunset Beach. Follow it one hundred yards west of 'Ehukai Beach Park and you will reach an area where winter swells rise steeply as they climb a shallow reef about 150 feet offshore. Each monstrous wave breaks almost at once, curling from one end to the other into a tube known as the Banzai Pipeline. Surfers did not successfully ride these famous waves until the 1960s.

HI 83 passes Pūpūkea Beach Park about 3 miles west of Sunset Beach, not far from the Banzai Pipeline. The shoreline is rocky with little sand. Three Tables at the southwest end of the beach park and Shark's Cove at the northeast end are popular swimming areas. Some people regard Shark's Cove, with an assortment of tide pools, as one of the prettiest stretches of shoreline on the island. Concentric growth rings mark coral heads several feet across in the old reef. Pebbles of white coral are mixed in with the beach gravels. A natural breakwater of ancient reef rock partly protects the inlet at Three Tables, which are named for flat-topped pieces of the reef that stand above modern sea level.

Waimea Valley

Waimea Bay occupies the flooded mouth of this sunken river valley. A sand bar and beach form the shoreline across the mouth of the valley. At low water, Waimea Stream cannot maintain a channel across the beach and merely soaks through the sand en route to the bay.

Waimea Valley is typical of hundreds of valleys and gulches along the northern and western flanks of Ko'olau Volcano. Almost nothing of the original surface of the shield volcano survives, but you can see evidence of the ancient slope in the general gradient of the divides between streams. Erosion has not progressed to the point of carving deep canyons or opening broad amphitheaters at the heads of the valleys, as on the Windward Coast. Deposits of coarse gravel, boulders, and other sediments lie beneath the flat floor of lower Waimea Valley. They fill the floor of the canyon that was eroded here when sea level was much lower around O'ahu.

At Waimea Valley, turn inland on Waimea Valley Road to see Waimea Falls, which drops 80 feet across a resistant ledge of lava that crosses the valley. The falls, reached at the end of a short walk through Waimea Falls Park, are spectacular when the stream is full, but they dwindle to a trickle during dry weather. At the entrance to the park, outside the parking area, examine the cliff face to see the massive cores of 'a'ā lava flows. The lavas exposed in Waimea Valley are basalt flows erupted during the shield-building stage of Ko'olau Volcano.

Waimea Stream is slowly filling Waimea Bay with sediment, patiently converting it into a marshy coastal flatland. Although most of the giant surf breaks for which the bay is famous are out beyond the point, the inshore surf can still break as high as 10 feet against the calcareous sand of Waimea Bay Beach. A strong rip current runs down the middle of the bay. There are safer places nearby to swim.

Past Waimea, slopes become gentler and the Ko'olau Range crest is farther from the highway. Leeward conditions are somewhat drier, and the flanks of the old volcano have been more stable than the Windward side throughout time. The overall slope angle of the western side of the Ko'olau Range is similar to that of the original volcanic shield.

HI 930 TO KA'ENA POINT

HI 930 heads west from Hale'iwa across a plain once planted with sugarcane. The Wai'anae Range, to the south, slopes steeply seaward from its highest summit, Mt. Ka'ala. Like the west coast, the northern shore of the Wai'anae Range owes its straight trend and steep palis, or cliffs, to a massive undersea landslide. The Ka'ena debris avalanche probably took place when Wai'anae Volcano was in the waning stage of its activity, though no one knows for sure.

The highway skirts Mokulē'ia Beach to end at the foot of the pali, about 9 miles west of Hale'iwa. You may walk past a gate to reach Ka'ena Point, O'ahu's sharp northwestern cape, about 2.5 miles away. Pali faces nearby expose countless thin lava flows, mostly 'a'ā, from the shield-building stage of Wai'anae Volcano. At the shore, explore the 4,000- to 3,000-year-old elevated reef shelf, best seen where erosion in inlets has exposed cross sections. In places, the now dead reef grew atop rounded boulders of older basalt and related rubble, showing that the reef was able to thrive in a rough surf zone. The reef lived for centuries before rapidly falling sea level left it high and dry. The drop in sea level was due to a climate-related loss of seawater, though slow tectonic uplift may have also played a role. A similar high-stand reef terrace and wave-cut bench lines many shorelines on Kaua'i, 90 miles to the west.

An ancient reef terrace, several thousand years old, lies atop the rubble of a surf zone active just before the reef cap formed, at the western end of HI 930.

Fossil coral heads appear in many places in the elevated reef terrace on the way to Ka'ena Point.

HI 93 VIA H-1 (FARRINGTON HIGHWAY)
Waipahu—Wai'anae—Keawa'ula Bay Beach
20 miles

West of Pearl Harbor, H-1 heads toward the southern end of the Wai'anae Range. As much as 1,000 feet of sediments deposited on hard basalt bedrock lie beneath the wide coastal lowland south of the freeway. This substantial accumulation includes sand and gravel laid down in streams and beaches, silt and clay deposited in coastal lagoons, reef rock, and even a few beds of coal. It was all laid down as the island slowly sank as it moved away from the Hawaiian hot spot.

Just west of Waipahu, Fort Weaver Road (HI 76) leads south 5 miles to 'Ewa Beach, crossing the coastal plain and passing through the community of 'Ewa Gentry. Like almost all the beaches on O'ahu, the beaches along the 'Ewa coast consist of soft, calcareous sand washed in from the fringing reef. These beaches are also fine swimming areas with continuous, pleasantly warm trade winds.

As the H-1 freeway approaches the Wai'anae Range, look for several eroded cinder cones upslope covered with vegetation. These erupted during the Kolekole phase of volcanism, the last gasp of Wai'anae Volcano. The reddish cone directly ahead as you pass the exit to the West O'ahu UH campus is Pu'u Makakilo. The H-1 freeway passes around its southern base. Not far past this point, the freeway narrows and becomes HI 93, the Farrington Highway, which follows the west coast of O'ahu. It skirts the eroded caldera of Wai'anae and ends in the volcano's eroded northwest rift zone near the island's northwest cape.

Frequently heavy traffic makes a turn left to the shore difficult before reaching Nānākuli, where a traffic light allows access to Kalaniana'ole Beach Park, with white, reef-derived beach sands. Stacked lava flows in hillsides lining Nānākuli Stream valley show the rapid growth of Wai'anae Volcano during its tholeiitic shield stage of growth. An erosional remnant of these flows also crops out as steeply pyramidal Mā'ili Point up ahead.

Mākaha and Wai'anae in the Lualualei Caldera

Just past Mā'ili Point the interior opens into the wide plain of Lualualei caldera. Rimmed with towering cliffs and steep hills, the caldera is between 4 and 5 miles across and 11 miles long. The highest point in the Wai'anae Range, 4,025-foot-tall Mt. Ka'ala, looms at the northern end. You can drive inland a short distance on Pa'akea Road and other public streets to get a better view of the basin, which is largely a closed military installation.

Both Wai'anae Valley and Kuwale Roads take you close to the foot of Mauna Kuwale, a small but conspicuous knob near the northern edge of the caldera. (Be sure to respect private properties next to these public streets). Mauna Kūwale is an extraordinary feature, perched atop thick, reddened flows of rhyodacite, a high-silica lava. Rhyodacite is not found on most oceanic islands; it occurs primarily in continental areas. Some geologists think it is residue from crystallization of a basaltic magma chamber that stagnated for a long time toward the end of Wai'anae's shield growth.

Lahilahi Point, between Wai'anae and Mākaha, is a small, steep-sided peak, Mauna Lahilahi. This erosional remnant of basalt, shaped like a giant blade 230 feet high, separates two small bays. The southern beach is Mauna Lahilahi Beach Park.

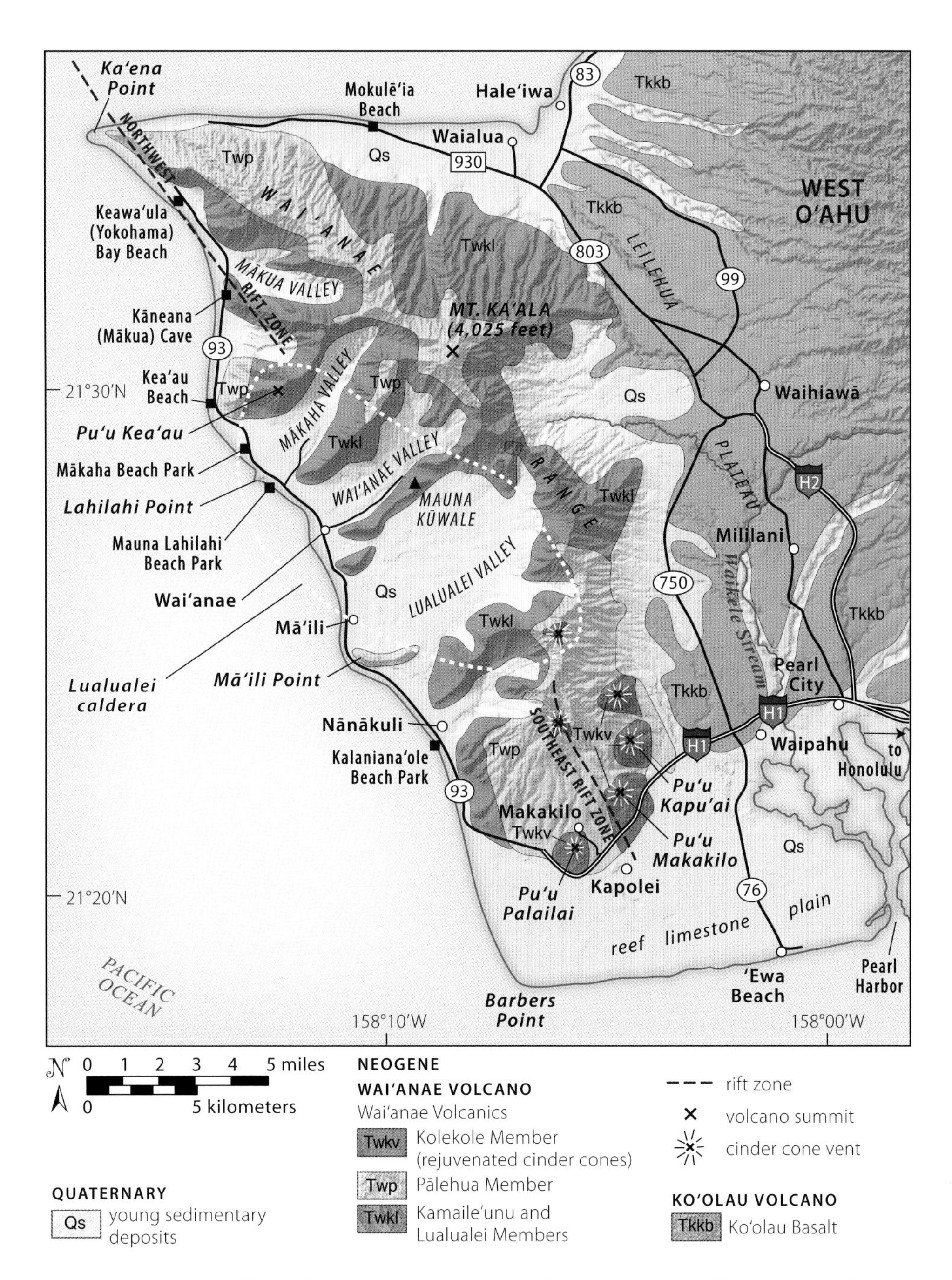

Geology along HI 93 as it follows the base of the Wai'anae Range up the Wai'anae coastline.

Mauna Kūwale near the northern edge of Lualualei caldera. The slopes consist of thick rhyodacite lava flows, rare in Hawai'i. Mt. Ka'ala and adjoining shaded ridges loom in the background.

Calcareous sand washes onto the beach with the gentle waves of summer and is carried away with the heavy surf of winter. A low, wave-cut bench around the point is cut into basalt flows and capped with white fossil reef and layered beach rock. Reef exposures are fairly well preserved, including some staghorn coral. You can reach Papaoneone Beach on the north side of the peak via Moua Street. The water deepens near shore, with a strong longshore current.

Across Farrington Highway from Mauna Lahilahi Beach Park, Mākaha Valley Road leads inland to a country club and condominium complex. The towering cliff face of Pu'u Kea'au, 2,650 feet high, rises northwest of the country club. Countless lava flows are stacked in the slopes. Those nearest the ocean are thin and dip seaward, but those inland are thick and nearly horizontal. They are separated by a fault where they come together, though slope sediment obscures the exact location of the fracture line.

Harold Stearns correctly concluded years ago that this fault was the rim of Lualualei caldera. This site may record the filling and overflowing of the caldera near the end of volcanic activity. If so, the inland flows would indeed be horizontal and thick because they ponded inside the caldera, and the seaward flows would have overflowed the rim of the caldera to pour down the flank of ancient Wai'anae shield volcano.

Mākaha Beach Park, near the mouth of Mākaha Valley, is one of the world's classiest surfing locales. But when the surf is high, strong longshore currents merge into a dangerous riptide at about the middle of the beach. The current flows out to sea along a former channel of Mākaha Stream, which cut across the beach when sea level was lower.

The ridge at the northern side of Mākaha Valley from the area of Wai'anae Intermediate School. White bar on left shows seaward-dipping outer Wai'anae shield flows. The bar on the right shows thicker, flat-lying caldera-filling flows (including the Kamaile'unu Member). Dashed line marks the approximate ancient buried caldera wall.

Kea'au Beach to Keawa'ula Bay Beach

North of Mākaha, the highway passes Kea'au Beach Park, with its 2 miles of ancient reef terrace. Dangerous offshore currents run along this section of coast. Farther north, Farrington Highway threads its way between the ocean and the steep western ridges of the Wai'anae Range. These ridges are in the northwest rift zone of the old shield volcano. Many basalt dikes slice through the dark flows of basalt that built the main part of the volcano. The tops of some of the ridges carry a veneer of pale alkalic basalts that erupted during the later stages of volcanic activity. Like the older flows, these dip gently seaward.

While large portions of the western flank of Wai'anae Volcano slid into the ocean, this part did not collapse catastrophically. Sonar surveys of the ocean floor show that a few large fragments are still nearby, indicating that they slid piecemeal in a series of slow slumps.

Kāneana (Mākua) Cave, the Cave of the God Kāne, is south of the mouth of Mākua Valley. Waves carved out this enormous sea cave when sea level stood about 100 feet higher along this shore, several hundred thousand years ago. Geologists refer to this as the Ka'ena high stand of sea level. Ocean waves exploited a swarm of fractures to open a passage about 450 feet long and 15 to 20 feet high. The coastal platform shoreward of the cave is a reef terrace that records another high stand in sea level, later and lower than the Ka'ena stand.

Many lighter-gray dikes, vertical stripes 2 to 5 feet wide in the nearby cliffs, show that this area lies within the Wai'anae northwest rift zone. A few dikes may be inspected up close in the walls of Kāneana Cave. If you look carefully, you will see that the edges of the dikes have darker border zones. Rising basalt magma cooled against the older rock to become finer grained and darker. The ancient rift zone stretches nearly 70 miles from Lualualei caldera to the foot of the Wai'anae shield

A swarm of vertical to steeply tilted dikes cut horizontal flows and mark the ancient northwest rift zone of Wai'anae Volcano along the highway near Mākua Valley.

Kāneana Cave originated as a sea cave, not by activity in a lava flow. Horizontal cooling fractures in dikes are visible in the wall overlying the entrance and are especially prominent along the left side.

volcano, which lies submerged under 2.5 miles of water. Only about 15 miles of it are above sea level today.

Just north of Kāneana Cave, look inland to see a beautiful landscape with a broad amphitheater at the head of Mākua Valley. Several basalt dikes of the northwest rift zone cut through the southern slopes. A long beach of beige reef sand stretches between rocky points at the mouth of the valley.

The highway ends at Keawa'ula Bay Beach, sometimes called Yokohama Bay because of its popularity with Japanese fishermen. The dirt road beyond clings precariously to a narrow fringe along the base of the mountain front through a conservation area to Ka'ena Point, another 2 miles away.

The Keawa'ula shoreline is rocky. At the north end of the beach, you can walk along the crest of the 4,000- to 3,000-year-old raised reef terrace, the flats next to the shore. Look for several species of coral, along with mussel and periwinkle shells, especially in reef ledges in the small, protected coves. The partly cemented beach and dune sands covering sections of the terrace must also have accumulated when sea level stood slightly higher than now. Look for cross beds that formed as waves shaped and reshaped the ancient beach.

A raised 4,000- to 3,000-year-old reef terrace fringes the shoreline near Keawa'ula Bay Beach at the end of the Farrington Highway, a popular fishing area.

Molokai

Molokai is the oldest member of the Maui Nui group of Hawaiian Islands. Molokai, Maui, Lāna'i, and Kaho'olawe formed a single landmass during the latest ice age, when sea level was low enough to drain the narrow straits now separating them. The group is composed of at least six, and possibly seven, coalescing shield volcanoes.

Molokai has an area of 260 square miles; it is 38 miles long and only 10 miles wide. A bird's-eye view suggests a deformed peanut, with shield volcanoes at both ends. East Molokai is the younger and higher. West Molokai is a small, low volcano, long quiet by the time East Molokai grew against it.

WEST MOLOKAI VOLCANO

West Molokai built up above the waves around 1.9 million years ago. Like other Hawaiian volcanoes, it erupted fluid flows of basalt during the initial period of rapid growth, constructing a broad, gentle-profile shield. Rift zones developed along trends northwest and southwest from the summit. If the volcano ever had an east rift zone, any evidence of it either was buried beneath younger lavas of East Molokai or went to the deep ocean floor in a landslide, perhaps together with the summit and caldera of West Molokai.

As West Molokai drifted west off the Hawaiian hot spot, the composition of lavas in it changed from the ordinary tholeiitic basalts of the main stage of shield volcanism to the more exotic alkalic basalts of late-stage activity. The late eruptions added sixteen cinder and lava cones to the landscape and several alkalic lava flows. Most of these eruptions were in the northwest rift zone. West Molokai shows no rejuvenated-stage volcanism.

A high submarine platform stretches about 40 miles southwest from West Molokai and culminates in Penguin (Penquin) Bank, which covers about 380 square miles and lies at a water depth of only 170 feet. The flat top on the platform may be from a combination of wave erosion and reef growth when sea level was lower, presumably during one or more recent ice ages. Some geologists speculate that Penguin Bank is an offshore extension of the West Molokai's southwest rift zone. Others suggest it may be a separate, submerged shield volcano topographically linked to West Molokai simply by submarine extension of the southwest rift zone.

During or shortly after the late stage of alkalic volcanism, the northeastern flank of West Molokai collapsed into the ocean, leaving a set of large slide scarps still partly exposed where the volcano abuts Ho'olehua Saddle (the broad, flat area between Molokai's two volcanoes) and along the north shore. Flows from neighboring East Molokai built up against these scarps, showing that it is a much younger volcano.

West Molokai is dry. It is too low to snatch moisture from the clouds, and East Molokai shelters it from the big rains that come in on the northeast trade winds. Stream erosion has minimally modified the volcano since it went out of business.

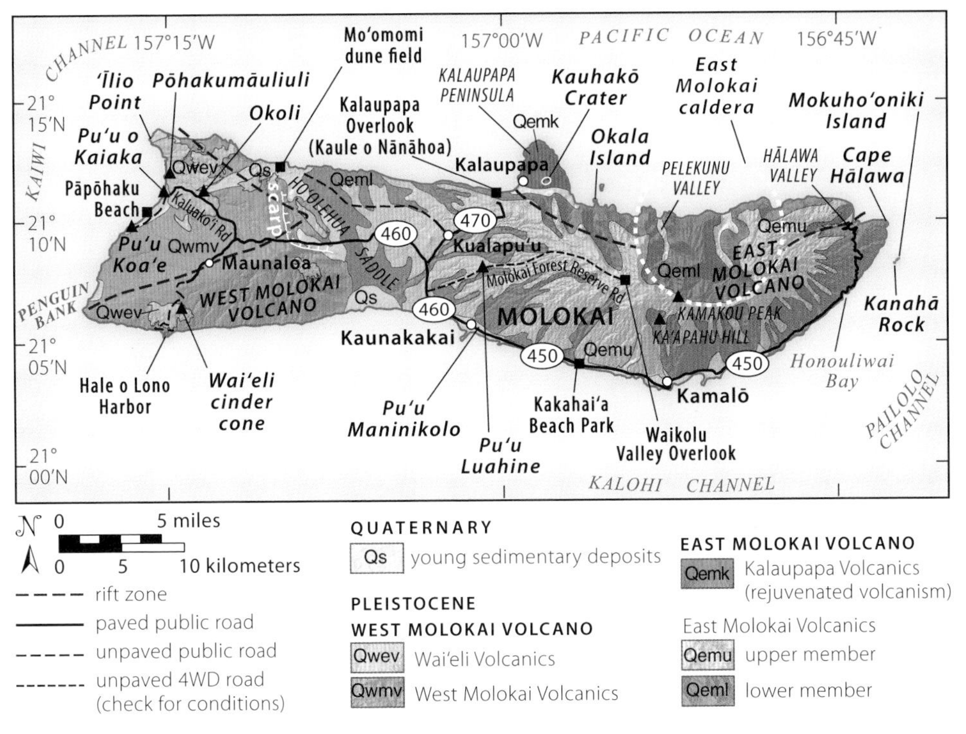

Geologic map of Molokai.

EAST MOLOKAI VOLCANO

Age dating tests show that East Molokai Volcano was building above sea level as early as 1.75 million years ago. As East Molokai grew, lava flows from it piled up against the flank of West Molokai, building the broad, nearly level Hoʻolehua Saddle between them. Several feet of red laterite soil (intensely leached oxisol) had developed on the lavas of West Molokai before the eruptions from East Molokai covered them, indicating that the West Molokai Volcano was thoroughly dead for some time before East Molokai reached its full size.

East Molokai had two rift zones: one extended west of the summit and another reached eastward. A caldera 5 to 6 miles across opened in the summit, which may once have stood 11,000 feet above sea level.

In many places, a prominent laterite paleosol (ancient, buried soil bed) separates the lower lavas of the shield-building stage from the alkalic lavas of postshield activity. In other places, notably in the western rift zone, the two kinds of flows are interbedded. Apparently, the change from one type of volcanism to the other followed a long pause in activity in some areas, but not in others.

Radiometric age dates on the lava flows show that the change from shield to postshield volcanism was around 1.5 million years ago. The postshield eruptions built at least a dozen cinder cones and domes and covered the shield with lava flows of alkalic basalt and, in a few places, trachyte lava. Many streams have since cut through this veneer of postshield lava, exposing the shield basalts beneath.

WAILAU DEBRIS AVALANCHE

US Geological Survey geologists Robin Holcomb, Bruce West, and others think that as late-stage volcanism waned, the northern flank of East Molokai broke approximately along its two rift zones. The northern half of the volcano dropped into the ocean in the Wailau debris avalanche. It must have been moving fast because it shattered into fragments that spread nearly 100 miles north across the Hawaiian Deep in a strip 25 miles wide.

The Wailau avalanche is the third-largest landslide identified in Hawaiian waters, after Kaua'i's Nā Pali slide and Ō'ahu's Nu'uanu slide. James Moore of the US Geological Survey estimated that the volume of the Wailau slide corresponds to the amount of lava the Hawaiian hot spot would erupt in 10,000 years at its current rate.

The Wailau slide split the caldera of East Molokai in half. The scar it left behind is the towering sea cliff along the island's north shore, the highest and quite possibly the most spectacular shoreline cliff in the world. It rises almost vertically 3,700 feet out of the sea, with one waterfall plunging 1,750 feet. Islets and sea stacks along the base show that waves have eroded the cliff at least a short distance inland.

Large stream valleys have cut across the cliff face to a depth of nearly 6,000 feet below sea level. They were eroded when the island stood much higher and became submerged as it sank. The great cliff stood 6,000 feet higher back then, an astonishing wall towering above the ocean.

REJUVENATED VOLCANIC ACTIVITY

Between 570,000 and 340,000 years ago, long after the Wailau avalanche, a series of rejuvenated eruptions built a large volcanic shield—now the Kalaupapa Peninsula—against the western base of the East Molokai sea cliff. The mostly alkalic basalt flows of the 2.5-mile-long peninsula are ropy, inflated pāhoehoe with glassy surfaces.

Looking east across Mo'omomi Bay toward the north shore of East Molokai. The towering sea cliffs left by the Wailau slide show up in profile, with the low, broad Kalaupapa Peninsula stretching on the skyline to the left.

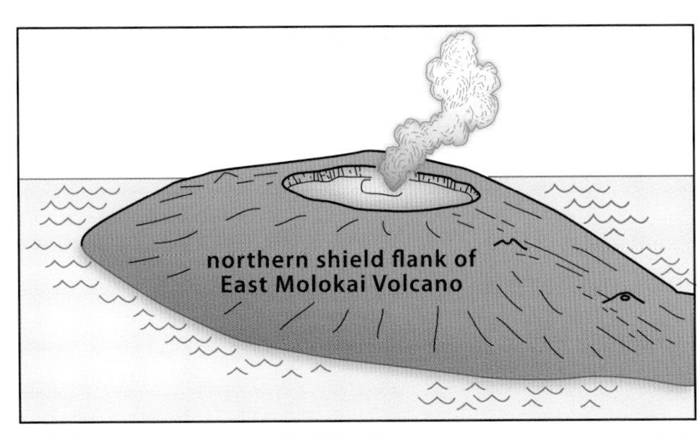

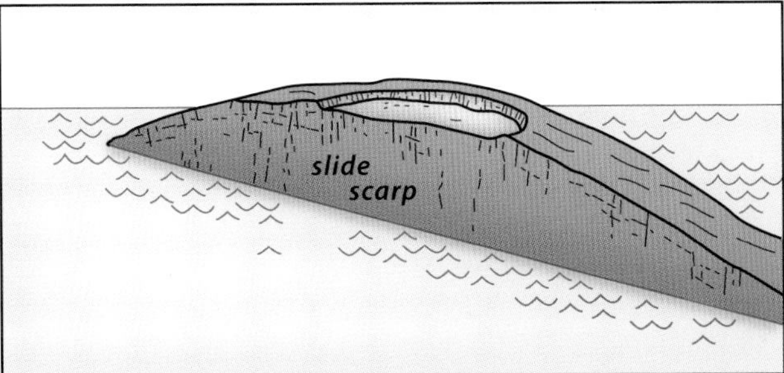

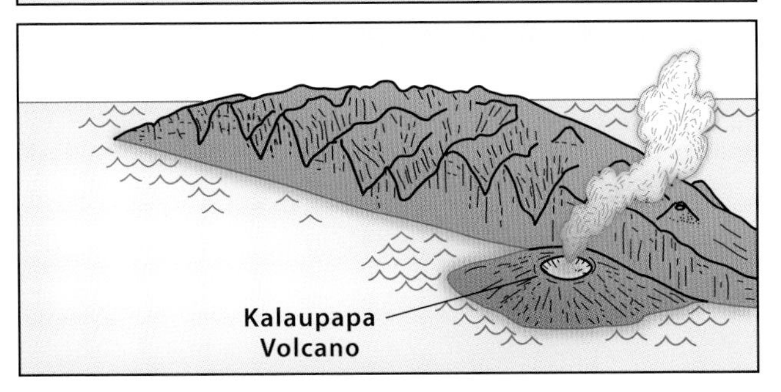

After half of the East Molokai Volcano slid away in a debris avalanche, Kalaupapa Volcano grew at the base of the slide, forming the Kalaupapa Peninsula.

Kauhakō Crater, a combination explosion and collapse pit, is about 1,000 feet wide and lies at the summit of the Kalaupapa shield, nearly 500 feet above sea level. The crater appears to be the main vent for the shield, which is also called Puʻu ʻUao. Groundwater and surface runoff maintain a stratified, algae-rich lake at the bottom. At some 800 feet deep and only 160 feet across, Kauhakō has the largest depth-to-area ratio for any lake on Earth. Very rare turnover (or drying up) of the lake has been known to release hydrogen sulfide gas from its anoxic depths. A huge lava channel and cave system extend north from Kauhakō Crater along the length of the peninsula. Five smaller vents, apparently rootless—meaning they come from surface lava tubes rather than deeper dikes—lie along this trend. Pāhoehoe lava flows spilling from this channel and the vents built the surrounding land.

Another eruption during the rejuvenated phase of volcanic activity built a small ash and lava cone near the eastern cape of Molokai. It has since eroded into tiny Kanahā Rock and larger Mokuhoʻoniki Island.

HI 450 (KAMEHAMEHA V HIGHWAY)
KAUNAKAKAI—HĀLAWA VALLEY
27 miles

Kaunakakai, the main seaport and population center of Molokai, lies near the middle of the south coast. The center of town is close to sea level; newer residential areas are creeping up the gentle flanks of East Molokai Volcano. Pu'u Maninikolo, a low, rusty cinder cone on the northeastern edge of town, has been partly excavated for road gravel.

Much of Kaunakakai is built on deposits from an ancient giant tsunami, perhaps the one generated by the collapse of the 'Alika-2 debris avalanche off the west coast of the Big Island 127,000 years ago. Development has covered up these deposits almost entirely, however, and there is no quick and easy way to see them.

HI 450 follows the dry southern coast of Molokai from Kaunakakai to the eastern cape of the island. For the first half of the distance it stays close to sea level, passing wetlands, scrub forest, and numerous fishponds and shallow bays sheltered behind a fringing reef. Many of the coastal rocks on Molokai record fluctuations in sea level, as do those on the other older Hawaiian Islands. Watch for the roadcut along the north side of HI 450 about 1 mile east of central Kaunakakai. It exposes a lithified coral sand deposit—limestone—3 or 4 feet thick with a flat base on rubbly basalt. The limestone is mostly darkened from weathering but shows white where freshly broken. It originated as coastal dunes when sea level stood much lower, possibly 18,000 years ago at the climax of the last ice age. Offshore reefs at the time were exposed high and dry because of dropped sea level as glaciers grew worldwide. The reefs died, broke down, and shed sand carried inland by onshore winds that may have

Layers in dunes, turned to hardened limestone on HI 450, about 1 mile east of Kaunakakai. These features are a legacy of the last ice age.

been much stronger than they are today. You can see some dune layering preserved in the rock near the west side of this outcrop.

The roadside bedrock exposed for the first 4 miles east of Kaunakakai is alkalic basalt that erupted late in the volcanic life of East Molokai. If you look high on the slope inland from Kakahai'a Beach Park, you can see a subdued group of young cinder cones profiled against the sky. They erupted during postshield volcanic activity. There are also nice views offshore to Maui and Lāna`i.

At milepost 8, a roadcut on the north side of the highway exposes glassy, dense alkalic basalt with conspicuous flow banding. The flow erupted from a vent a short distance upslope. The basalt contains widely scattered, small, angular gabbroic xenoliths, probably fragments scooped from crystallized portions of the magma chamber below. The xenoliths stand out as light-colored angular chips and are mostly just a fraction of an inch across. Their olivine crystals are typically reddened from oxidation. A younger 'a'ā flow covers the eastern margin of the roadcut.

Outcrop of flow-banded lava at milepost 8 along HI 450.

Not far past milepost 9 is a wetland around the large Kamāhu'ehu'e Fishpond. From there you can look upslope to see Ka'āpahu Hill high on the mountainside. This prominent, steep, light-colored knob is a large dome of trachyte—a pale, evolved, alkalic lava—that oozed downslope more than 1 mile as an enormously thick lava flow. Trachyte lava is far more viscous than basalt lava, so it tends to form domes and thick flows. Streams have eroded the deep canyons flanking the trachytic mass. The ridgeline above Ka'āpahu culminates in Kamakou Peak, the highest point on Molokai, at 4,961 feet. A cluster of largely reddened (oxidized) younger cones is visible to the west of the trachyte dome.

At milepost 10, look back upslope for excellent views of a deep, amphitheater-headed valley. These scooped-out valleys form by the gradual dissolution of the

Old, eroded trachyte dome (blue arrow), Ka'āpahu Hill, and associated flow (black arrow) on the south flank of East Molokai Volcano.

basaltic lavas exposed to Earth's atmosphere combined with the headward erosion of waterfalls, one of which may be visible. Other impressive valleys loom above the highway. Fairly recent flood deposits may be present in many places along this part of HI 450.

Between mileposts 16 and 17 watch for views of fishponds along the coast and the fringing reef, keeping the waves breaking a couple hundred yards offshore. These features mark the gradual subsidence of Molokai into the sea, a process that is slowing in present geologic times as tectonic drift carries the island into an area of upward arching crust caused by the heavier weight of the bigger islands to the southeast.

Between 18 and 19 miles east of Kaunakakai, look south across the sea on a clear day to see the islands of Lāna'i and Kaho'olawe. Also watch inland in the upper half of roadcuts for a distinctive layer of bright-reddish paleosol full of rounded corestones of dark basalt. A few hundred yards east of milepost 19, a lava flow probably covering the same red soil layer shows spheroidal or concentric weathering. The palosol appears in many other roadcuts along this part of coast. Lavas below the ancient soil level are basalt that erupted during the main shield-building stage of volcanic activity. Lavas above it are alkalic basalt that erupted late in the life of the volcano. The paleosol must represent many thousands of years of weathering, indicating that quite a significant period of time elapsed between the end of the shield eruptions and the first appearance of alkalic lavas, at least on this part of Molokai.

Mokuho'oniki Island and Kanahā Rock come into view at milepost 20. They are remnants from one of two known eruptions on Molokai in the rejuvenated phase of volcanic activity.

A wide pullout on the ocean side of the road just past milepost 20 is a good place to park and take a closer look at the red paleosol and the lavas lying above and below it. At the top of the steep slope next to the road is a dense basalt layer sitting above a thinner vesicular but heavily weathered basalt flow. A baked zone may be seen where the younger lava came into contact with the older flow. You'll find few if any crystals in these dark alkalic rocks. Beneath the thin, lower basalt is the familiar red paleosol. It lies perched atop a paler-gray lava flow extending all the way down to the ocean, packed with countless white plagioclase crystals. Some crystals are as much as an inch long. In addition to plagioclase, dark, squarish crystals of augite, a pyroxene mineral, are abundant. Similar tholeiitic, crystal-rich basalt is common in other Hawaiian shield volcanoes and typically erupts during the waning phases of shield-building activity.

A red paleosol separates the overlying alkalic postshield lava from the underlying, crystal-rich late-shield lava near milepost 20 on HI 450.

Underneath the plagioclase-rich flow is a jumbled layer of slide rubble or a mudflow, which in turn overlies another massive basalt bed containing few crystals. The contacts between the plagioclase-rich flow, paleosol, and the mudflow continue about 50 yards seaward of the road, forming the rocky, roughly flat-topped point. The surface of the plagioclase-bearing flow by the water's edge shows a honeycomb texture known as tafoni, caused by salt corrosion of solid rock exposed for a long period of time to wave spray and sea mist.

Molokai's Eastern Cape Area

East of the big pullout near milepost 20, the road narrows and continues down through a section of massive basalt flows as it hugs the shore of Honouliwai Bay. The plagioclase-rich flow and reddened paleosol then show up again along the road, starting on the far, eastern side of the bay. At Honouliwai and farther east, picturesque little pocket beaches between rocky points of black basalt are beige coral sand eroded from the offshore fringing reef.

Between miles 22 and 26, watch for the closest views of Mokuho'oniki Island as the road climbs inland to grassy rangeland. It passes into a forest of ironwood, gums, and other trees as it rounds the eastern cape, entering the windward rain belt. Roadcuts in the forest expose intensely weathered soil with a much lusher plant cover than farther west. After mile 25, watch for the widening of the road right before it begins a steep descent into Hālawa Valley. Another very thick layer of late-shield-stage pinkish-gray crystal-rich lava outcrops along the road, so heavily weathered that it crumbles to the touch, though the minerals retain their shapes and colors. Rock weathered in place like this is called saprolite. It may take thousands of years still before this ancient lava completely transitions to soil.

Near mile 26, Hālawa Valley Lookout provides views north across Hālawa Bay and east to the tall sea cliff, Kapaliloa, reaching out to Cape Hālawa at the easternmost end of the island. The road finally twists down to the floor of Hālawa Valley, where it leads out to the mouth of Hālawa Stream. The river has created a boulder jetty—now capped inshore by overgrown sand dunes—that divides two separate gray sand beaches.

Hālawa Valley is the location of one of the earliest-known settlements in Hawaii. Archaeological sites, including terraces, temples, and irrigation works, are numerous across the valley floor. Several thousand people may have lived here in prehistoric times. Farmers continued to cultivate many kalo (taro) patches near the riverside until tsunamis in 1946 and 1957 wiped out their fields and a small village near the river mouth. Waves washed as far as 1.4 miles inland. Fortunately, no one died.

If the weather is clear, you can see two spectacular waterfalls about 2 miles inland. A foot trail leads to one falls, but though publicly accessible, a guide service may constrain access. (The cultural tours are certainly worthwhile in any case.)

View looking inland at Hālawa Valley with Moa'ula Falls visible at the head of the valley. This area is one of the richest archaeological sites in all of Hawai'i.

A close-up view of 250-foot-tall Moa'ula Falls in the Hālawa Stream, reached by footpath through jungle from the end of HI 450.

HI 460 (MAUNALOA ROAD)
KAUNAKAKAI—MAUNALOA
17 miles

HI 460 follows an inland route across Ho‘olehua Saddle and the low shield volcano of West Molokai. This country is more open than two centuries ago when scrubby woodland covered it. The land was cleared for cultivation and grazing. Wherever cattle graze an area faster than grass grows, they soon strip the ground bare. Rainwater falling on bare soil runs off the surface instead of soaking in. In many places, broad swaths of red laterite soil on West Molokai are now exposed in barren, eroded terrain—a sobering monument to land abuse.

West of Kaunakakai, the highway climbs gradually onto the wide saddle linking East and West Molokai. Outcrops of lava are sparse in this terrain of deep, windswept soils, but near the coast, as in a roadcut 1.4 miles west of Kaunakakai, basalt does crop out.

Just south of the junction with HI 470, you can look directly upslope about 3 miles to see Pu‘u Luahine, a prominent late-shield-stage cinder cone of East Molokai vintage. This cone has a distinct horseshoe shape formed when lava exited from the downslope side of the vent and carried the falling cinder and spatter away while the cone grew.

HI 460 reaches the flank of West Molokai near milepost 9, 4.7 miles west of the intersection with HI 470. The steep, rugged slopes ahead are eroded scarps at the headwall of a massive landslide that dropped much of the northeast flank of West Molokai into the ocean sometime between 2 and 1 million years ago. Most of the mass moved underwater, but the slide scar extended well above sea level; West Molokai was a much higher mountain at the time.

Watch for deep laterite soil exposed in roadcuts where the highway begins climbing around mileposts 10 to 11. A small cinder cone of late alkalic basalt perches on the crest of the scarp to the south. The projections from the northwest and southwest rift zones of West Molokai intersect here, indicating that the vanished ancient summit of the volcano must have stood nearby.

A large roadcut towers above both sides of the highway near milepost 12. The rocks show remarkably regular, horizontal color bands, each 1 to 2 feet thick. Most consist of pale-gray ledges, the solid interiors of thin ‘a‘ā flows, separated by zones of oxidized, more easily eroded flow clinker. The flows erupted late in the main shield-building stage of the volcano, possibly associated with the growth of a lava shield. They are thin because they were flowing down a steep slope. ‘A‘ā flows pouring across more gently sloping terrain tend to be much thicker—as much as tens of feet thick per flow. The trend of the small dike of dense, light-gray rock cutting across the lower layers in the roadcut indicates that this site lies within West Molokai's southwest rift zone.

Several prominent, yellowish-brown to red zones a few inches thick between flows are baked ash and flow rubble. Each successive lava flow heated and altered the underlying wet ash, converting the iron in the upper layer to red iron oxide, while leaving the less baked ash below yellow. One especially conspicuous red zone 10 to 15 feet high in the central and eastern part of the cut is an ancient soil, a paleosol as much as 1 foot thick. A wide, smooth-faced layer higher in the section appears to be a well-sorted tuff from a more explosive eruption.

How long did it take for this sequence of layers to accumulate? Based on observations of modern eruptions on the Big Island and elsewhere, stacks of lava flows up to several hundred feet total thickness can pile up within a few months to years. That may have happened here, too, but the presence of a paleosol shows that there was a pause in volcanic activity that may have lasted thousands of years. There is no telling how much later the dike intruded these layers, but West Molokai must still have been an active volcano when it did.

Thin lava flows in a HI 460 roadcut at West Molokai. White arrow points to paleosol layer. Yellow arrow highlights an ash bed. Red arrow shows location of dike near roadside.

Spheroidal corestones weathering from basalt in a lateritic soil bed near milepost 13 along HI 460.

At around milepost 13, you'll see big roadcuts in rubbly black basalt layers that have deeply weathered to reddish-brown laterite and rounded corestones. Spires of less weathered basalt stick up into the gradually thickening lateritic soil beds.

Farther west, HI 460 passes onto a broad tableland that slopes gently seaward. Between miles 15 and 16 is a pullout with excellent views of O'ahu on a clear day, 30 miles across Kaiwi Channel (Maunaloa town is at milepost 16.5). If the island is visible, look for the iconic shape of Diamond Head (Lē'ahi), a rejuvenated tuff cone on the left side of the island. Other cones of the related Honolulu volcanic series may also be visible.

Maunaloa to Hale o Lono Harbor

Maunaloa Road/HI 460 turns sharply to the left as it enters the old pineapple plantation town of Maunaloa, near its end around milepost 16.5. After making this elbow turn, take the first right onto Mokio St. As Mokio Street leaves Maunaloa, it becomes a well-compacted dirt road skirting the edge of Molokai Ranch as it descends 5 miles to the southwest coast. The road traverses an arid savannah highlighted by red lateritic soil beds with scattered rounded corestones where the soil has washed away. Bits of black plastic littering the road are remnants from the pineapple days, when the plastic was used to prevent weed growth in the fields. Eventually the road begins a notably steeper descent and crosses onto grass-covered, postshield alkalic lava flows from Wai'eli, a vent to the east.

At the coast you'll find several shaded pullouts next to beaches of white sand sitting on cemented calcareous sand layers, or beach rock. Drive another half mile east along the coast to the old harbor and look upslope at the jutting cliffs nearby.

Severe soil erosion and corestones along the road to Hale o Lono. The fence posts here now hang above ground; the lateritic soil in which they were placed has disappeared.

Beach rock exposed where active beach sand has washed away near Hale o Lono. It may have formed when sea level stood somewhat higher than at present or during a time of greater beach sand supply. Fresh beach rock could be forming beneath the active sand to the right.

The vertical upper part of the cliffs consists of postshield Wai'eli lava. The massive alkalic flow is very uneven in thickness, possibly having accumulated in gullies or depressions as it flowed downslope. The flow shows crude columnar jointing in its thickest parts. Flow-banding can also be seen, showing that this lava was highly viscous, stiffer than the fluid tholeiitic lavas that built most of Molokai.

Lower in the slope lie two spheroidally weathered grayish lava flows related to the earlier shield stage of West Molokai's history. Near the foot of the slope is a very flat contact with a basalt flow so deeply weathered that only corestones and oxidized soil remain at its upper section. This paleosol indicates a long gap in time between eruptions in the waning stages of shield volcanism.

The harbor at Hale o Lono is enclosed by two artificial jetties. A ruined pier lies close to the road at harbor's edge. Most of the rock and sand that make up the jetties and provided support for the pier came from mining elsewhere on West Molokai. Engineers delivered it directly to the shore via large chutes positioned atop the concrete platforms still visible on the cliff above the harbor. Some rocky material also ended up being shipped to O'ahu as fill for the runways at Honolulu International Airport. The Hale o Lono Harbor remains active today, a starting point for canoe races.

The road ends on the eastern jetty, where you can enjoy a sweeping view of the coastlines of three islands: eastern Molokai, West Maui, and Lāna'i. The word *Molokai* means "turning" or "swirling ocean," perhaps in reference to eddies in offshore currents along this stretch of shore, certainly a challenge for canoers. The

In the cliff above Hale o Lono Harbor; a thick postshield alkalic flow overlies thinner, lighter-gray flows of shield-stage vintage.

eastern side of the jetty, facing the sea, is very rocky. Basalt rubble, sand, and coral bits cover parts of the parking area here, swept across the top of the jetty by occasional big storm swells.

KALUAKO'I ROAD
HI 460—Pāpōhaku Beach

6 miles

From HI 460, Kaluako'i Road leads to the shore just south of 'Īlio Point, the northwestern cape of Molokai. Trees along the road lean sharply to the southwest, parallel to the prevailing trade wind direction. Look to the north as you drive the first few miles of this road to see an open, deeply eroded landscape of red lateritic soil and white sands, part of the Mo'omomi dune field extending 3 miles inland from the north coast. Dune sands blew all the way across this part of Molokai toward the close of the last ice age, supplying much of the sand making up the western shore's beautiful beaches.

Postshield cinder cones and lava shields composed of basalt that erupted along the northwest rift zone lie north of the road and around Kaluako'i Resort. Okoli, a

North of Kaluako'i Road is red scrub desert with white calcareous sand streaks: the western trace of the Mo'omomi dune field.

prominent dark-red cone, stands next to the road 2.6 miles west from HI 460. The east and north flanks have been excavated for road gravel.

Pōhakumāuliuli

The first right turn as you approach the shoreline is Kakaako Road/Lio Place to Paniolo Hale. Follow this paved route 0.4 mile to its end near the condominiums, where you'll find beach parking. An additional short (0.3 mile) walk down an unmaintained dirt road takes you to a small, secluded white sand beach. A gully at the northern end of the beach has cut a channel into the thick red soil, exposing a couple of layers of coarser flood debris, much of it containing cobbles already rounded from spheroidal weathering.

Walk along the rocky headland on the north side of the beach; the rocks along here show a honeycomb texture (tafoni) formed by salt weathering. Salt spray soaks into the surface of the rock, then little salt crystals form as it dries. The crystals growing in the rock pry little chips from the surface. More salt spray concentrates in the depressions, localizing the weathering and deepening the holes.

Pick your way, mindful of the tide and surf, across the headland north of the beach to discover another, larger white sand beach on the other side. Here, too, is a spectacular view of a large cinder cone named Pōhakumāuliuli, the western flank of which has been removed by wave erosion. Layers of cinder on the north side of the eroded cone dip north; a small exposure of cinder near the south side dips south; and lava flows are thicker and lie flatter in between. Offsets by faults or slumping and denser flow lava or dikes are also visible inside the cone. The old crater still lies largely intact inland, partly buried by sediment out of sight from the beach. Near the headland, outcrops of ancient, hardened dune sand also protrude, developed long after the cone was active. Pōhakumāuliuli did not form next to the shore originally. Tectonic subsidence of West Molokai dropped it to sea level where the ocean could eat it away, probably hundreds of thousands of years after it erupted.

Flood deposit layer in lateritic soil beds at the mouth of the gully near Paniolo Hale beach.

Pōhakumāuliuli cinder cone just north of Kaluako'i resort. Wave erosion has cut open its interior, revealing strata and dikes likely emplaced by the magma that built the cone. Layered limestone beach rock also outcrops along the active sandy beach.

Kaiaka Rock

After passing the resort on Kaluako'i Road, watch for the Kaiaka Rock sign and turn right onto Kaiaka Road. You will park on the ancient postshield cinder cone Pu'u o Kaiaka. Follow the trail down to the white sand beach below where you can see exposures of calcareous beach rock etched by dissolution. To the left as the trail reaches the beach is one of the most accessible exposures of a volcanic cone on the island—a 50-foot-high wave-cut cliff containing landward-dipping layers of red to black cinder, spatter, and lava bombs. These are interspersed with denser

Wave-cut face of Pu'u o Kaiaka ("Sea Shadow Hill") cinder cone. A few darker, denser layers of lava—mostly welded spatter—stand out from the landward-dipping layers of highly oxidized cinder.

Dense dike rock cutting cinder and spatter in a boulder at Pu'u o Kaiaka.

layers where the rain of spatter fell fast and hot enough to merge and move as short-traveling lava flows. The layers are thicker and more prominent toward the top of the section, and even seem to thicken toward the sea, implying that the center of the cone may have been just offshore but has since been eroded away. Another trail to the south from the parking area leads to the northern end of Pāpōhaku Beach and a sweeping coastal panorama.

Pāpōhaku Beach and Farther South

Kaluako‘i Road leads to several access points and the county facilities for 2-mile-long Pāpōhaku Beach, one of the longest calcareous sand beaches in Hawaii. Without a fringing reef to provide a local source of sand, it seems likely that the sand is eroded from local dune rock. Sands related to the Mo‘omomi Dunes several miles to the northeast probably contributed much of the original sand supply. Until the 1970s, this secluded beach was mined for sand to help build the Hale o Lono pier and replenish Waikīkī beaches on O‘ahu.

The lack of an offshore reef also leaves the coast with nothing to absorb the force of the ocean waves. The offshore slope is very steep. Riptides are common because waves do not meet shallow water until they come crashing onto the beach. Swimming is treacherous here.

The rocky headland at the south end of Pāpōhaku Beach showcases limestone weathering. You can drive there by continuing about 2.2 miles farther south on Kaluako‘i Road, then turn right onto Pohakuloa Road, then right again onto Kulawai Place, and then left on Kaula Ili Street. Just before Kaula Ili dead-ends, a short side road accesses public parking at the edge of a bluff. Looking north from here, you

Ancient dune rock, limestone, slowly dissolving into stone forest structure near Pu‘u Koa‘e, just south of Pāpōhaku Beach.

will see the small hill of Pu'u Koa'e, another shoreline vent cone but largely covered with white calcareous sand on an underlying platform of ancient dune limestone. A short walk along the bluff line (take care of the edge!) reveals fantastic scenery of dissolved dune limestone, both from sea sprays and the atmosphere. In places the corroded rock is so jagged that it sticks out like a forest of spears. In China, similar larger structures in limestone are called *shilin*, or "stone forest." You may be able to make out the original, finely spaced layering of the dunes preserved in the limestone as it is slowly dissolving away.

Farther down the road, public access routes spurring off the main road lead to several little pocket beaches and rocky coves. The southernmost beach, Kapukahehu (or Dixie Maru), and the next beach south of it, accessible only by trail, are white sand beaches popular with surfers. The farther south we travel from the blustery northwest cape, the less sand there is and the rockier the coast becomes.

On clear days, you can see neighboring O'ahu from the shoreline, 25 miles across the Kaiwi Channel. O'ahu stands on the same general submarine platform as Molokai, Maui, Lāna'i, and Kaho'olawe, but the Kaiwi Channel is deeper than the channels between the other islands. O'ahu may once have been linked to Molokai, but tectonic subsidence has long since separated them.

HI 470 (KALA'E HIGHWAY)
HI 460—KALAUPAPA OVERLOOK

6 miles

Branching off HI 460 (Maunaloa Highway) in Ho'olehua Saddle, HI 470 winds through the Molokai upland past the town of Kualapu'u (start of the Farrington Avenue) into dense ironwood forest and a spectacular lookout on the rim of the big sea cliff southwest of the Kalaupapa Peninsula. Just before reaching the overlook, the road intersects the start of the Kukuiohāpu'u Trail, which descends 1,600 feet to the west coast of the peninsula. You have the option of walking or riding a mule down this roughly 2-mile-long (one-way) track. Either way, a paid permit is required to do so. Knowledgeable guides are provided who can introduce you to the rich history of the national historical park below. A more expensive option is to fly into Kalaupapa Airfield at the northern tip of the peninsula for a visit.

At its highest point, Kalaupapa (meaning "flat plain") Peninsula rises about 500 feet above sea level. It originated as a lava shield that grew atop giant blocks of landslide material sometime between 570,000 and 340,000 years ago. The looming cliff at the southern base of the peninsula, the scarp of the Wailau debris avalanche, rises 2,000 to 3,000 feet and opens into two deep valleys that largely eroded before the peninsula formed. Damien Road leads from Kalaupapa settlement to the southwest shore of the peninsula, affording a breathtaking view of the even taller cliffs beyond, all left by the giant debris avalanche late in the history of East Molokai volcano. These cliffs stretch for nearly 20 miles and are the tallest sea cliffs in the world. They are composed of countless mostly tholeiitic lava flows.

Another short trail from the Kalaupapa Overlook parking area leads to Phallic Rock (Kaule o Nānāhoa), a cultural site set in a cluster of basalt outcrops rounded by weathering into unusual shapes.

Kalaupapa Peninsula, 1,600 feet below, viewed from the overlook at the end of HI 470. The small shield volcano making up this isolated peninsula grew during a spasm of rejuvenated volcanism long after the northern flank of East Molokai Volcano collapsed into the sea.

Fantastical rock forms appear in deeply weathered lava on the trail to Phallic Rock, next to Kalaupapa Overlook.

HI 480 (FARRINGTON AVENUE)
Kualapu'u—Mo'omomi Sand Dunes

8.4 miles

Beginning as a turnoff from HI 470, Farrington Avenue is the main road through the town of Kualapu'u. If you follow it westward out of town, the avenue eventually becomes a dirt road leading downslope 8.4 miles to the Hawaiian Home Lands Recreation Center at Mo'omomi Bay. A four-wheel-drive or high-clearance vehicle is strongly recommended for the unpaved part, and it should be avoided entirely when muddy. But the visit is certainly worthwhile. A walk westward for a little less than 1 mile from the recreation center takes you into the Mo'omomi Nature Conservancy Reserve, an area of active and ancient, hardened sand dunes, pocket beaches, and spectacular coastal scenery.

The main part of the Mo'omomi Dunes probably formed during the latest ice age, when sea level was a few hundred feet lower than now. Reefs presently submerged offshore were dry and feeding sand into the wind. Since the ice age, precipitation of dissolved calcite from groundwater in the buried lower levels of the sand dunes has converted them into hard, finely layered limestone. Subsequent shifting of the loose sand cover has exposed these hard base rocks in many places. Shifting sands have also led to the preservation of bird fossils of about thirty species, ten of them now extinct. These include a fossil ibis, giant sea eagle, and a long-legged (stilt) owl. If you happen to discover any fossils while exploring, please leave them in place. This area is still being researched. Also, please be sure to stay on paths, open ground, or

Eroded layered ancient dune rock, probably no younger than 20,000 to 15,000 years at Mo'omomi Dunes on Molokai's windblown northwest coast.

rock to avoid disturbing nesting wedge-tailed shearwater burrows and stepping on rare or endangered plants. Close all gates behind you; recent raids by wild dogs have killed a number of birds.

A hike west along the shore quickly takes you to an area of intensely weathered basalt with abundant rounded corestones littering the ground or in the process of forming as the enclosing lava slowly weathers away. Weathering around some in-place stones has formed concentric bands accented by rust-red oxidation on the surface. Farther west, you'll cross active dunes and reach a wide sandy beach that has sealed off a small stream mouth, creating an estuarine pond. At the far western end of this beach is a spectacular outcrop of hardened, white limestone, jagged from dissolving along the fine layers of cemented dune sand that compose the rock. Reaching this point, you may notice a lot of litter (plastic, fish nets, and other flotsam) cast upon the beach. Volunteers periodically clean this up, but here you are only a few hundred miles south of the Great Pacific Garbage Patch, and the task of keeping these shores trash-free is endless.

West of the limestone, climb upslope to a gate in the reserve fencing admitting you to a headland with more deeply corroded limestone. In places here you'll discover the hardened, calcareous molds of numerous tree roots—tubes as much as an inch across that lace across the ground like veins in a giant hand—all that remains of an ancient forest that once grew here. The dunes have since shifted, revealing the old root traces. Perhaps moving sand killed the trees, or overgrazing, or a combination of both. Erosion has since exposed them. The main shearwater nesting area is just past this point, so we suggest turning around here.

A field of corestones on the way to Mo'omomi Dunes (background).

Oxidation bands enclosing spheroidal core rocks in deeply weathered basalt, on the way to Mo'omomi Dunes.

Tubes of calcareous rock show the location of ancient tree roots at a headland in the Mo'omomi dune field. The ancient forest here was likely smothered by shifting sands centuries ago.

MOLOKAI FOREST RESERVE ROAD
HI 460—WAIKOLU VALLEY LOOKOUT

16.2 miles

This rugged dirt road requires a high-clearance vehicle. If you continue past Waikolu Valley Lookout another couple of miles on the Kamakou Nature Conservancy road to the Pēpē'opae Bog Trail, you will also definitely need four-wheel drive. Avoid going during wet weather conditions, which generally includes all morning hours. Dedicate four to six hours for the complete round-trip.

Despite this forewarning, a drive up the Molokai Forest Reserve (or Maunahui) Road can certainly be worthwhile. The road begins at the Homelani Cemetery turnoff from HI 460, 3.5 miles west of Kaunakakai. It gradually ascends through three forest belts: grazed dryland kiawe shrubs at lower elevations (cemetery and surrounding ranchlands), an intermediate belt of imported forest trees (ironwood, eucalyptus, lodgepole pine) in the Molokai Forest Reserve, then finally native cloud rain forest in the Kamakou Preserve. Small reservoirs important to island water supply cling to the slopes in a few available flat areas within the forest reserve. You may see pipes related to this reservoir system along the road in places. Water feeds to a main reservoir in Ho'olehua Saddle. At around 8.8 miles the road passes a pit, a Lua Moku 'Iliahi, dug into the ground during the sandalwood export frenzy of the early 1800s.

Waikolu Valley from the overlook at 4,200 feet elevation on Molokai Forest Reserve Road. This steep coastline originated when the northern flank of East Molokai Volcano slid into the sea sometime around 1.5 million years ago.

These pits were used to measure enough harvested sandalwood to fit a ship's hold before the logs were carried to the coast. As you near Waikolu Valley Lookout, note the light-colored lavas cropping out in roadcuts. These are trachyte, the alkalic lava of the postshield stage of volcanism—East Molokai's last (volcanic) gasp.

The culminating attraction, Waikolu Valley Lookout and picnic area, provides views of the 3,000-foot-deep Waikolu Valley and the north shore of the island 3 miles away. Dizzying valley walls are laced with waterfall runnels and pools. Okala Island, an offshore sea stack, may be visible just past the surf line far below. This overlook is one of the finest places to get a sense of the magnitude of the great Wailau landslide and what it did to ancient Molokai. Viewing conditions are generally best late in the day after the clouds begin to dissipate.

The Pēpē'opae Bog (Boardwalk) Trail starts at the end of drivable road in the Kamakou Preserve, 2.2 miles past Waikolu Valley Lookout. The light-gray and white rocks around the trailhead are all trachyte, but you are not apt to see much rock after you start hiking. The narrow boardwalk crosses dense native forest and mountaintop bog lands and gradually ascends to the summit of Kamakou Peak, nearly 5,000 feet tall—the highest point on the island. It leads to another overlook from which you can look far down into Pelekunu Valley, also opening out onto Molokai's cliffy north shore. Keep to the planks; the surrounding earth is perpetually muddy. The path closely follows the rim of East Molokai's ancient, lava-filled caldera. Count yourself lucky if you aren't in mist and cloud as you near the top.

Lāna‘i

Lāna‘i, the smallest of the easily visited Hawaiian Islands, is 18 miles long and 15 miles wide, encompassing only 141 square miles with less than 50 miles of shoreline. The island nestles in the dry rain shadow of neighboring Maui, leaving it without a wet windward side. Its single shield volcano first appeared above the waves around 1.5 million years ago. The ancient volcano continued to erupt for the next quarter million years or so but in its primacy never rose more than about 4,500 feet above sea level. The island has since lost much elevation, not so much because of erosion—which certainly has been considerable—but due to tectonic subsidence of more than 1,000 feet since the volcano last erupted.

Lāna‘i has an unusual volcanic history as far as Hawaiian volcanoes go. Like each of its neighbors, it grew by erupting countless tholeiitic basalt flows. But as it aged, it did not experience the typical late stage of alkalic eruptive activity—and has

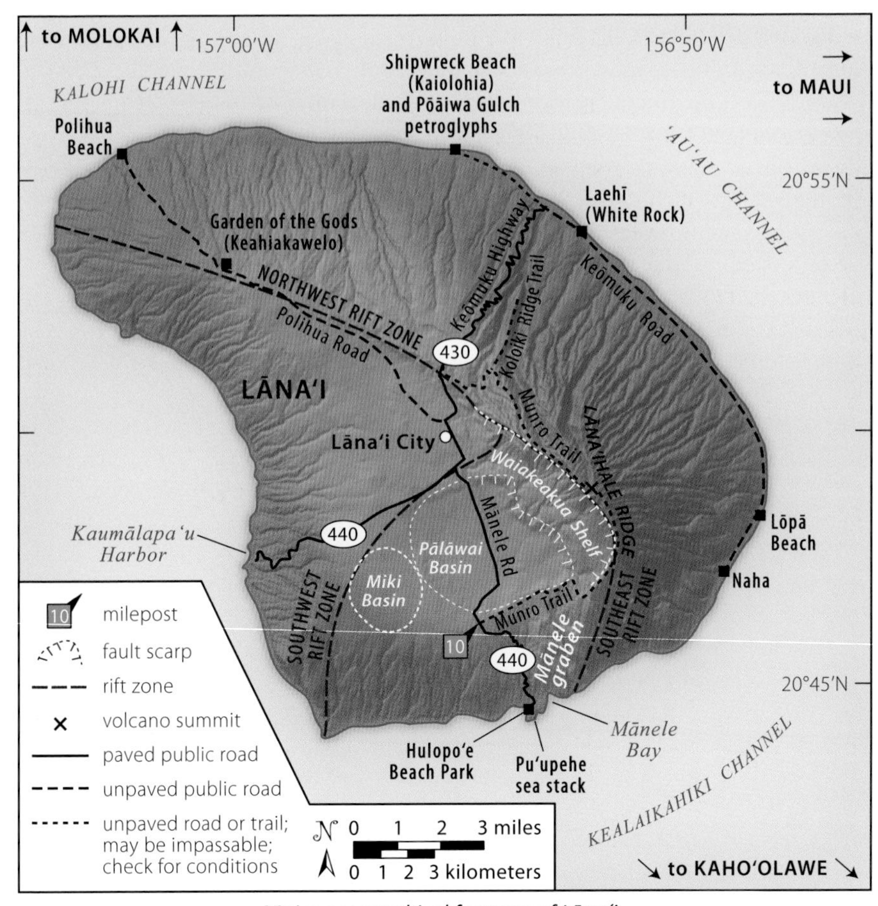

Major geographical features of Lāna‘i.

never experienced rejuvenated volcanism. Still, deep well drilling by the University of Hawai'i to more than 3,150 feet, just a few miles southwest of Lāna'i City, has revealed the presence of groundwater at 150 degrees Fahrenheit, indicating that the old volcano's core remains hot.

Three rift zones appear to radiate from the south-central part of the island. The northwest rift zone is the longest and was probably the most active; it built the island's prominent northwestern bulge. Several dikes appear in the walls of shoreline cliffs at the northwestern cape, where the rift zone meets the sea. Inland of the cape, only a few vents have survived erosion.

The southwest rift zone enters the ocean at the island's southwestern cape. One greatly eroded cinder cone still rises in the rift zone. You can also see an eroded lava shield and a great swarm of dikes exposed in the sea cliffs. Large fissures that have opened along some of these dikes show that island slopes have grown weak in the area of the cape.

The third and most recently active rift zone extends southeast, where numerous small cones and craters, subdued by aging, survive. A network of faults trending generally southeast along the axis of the rift zone border a sunken strip of land, the Mānele graben. A great swarm of rift zone dikes is exposed in the shoreline cliffs just east of Mānele Harbor.

Most Hawaiian geologists have identified the Pālāwai Basin south of Lāna'i City as the old summit caldera of Lāna'i. They have interpreted Miki Basin, adjoining the western rim of Pālāwai Basin, as a major, largely buried pit crater linking the basin with the southwest rift zone. The deep drilling mentioned above reveals shattered rocks, dikes, and the intensive filling of vesicles with zeolite crystals, all characteristic of a once-active caldera. The stack of thick, horizontal lava flows that form a wide, flat shelf (Waiakeakua) east of Pālāwai Basin, however, look like the kind that erupt onto a caldera floor. If so, the basin may simply have been a centralized subsidence area inside a larger depression, much like the collapsed inner floor of Kīlauea caldera on the Big Island today.

Faults that follow the northwest rift zone and continue south to the Mānele graben seem to strike across Pālāwai Basin, but younger sediment mostly obscures these faults in the basin. Whereas the faults bordering the eastern margin of Pālāwai Basin could at least partly be explained as evidence of caldera collapse, faults continuing to the north and south of the basin must be tectonic in origin and might relate to a mega-landslide that tore the southwestern side of Lāna'i sometime toward the end of, or soon following, volcanic activity. Sonar maps of the ocean floor reveal a broken-up mass of shield fragments stretching nearly 70 miles from the southwest coast of Lāna'i and spreading into two huge lobes where it enters the Hawaiian Deep. Geologists call this the Clark slide. It must have moved very fast to have traveled so far. This catastrophic event could also explain the towering sea cliffs that rise to 1,000 feet above the waves. The cliffs rim much of the western coast of the island.

The natural landscape of Lāna'i today is utterly different from what it was 150 years ago, and the transformation is due entirely to human activity. When early Hawaiians first arrived here, Lāna'i was completely covered with forest vegetation scarcely seen today. The dominant herbivores, giant flightless ducks (moa-nalo), prowled the woodlands. Hawaiians modified the environment as they killed off the moa and began cultivating, but the biggest changes began after the arrival of

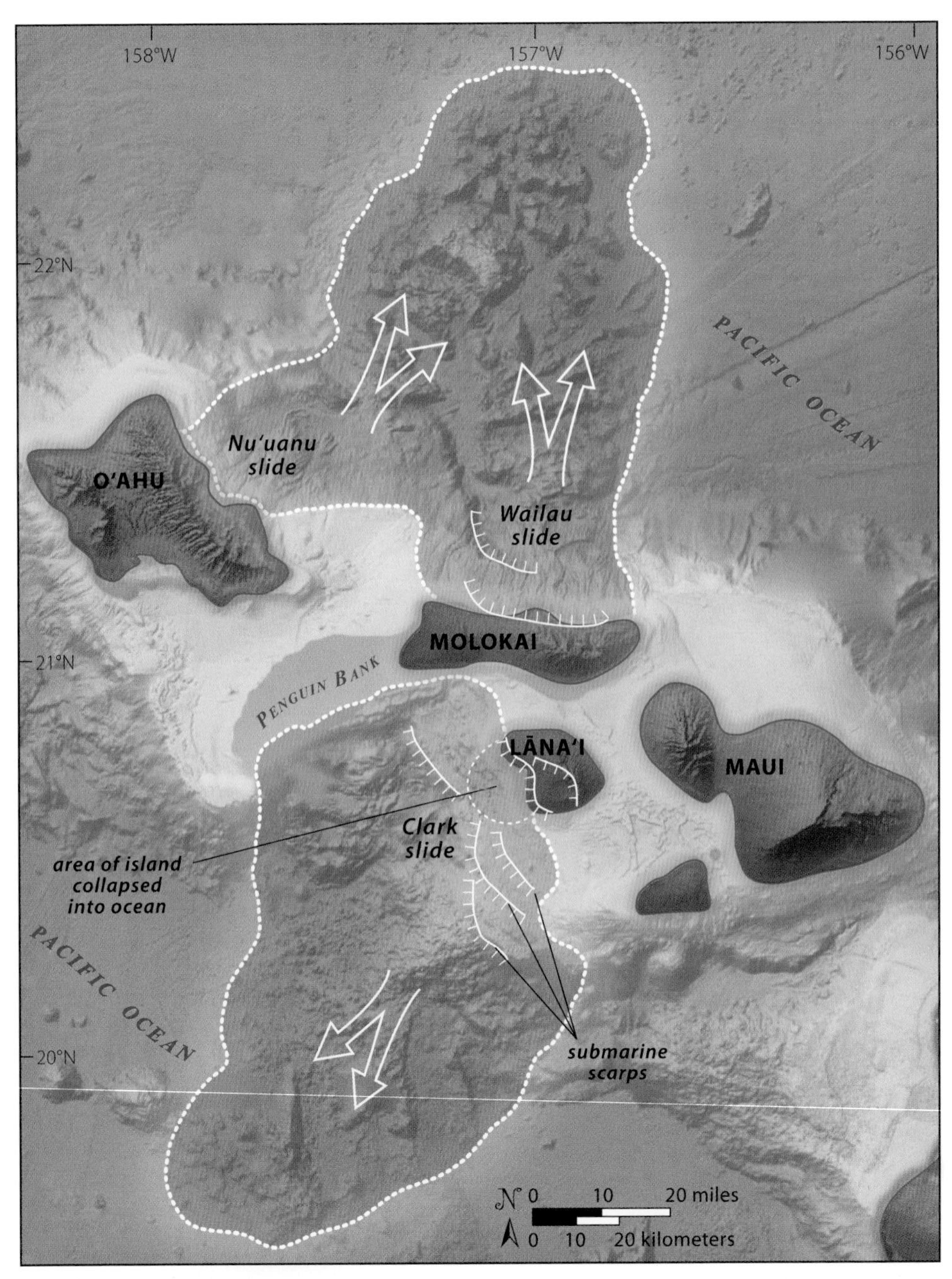

The giant Clark submarine landslide removed the southwest side of Lāna'i, sometime between 1.5 and 1 million years ago.

Europeans and Americans. In the mid-nineteenth century the first ranching of sheep and goats started, while the Hawaiian population plummeted owing to diseases and emigration. By 1901, only about 125 people lived on Lāna'i, while tens of thousands of sheep and cattle roamed essentially free, stripping the forests. Sugar cane came soon after 1900; and in 1920, Charles Dole purchased most of the island for pineapple production. Conversion to a dry, open landscape accelerated. Streams and springs that once flowed year-round on Lāna'i's rugged east coast disappeared, and vast amounts of soil sediment added new land to the island in the form of coastal flats hundreds of feet wide. The outpouring of sediment filled ancient fishponds and buried nearshore corals and reef flats. Kiawe, ironwoods, and other deliberately introduced trees replaced native species in order to provide shade and help arrest the loss of fertile soil. Today, conservation efforts are underway to reduce or heal these impacts, but Lāna'i has forever changed. More than any other for the roadside geologist, it is the Island of Erosion.

HI 440 (MĀNELE ROAD)
LĀNA'I CITY—MĀNELE HARBOR

8 miles

A good paved road leads to the Mānele resort area from an intersection with the Kaumalapau Highway at the southern end of Lāna'i City. This southern branch of HI 440 is the most traveled route for visitors on the island.

Leaving the Lāna'i City area, the road to Mānele winds down through low forest to enter the Pālāwai Basin, site of Lāna'i's long-extinct caldera. For the next 3 miles, the road crosses the roughly 350-foot-deep floor of Pālāwai Basin, flanked on both sides by stately rows of araucaria pines. About 1.5 to 3 miles east of the road, the ragged skyline of Lāna'ihale Ridge rises to an elevation of 3,370 feet, the island's highest point. The stream-cut bench flanking the ridge directly above the basin is a down-dropped fault block, Waiakeakua ("Spring of the Ghosts"), probably related to caldera collapse near the close of Lāna'i's volcanic activity hundreds of thousands of years ago. The bench rises about 500 feet above the basin floor.

Exiting the basin at mile 3.7, turn left at the intersection to follow the Mānele Road the rest of the way down to the sea. (The route straight ahead at this intersection is reserved primarily for construction vehicles.)

At the divide separating Pālāwai Basin from the southern slope of Lāna'i, the neighboring island of Kaho'olawe, across Kealaikahiki Channel, comes into view. Roadcuts reveal dark lava blotched white in many places with caliche (calcium carbonate) resulting from weathering in the dry, hot climate. Reaching the shore, the road winds to an end at Hulopo'e Beach Park, a fine, wide carbonate sand beach with picnic tables and restrooms—the best accessible swimming place on the island. Walk west along the beach toward the Mānele Bay resort. If you look carefully along the rocky shore just past the end of the beach, you can see a blowhole spraying from time to time. A small stone platform (heiau)—the site of the ancient Kapiha'ā temple—rises in the grass just inland from the end of the beach. Tread respectfully because the entire area is the site of a prehistoric Hawaiian village.

You'll also come upon a rough, shallow, dry streambed containing an extraordinary deposit. Walk along this channel to observe embankments of black basalt

MUNRO TRAIL

The 12.8-mile-long Munro Trail follows the spine of the island's mountainous central ridge, along the way crossing Lāna'i's highest summit—Lāna'ihale (3,370 feet). The trail is occasionally open to adventuresome four-wheel and ATV drivers but could well be closed for much of its length owing to washouts, mud, and other maintenance issues. Inquire locally with rental agencies or the police department about accessibility.

The south end, far more favorable for hikers bound for Lāna'ihale than the north end, is reached by taking Munro Trail Road from HI 440 at the southern rim of Pālāwai Basin (at milepost 10). Follow the road about 3.6 miles east into the foothills, where you can park (avoiding a much steeper and trickier route up ahead) next to the Naha Overlook picnic site and Pu'u o Manu Heiau Trailhead. From this location at an elevation of 1,900 feet, it is a strenuous but rewarding 5-mile round-trip hike to the summit, where you may enjoy sweeping views of Pālāwai and Miki Basins, the southwest rift zone, and Lāna'i City. This approach also allows you to see at close hand the mile-wide, fault-bounded bench of Waiakeakua, lying between Pālāwai Basin and Lāna'ihale ridge.

Even if you choose not to hike past the Naha Overlook, the view from here is certainly one of the most impressive on Lāna'i, providing a panorama of neighboring seas and islands, including West and East Maui, Kaho'olawe, and even the Island of Hawai'i, where on clear days both Mauna Kea and Mauna Loa are visible—the latter nearly 100 miles away. If you look carefully at the slope of West Maui, you may spot the great battery of windmills that produces much of Maui's electricity.

Whether you take Munro Trail from its northern or southern terminus, you'll experience a striking transition in climate and vegetation as you climb higher, finally entering a cool, drippy cloud forest at the top. Here are some of the only native trees and plants left on the island—the widespread false staghorn fern (uluhe), pa'iniu, koa, and 'ōhi'a among them. Mostly, though, the woods consist of mixed blue gums, ironwoods, strawberry guava, silky oaks, araucaria and true pines—non-natives introduced early in the last century to protect water resources and reduce the alarming rates of soil erosion. The nineteenth-century conservationist George Perkins Marsh introduced the basic ideas behind this effort. While soil erosion still appears dramatically in many places along Munro Trail, perhaps now the overall loss rate is much less.

On Munro Trail near the summit of Lāna'ihale. A mix of native understory and imported forest trees crowds the surrounding landscape.

cobbles mixed with buff-colored reef limestone. This mishmash is the rubble left from a giant tsunami that battered the southern side of Lāna‘i around 127,000 years ago, based upon radiometric dating of the old limestone chunks. The probable cause was the youngest of three great ‘Alika landslides from Mauna Loa's western flank on the Island of Hawai‘i, 90 miles to the south. The titanic sea waves washed inland to as much as 1,000 feet above present-day sea level, tumbling and rounding rocky debris torn from living reefs and square miles of older lava flows. The catastrophe also certainly stripped lower slopes of the island bare of vegetation.

A deposit of mixed basalt and coralline limestone deposits, left by a gigantic tsunami more than 100,000 years ago, lines a gulch near the Mānele Bay resort.

A blowhole erupts frequently just past the west end of Hulopo‘e Beach.

At the east end of Hulopo'e Beach, a trail follows the rocky shoreline out to a point, offering views of an enormous, 125-foot-tall sea stack (Pu'upehe), wave-cut terraces, and a red sea cliff composed of countless thin, piled 'a'ā flows. You can follow the trail a short distance farther along to the highest point at the eastern rim of a small bay ("Shark's Bay"), directly overlooking the sea stack. It consists of the same red 'a'ā flows as the adjacent cliffs, and no doubt at one time was linked to the land. The extraordinary thinness of the individual 'a'ā flows can be attributed to frequent rapid outpourings of lava on moderate to steep slopes. Atop the sea stack stands an ancient heiau, an amazing testament to the construction skills of early Hawaiians and the sacredness with which they regarded this site.

Where was the source of all these lava flows? Some geologists think that they erupted from the double-crested Pu'umāhanalua cones close to the rim of Pālāwai Basin. But the low hill just inland from the sea stack, Pu'u Mānele, possibly a lava shield, might be the actual source. Despite its semicircular shape, Shark's Bay is not a water-filled volcanic crater. The lava flows exposed in the cliffs do not incline away from the bay's shoreline all around it—the expected pattern for crater overflows.

HI 430 (KEŌMOKU HIGHWAY)
LĀNA'I CITY—SHIPWRECK BEACH AND KEŌMOKU
7 miles

Begin this route by zeroing your odometer where the entrance drive to Kō'ele Lodge meets the highway heading north out of Lāna'i City. The road gradually rises to a divide, at around the 1.0-mile point, separating the west Lāna'i tableland from the northern slope of the island. Just past the Cemetery Road intersection, West Maui can be seen rising from the sea across 10-mile-wide 'Au'au Channel. A short distance past the Cemetery Road intersection, look for a wide pullout to the right—directly across the highway from a roadcut of deeply weathered lava flows. The transition from moderately weathered basalt at the base of the outcrop to a full bed of reddish soil at the top is well displayed here. Infiltrating water percolating through these rocks over thousands of years is gradually leaching them of most of their original constituents,

KOLOIKI RIDGE TRAIL

The Koloiki Ridge Trail branches off the Munro Trail near its northern end (see the previous sidetrip guide for the southern end of the Munro Trail). The north end of Munro Trail is reached just past the Lāna'i Cemetery entrance, a short distance north of Ko'ele Lodge. There is a parking area next to the cemetery for hikers. Follow the Munro Trail southeast from the cemetery about 2 miles (one way) through heavily wooded, fast-eroding Hulopo'e Gulch to reach the posted Koloiki Ridge trailhead.

Koloiki Ridge Trail itself is a gentle half-mile (one way) walk from Munro Trail onto a narrow ridge crest lying between two large canyons: Nai'o Gulch to the west and Maunalei Gulch to the east. The view of Maunalei Gulch off the end of the trail (just past the grove of araucarias) is truly breathtaking, showing many of the landforms characteristic of steep-walled erosional valleys forming in tropical settings. These include cliffy headwalls and

Looking toward the cloud-swept head of Maunalei Gulch from the end of the Koloiki Ridge.

Island highlands with high stands of araucaria trees along the Koloiki Ridge Trail.

close-spaced, sharp drainage grooves in flanking slopes. Intensive chemical weathering, accelerated in the warm, moist climate, accounts for this exotic topography. The old flows are dissolving as well as being scraped or toppled away.

Looking across Nai'o Gulch, you can see a cross section of the island's lateritic soil cover gradually diminishing downslope where the terrain is steeper and historical grazing was more intensive. Red streaks of soil-stained lava on the far slopes of the gulch show that soil stripping continues into recent times.

Saprolite capped by a layer of lateritic soil at a roadcut on HI 430 just past Cemetery Road and mile marker 1.0.

leaving a crumbly residue of clay minerals including smectite and kaolinite. Iron oxide imparts a red color to the soil and results from an extreme degree of leaching. The resulting red soil beds are called laterites. Where the ghostly outlines of original flow structure can still be seen in the soil, geologists call such material saprolite.

A pullout at around mile 2.2 provides a sweeping view of East Molokai and West Maui, with 8-mile-wide Pailolo Channel separating them. West Molokai comes into view as you continue a bit farther downslope. Between 2.5 and 3.0 miles, dense thickets of non-native, fast-growing ironwood trees have been planted to help hold fragile soil in place. Farther downslope, though, it is too late; the soil cover is mostly lost, exposing jagged bedrock lava.

Knobs and pillars of weathering lava jut from rapidly eroding ground across much of the lower mountainside between miles 3.1 and 4.5. Originally, these prominences were fully covered with soil, as seen in the roadcut at mile 1.0 and suggested by many other roadcuts in this area. The terrain is not without its mystifying beauty, but it is certainly a landscape of loss.

HI 430 reaches the coast at a T-junction with two dirt roads in a dense forest of non-native kiawe trees. The road to the left leads a few miles to Shipwreck Beach; the rightward route follows the island's northeastern shore a dozen miles past numerous historical sites, small beaches, and other coastal features. Both routes are described below.

Deep soil erosion has exposed numerous knobs of resistant bedrock lava at the lower end of HI 430.

Shipwreck Beach Road

This 1.4-mile-long sandy but compact road ends at a pullout overlooking the coast of Kalohi Channel with Molokai on the other side, only 10 miles away. Walk a few hundred feet farther along the shore to the west and you'll soon see the wreck of the YOGN-42, a ferro-cement tanker built in 1943 that ran aground on nearby Kahā'ulehale reef in the mid-1950s. Also at road's end is the trailhead to a small petroglyph site in Pōāiwa Gulch, a short distance away.

Keōmoku Road

This 24-mile round-trip dirt road is mostly well compacted and easily passable to vehicles, though there may be patches of soft sand to negotiate along the way, and it can become quite muddy during periods of rain. Four-wheel-drive vehicles and light trucks ordinarily have no problems. Geologic attractions are sparse, but the route ties in closely to the story of soil erosion and ecological transformation. Zero your odometer at the intersection with HI 430.

The most interesting geologic feature along the Keōmoku Road is encountered only 1 mile from the start, at Laehī ("White Rock"), a wide strip of rough limestone that formed thousands of years ago—probably during the latest ice age—when winds blew a large patch of exposed reef sand inland as far as 2 miles, building dunes that later consolidated into solid rock. The original planar layering of the dunes can be seen in many places. Chemical attack from rainfall and humidity has also given the

Cemented reef sand that was once dunes at Laehī on Keōmoku Road.

limestone an extremely jagged, corroded appearance. You need good boots to walk far across this surface. (Geologists call it "tear-pants structure"!) You can reach the point of land formed by the limestone by following the road downslope to where it bends at the coast, then walk the beach back west a short distance.

At mile 5.4, the road reaches Ka Lanakila Church, in use from 1903 until 1951, in the heart of the abandoned sugar plantation town of Keōmoku, once the population and commercial center of Lāna'i (1899–1920). Introduced plantings included drought-tolerant kiawe (mesquite) trees to provide shade in a hot, deforested landscape. With the collapse of the sugar enterprise, Keōmoku withered. The kiawe trees spread, eventually creating the dense forest you presently see enclosing the road. Meanwhile, sediment continued to pour into the ocean from accelerated soil erosion. In some places, the shoreline shifted seaward as much as 400 feet! A measure of this change can be appreciated by taking the historical trail across the street from the church. The remains of a boat used to bring supplies from Maui to Keōmoku can be seen here—currently stranded far inland.

At mile 9.4, the road skirts the shore at Lōpā Beach, a narrow strand of coralline sand. Looking right (south) down past the end of the beach, you see a heavily vegetated point that was once an ancient Hawaiian fishpond called Loko Lōpā. Now a bird refuge, it has largely filled with sediment, providing a platform for the trees you see growing there.

Reaching mile 9.6, the road crosses an 'a'ā flow possibly related to one of several vents a short distance upslope. The road then abruptly ends at Naha (mile 11.9), next to a bank of mixed pāhoehoe and 'a'ā flows, also related (as at mile 9.6) to one of Lāna'i's latest eruptions.

POLIHUA ROAD
LĀNA'I CITY—KEAHIAKAWELO (GARDEN OF THE GODS)

6 miles

This 12-mile round-trip route begins as the first turnoff to the left a few hundred feet north of Ko'ele Lodge on HI 430. Look for the sign directing you to Keahiakawelo, and turn there. Zero your odometer at this intersection.

The first few miles of Polihua Road cross open tableland with rich, red lateritic soil. Bits of black plastic appearing in the roadbed relate to weed mat set down during pineapple plantation days. It cannot be easily removed and may be found throughout broad swaths of the landscape. The road then enters a fragment of moderately moist forest on the 1,800-foot-tall upland of Kanepu'u. At about mile 6, you'll find a short (ten-minute) nature trail that leads visitors through a fenced enclosure near the northwestern edge of the forest. Interpretive signs give an excellent introduction to some important native plants and the challenges facing them.

A few hundred feet past this point you'll reach Keahiakawelo Heritage Site and the Garden of the Gods. The strange landscape, among Hawai'i's most astonishing, is a product of erosion. It is certainly fantastic—countless rounded corestones rest on smooth multicolored beds of residual soil in a rolling, otherworldly terrain. The great number of rounded rocks clustered atop the present-day soft, rounded soil bed gradually accumulated from many levels in the originally much deeper soil cover. The rocks are too heavy for water to transport and became concentrated as the great

At Keahiakawelo, Garden of the Gods, corestones from a residual soil rest on an eroded landscape.

thicknesses of finer soil material around them washed away. As a bonus, the view looking north to Molokai is outstanding; practically the whole length of the neighboring island is in line of sight. Occasionally, even the profile of O'ahu, over 50 miles to the west, may be glimpsed.

Beyond Keahiakawelo, the Polihua Road winds downslope to the coast at Polihua Beach—a long, beautiful strand of coralline sands, and a turtle refuge right across the channel from West Molokai. When the roadbed is dry, the drive is generally fine for four-wheel-drive vehicles; but road conditions may deteriorate significantly past Keahiakawelo. (Keep in mind that you need to drive back up from down below.)

HI 440
Lāna'i City—Kaumālapa'u Harbor
7 miles

The 7.4-mile drive from Lāna'i City to Kaumālapa'u Harbor, on the west coast, skirts the nearly flat northern rim of Pālāwai and Miki Basins. Just past the airport the highway begins its nearly 2,000-foot descent to the sea. Between highway mileposts 5 and 4, exposures of lateritic soil embedded with old residual corestones become apparent. The underlying lava flows here begin to crop out in roadcuts farther downhill. Here and there, pinnacles of resistant bedrock project through the soil, which ranges from a few feet to more than 10 feet thick. These buried rock pinnacles resemble the ones exposed by erosion along Keōmoku Highway (HI 430), north of Lāna'i City.

Nearing the harbor, HI 440 winds steeply past roadcuts that expose numerous thin, weathering pāhoehoe flows. Most of these lavas erupted in the southwest rift zone, a mile or so to the south. The road's end at Kaumālapa'u offers a spectacular view along the precipitous west coast of Lāna'i. The impressive cliffs continue around the southern coast of the island. The cliffs probably began as a headwall scarp from the Clark slide. Waves have been modifying them ever since.

Waves attack a cliff by first eroding a notch at sea level, primarily during the heavy winter storm season. The notch undermines the cliff face, and eventually it collapses, forming a new face a short distance inland. The waves then work the rubble seaward across the wave-cut bench at the foot of the cliff. Look for prominent wave-cut benches at tide level extending from points to the north and at the shore directly below the end of the road.

No cliffs of any origin face the narrow channels along the northern and eastern shoreline of Lāna'i. That side of the island has not experienced large landslides, nor is it exposed to heavy surf, even in winter months.

Maui

Maui is the second-youngest of the main Hawaiian Islands. It consists of two large volcanoes, West Maui and Haleakalā, linked by the 7-mile-wide Isthmus of Maui (or the Valley). West Maui seems to be extinct. Haleakalā erupted perhaps as recently as about AD 1600 and, though in decline, is almost certainly capable of erupting again. The landscapes of Maui range from subalpine volcanic terrain to lush tropical valleys and include some of the finest beaches in Hawai'i.

Maui once belonged to a much larger island—possibly bigger than today's Big Island. Geologists call it Maui Nui, or Great Maui. At its prime around 1 million years ago, Maui Nui included the islands of Maui, Molokai, Lāna'i, and Kaho'olawe, plus what is now the shallow straits between these islands and a large portion of Penguin Bank to the northwest.

As the oceanic crust sagged under the weight of Maui Nui, the sea flooded the saddles separating its volcanoes, isolating volcanic summits to form today's much smaller islands. During the ice age around 20,000 years ago, when sea level stood 300

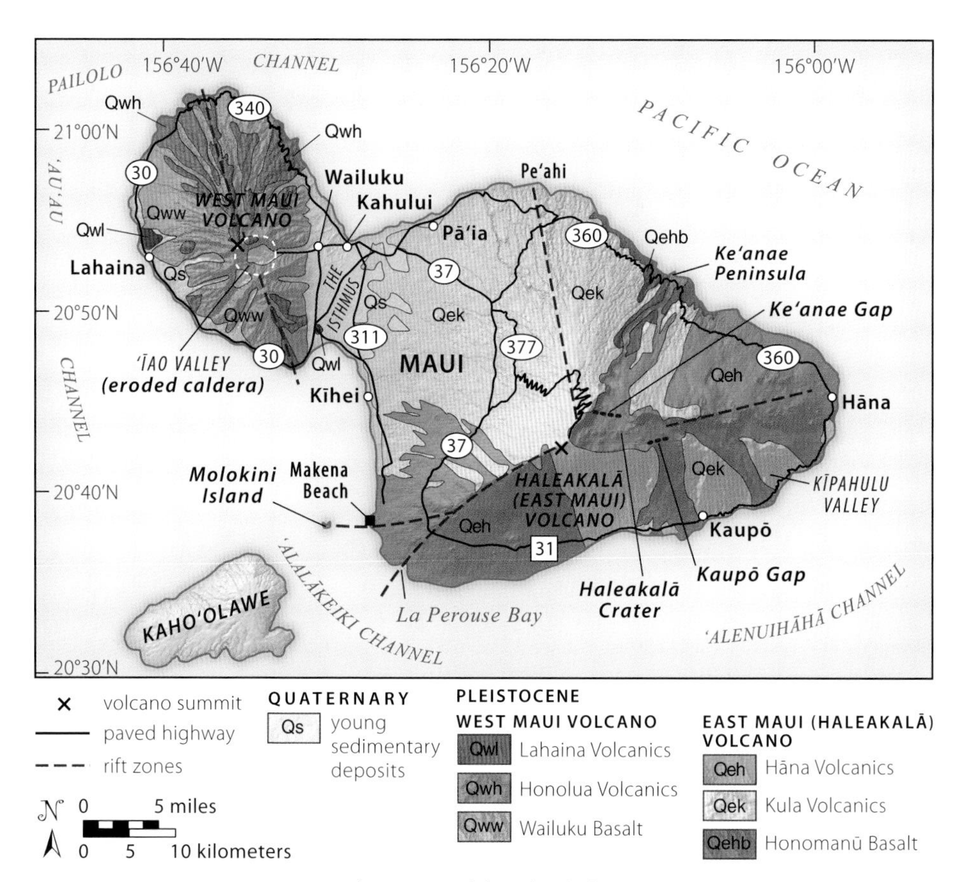

Geologic map of the Island of Maui.

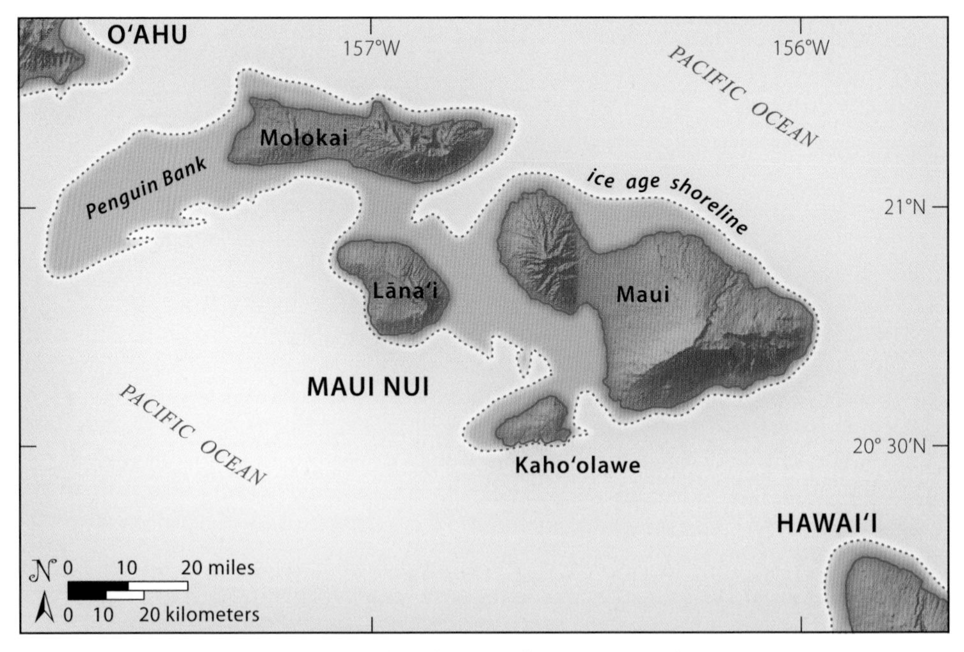

Maui was part of a larger island from 1 million to as recently as 20,000 years ago, at the climax of the last ice age.

feet lower than it is today, Maui, Molokai, Lāna'i, and Kaho'olawe briefly merged once more to resurrect the landmass of Maui Nui. But not for long; by around 10,000 years ago, with rising seas, our modern shorelines came into being.

WEST MAUI VOLCANO

The older of Maui's two volcanoes, West Maui, rises 5,788 feet above sea level. It is 18 miles long and 15 miles wide. The oldest lava exposed above sea level on West Maui erupted about 2 million years ago, so the island must have been above sea level before then.

Thin flows of mainly pāhoehoe lava accumulated to build the young shield of West Maui. It was complete by 1.3 million years ago, with a caldera at the summit about 2 miles in diameter. The lavas erupted during this main stage of growth are the Wailuku Basalt.

West Maui's young shield developed two well-defined rift zones that extend northwest and south of its summit. Later, as the younger bulk of Haleakalā began growing up against its eastern flank, West Maui adjusted by forming two new, less well-defined rift zones, trending northeast and southwest. For unclear reasons, but perhaps due to the deep surrounding ocean that then existed, the slopes of the growing shield became unusually steep—as much as 10 to 20 degrees.

In the typical way of Hawaiian volcanoes, the chemical composition of West Maui's lavas changed as volcanic activity declined. The frequent and generally mild eruptions of tholeiitic basalt in the early stage of shield building gave way to more explosive eruptions of alkali basalt and trachyte during late-stage volcanism, beginning around 1.3 million years ago. The new cinder cones and domes made the originally smooth profile of West Maui's shield look rough. A new cone or dome would appear on average once every 6,000 to 7,500 years; a very low eruption rate compared to the vigorous activity that took place during shield growth.

The postshield rocks are the Honolua Volcanics. They are not nearly as dark as the Wailuku Basalts, and form much thicker flows, individually up to 150 feet thick. If the weather is clear and the lighting is right, you can easily distinguish the Wailuku and Honolua rocks in the walls of the deep interior valleys on West Maui. The lower slopes are composed of the dark Wailuku Basalt, and the light Honolua flows cover it, like pale frosting on a dark cake.

Surprisingly, West Maui's wet northeastern slopes are considerably less eroded than the dry southwestern slopes, even though rainfall and erosion rates are greater on the northeastern slopes. The younger Honolua Volcanics did not cover the dry southwestern slopes, so the Wailuku Basalt there has been exposed to erosion far longer than the younger terrain on the wet northeastern side.

The youngest Honolua lavas erupted about 1.1 million years ago, and a long period of erosion and deposition of shoreline sediments began transforming the appearance of West Maui. Then, sometime between 600,000 and 300,000 years ago, rejuvenated volcanic activity built four cinder cones of highly alkalic basalt along the southwestern shore, between Lahaina and the Isthmus. Two of these cones diverted prominent streams, so erosion of the older volcanic rocks must have been well underway when they erupted. These latest and poorly dated volcanic rocks are the Lahaina Volcanics. No one can tell whether those will be the last eruptions on West Maui. A volcano that last erupted rejuvenated-stage rocks hundreds of thousands of years ago may be truly dead.

Alluvial fans extending seaward from the mouths of valleys on West Maui created the narrow, gently sloping plain along the mountain's western coast. The growing fans rest on the floor of the shallow channel separating Maui and Lāna'i. In places, as much as 200 feet of fan sediment has accumulated since the Honolua Volcanics erupted more than 1 million years ago.

HALEAKALĀ VOLCANO

Haleakalā means House of the Sun. In the tradition of early Hawai'i, the demigod Maui used a net to snare the sun as it crossed the sky above Haleakalā, slowing it down so that his mother, Hina, could have time to dry her kapa cloth.

Haleakalā, 33 miles long and more than 20 miles wide, first rose above sea level around 1.1 million years ago. At 10,025 feet, the summit is high enough to catch some snow. Haleakalā is not as high as Mauna Kea and Mauna Loa on the Big Island, but it is older than they are and has sunk much farther into the ocean floor. It comprises about 75 percent of the landmass of Maui and is one of the largest volcanoes in the main Hawaiian Islands, though less than half the volume of Mauna Loa, Hawai'i's volcano record-holder.

In its prime, Haleakalā was a vast shield of olivine tholeiitic basalt lava flows. Geologists call these early Haleakalā lavas the Honomanū Basalt. They are analogous to the Wailuku Basalt on West Maui, but younger. A succession of calderas probably formed at the summit of the growing shield, but younger rocks filled and buried them. About 900,000 years ago, shield growth slowed and explosive eruptions began to produce more alkalic rocks as Haleakalā entered the late stage of its development. Those eruptions, producing the Kula Volcanics, have continued into recent times, the longest episode of postshield volcanic activity known in Hawai'i. Geologists lump Haleakalā's most recent lavas, erupted within the past 150,000 years, into the Hāna Volcanics, but the young lavas are just a continuation of older Kula volcanism. Haleakalā's postshield basalts do not show much chemical variation.

Haleakalā has three rift zones. The northwest rift zone appears to be dead, but the southwest and east rift zones are still potentially active. The volcano slopes gently seaward about 5 to 10 degrees along the northwest rift zone. In contrast, the volcano's southern flank slopes seaward at an angle of more than 20 degrees, steeply enough to suggest that sliding may have contributed to the angle. No evidence of a debris avalanche from Haleakalā has been found on the adjacent ocean floor, however. The steeper angle may result from Haleakalā tilting toward the Big Island because the huge landmass of the Big Island is sinking more rapidly than East Maui. That would also explain why ancient submerged reefs on the lower east rift zone of Haleakalā also tilt southward.

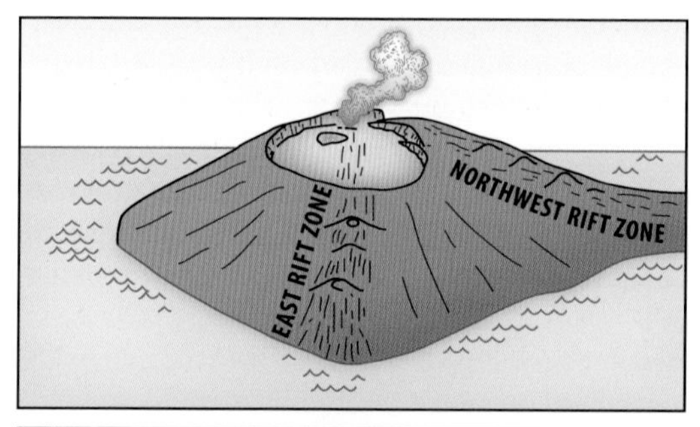

Haleakalā's changing summit over time. View looking west. The orange-brown rocks in the third frame are Hāna volcanic rocks including lava flows and pyroclastics related to numerous postshield cinder cones.

Lava pouring from the southwest and northwest rift zones of Haleakalā piled against the flank of West Maui, building the Isthmus that links the two volcanoes. Haleakalā then formed the broad, gently sloping lower flank along its western side—the beloved Maui Upcountry.

The western flanks of Haleakalā are too young and too dry to have been much eroded. River valleys deeply dissect the wet eastern side, where thousands of cascades and waterfalls tumble down steep valley walls. Gently sloping remnants of the original flank of the shield separate eight huge valleys shaped like giant amphitheaters. Four of them meet at their heads, in the area of what was probably the ancient caldera.

Ash, lava, and cinder erupted high in the Ke'anae and Kaupō Valleys during Hāna volcanism, burying the low divide that separates the heads of these two drainages. The tops of the sheer cliffs around other parts of the valley heads stood above the new volcanic field, enclosing a vast, cindery, subalpine basin 7.5 miles long and 2.5 miles wide. At first glance this spectacular basin resembles a giant crater—and indeed, many maps and publications refer to this area as Haleakalā Crater. But don't be deceived; it is primarily an erosional feature—though Ke'anae and Kaupō Streams may have eroded it from the weak, altered rocks of an older caldera.

Haleakalā has erupted around a dozen times in the past 1,500 years—the third-most-active volcano in Hawai'i (after Mauna Loa and Kīlauea on the Big Island) during that time frame. But Haleakalā's eruptions do not occur at regular intervals. Periods of volcanic unrest seem to be separated by quiet spells lasting 500 to 800 years. Local residents can take comfort that we now appear to be living in one of those quiet spells. The latest outburst of Haleakalā took place in the southwest rift zone just south of Kīhei, pouring a big ankaramite lava flow into the ocean. Geologists and historians used to think that this took place around the year 1790. More recent radiocarbon dating shows that this likely happened earlier, sometime between AD 1450 and 1630.

'ĪAO VALLEY ROAD
WAILUKU—'ĪAO NEEDLE

2.1 miles

See map on page 156.

Ka'ahumanu Avenue (HI 32) passes through Kahului to Wailuku, where it becomes Main Street. Continue straight past the intersection with HI 30 on Main Street about a half mile west, to the intersection with 'Īao Valley Road, which branches to the right. Zero your odometer at this intersection.

Along the first 0.2 mile of 'Īao Valley Road, roadcuts upslope reveal buff to orange-brown boulder conglomerate. Rocks embedded in the conglomerate are rounded and heavily weathered. You are looking at the interior of the ancient alluvial fan on which much of Wailuku stands. The fan developed hundreds of thousands of years ago when sea level stood much higher than it does now. Looking north (right side of road), you can see the gently sloping ancient fan surface. Modern-day 'Īao Stream (Wailuku River) has cut down through the no longer active fan, adjusting to present sea level.

ʻĪao Valley Road exposes the interior of an ancient alluvial fan, including numerous heavily weathered, rounded boulders and cobbles washed down by past floods from the mouth of the valley.

The road follows the stream into a narrow canyon with steep, fluted walls typical of tropical environments where intense weathering and gradual dissolving of rocks on slopes is widespread. Around mile 1.8 the vista ahead opens up. You are entering the 2.5-mile-wide ʻĪao Valley, site of West Maui's extinct caldera. The road ends at ʻĪao Valley State Monument. A ten-minute walk from the parking lot leads to the lookout for viewing ʻĪao Needle, one of Hawaiʻi's most famous landmarks.

When the caldera was still active, the greatly fractured lavas and intrusive rocks beneath its floor altered rapidly as volcanic gases and water heated by rising magmas percolated through them. The alteration made the rocks soft and punky—easy to weather and erode once eruptions ended and stream channels worked their way into the heart of the mountains. The surrounding unaltered lavas and dikes didn't erode so quickly, however, explaining why they now stand up as walls thousands of feet high enclosing today's ʻĪao Valley.

One of the products of lava alteration inside volcanoes is silica (SiO_2), usually in the form of chalcedony, a milky-white material deposited by volcanic solutions as they infiltrate cracks in the rock, cool, and evaporate, often long after eruption has taken place. Later erosion breaks up the silica and mixes it with other sediments throughout ʻĪao Valley. Stream action abrades and rounds the silica bits into smooth, white pebbles called moonstones that can be found on local beaches downstream from the old caldera.

The 1,200-foot-high ʻĪao Needle, or Kūkaʻemoku, is the traditional phallic stone of Kanaloa, Hawaiian god of the ocean. The Needle, viewed from the lookout at the

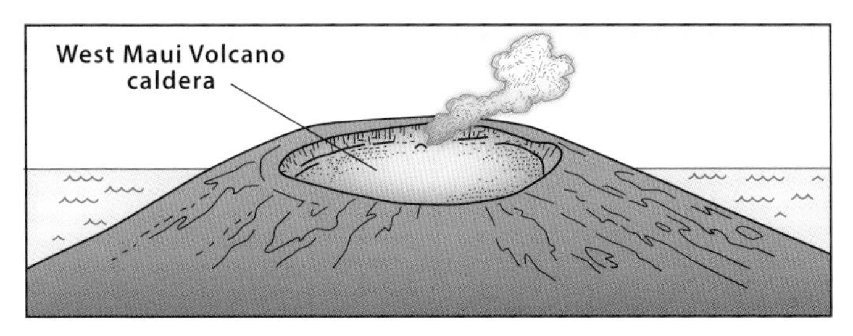

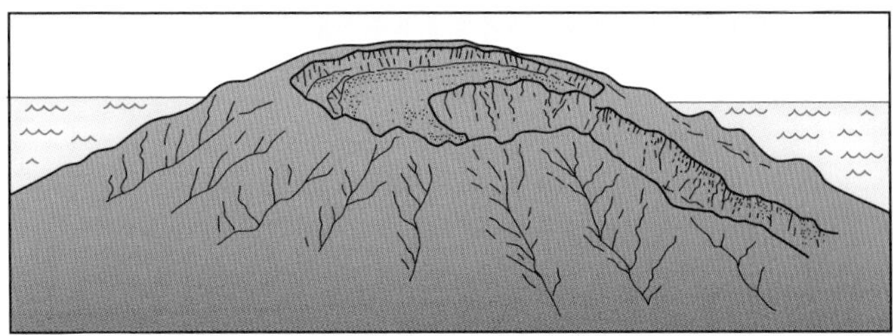

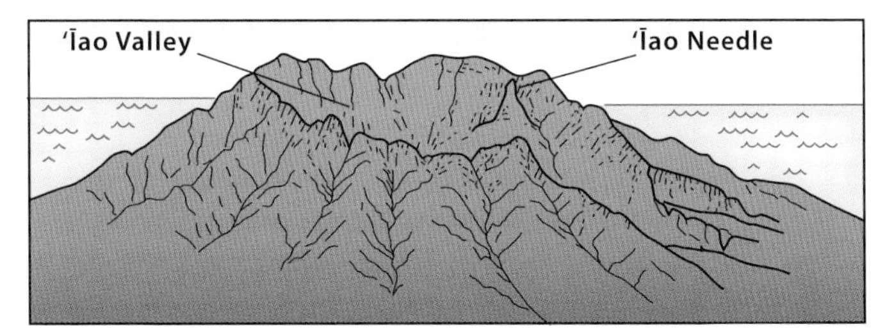

Erosion of West Maui and the origin of ʻĪao Needle.

A dike exposed by erosion forms the spine of the ridge that courses up the south wall of ʻĪao Valley opposite ʻĪao Needle.

end of the trail, is the sharp, downslope end of a narrow ridge composed of Wailuku Basalt intruded by dikes. The dikes strengthened the stack of flows, enabling them to resist erosion more effectively than the surrounding flows. Looking across the valley to the opposite (south) side, you can see a large dike snaking up the walls. A dense network of dikes exists throughout the former caldera area.

The flows making up 'Īao Needle all formed just outside the caldera, on the upper northeastern flank of West Maui's volcanic shield. But the rocks underlying the Needle lookout are steeply dipping rubbly debris shed into the caldera from its now buried, cliffy rims. In other words, the eroded edge of the old caldera lies somewhere between the lookout and the Needle. Unfortunately, dense vegetation makes it impossible to see this boundary at a glance across the landscape. Overall, the walls of 'Īao Valley enclose a somewhat wider area than the former caldera.

Magma welled up not only as numerous dikes beneath West Maui's summit, but also in bulbous masses that crystallized as bosses. One such boss crops out in the gorge just east of the Needle lookout, at about the same elevation. At the parking area, an outcrop of rubbly caldera-fill material is exposed near the entrance station. Watch for it as you exit.

'Īao Needle viewed from the viewpoint at 'Īao Valley State Monument.

HI 30
Kahului—Lahaina
25 miles

Kahului is at the northern edge of the Isthmus of Maui. From this bustling little port you can see towering Haleakalā to the southeast, with its youthful shield still intact, and the green, deeply eroded flanks of West Maui Volcano to the west, which are usually shrouded in clouds. Lava erupted from Haleakalā flowed against the eastern flank of the old West Maui Volcano to build most of the Isthmus of Maui. Alluvial fans and windblown dunes of calcareous sand, derived from exposed offshore reefs during the last ice age, have since blanketed most of the Isthmus. Deeper sand deposits have largely turned to solid rock as percolating groundwater precipitated calcite dissolved from closer to the surface.

Between Wailuku and the intersection with HI 310, near milepost 5, HI 30 skirts the eastern base of West Maui, which here is largely veneered with Honolua lavas. The small bumps on the flank of the volcano, on the horizon to the southwest, are eroded cinder cones, vents that erupted Honolua lavas along the south rift zone. The line of windmills marching high up the mountainside also marks the rift zone's trend.

Haleakalā dominates the view to the east along this section of highway. Its younger and gently sloping flank abuts West Maui's much steeper and more eroded flank. Gentle slopes at the base of the steep flank are alluvial fans that spread from the mouths of its valleys and until recent years supported vast fields of sugarcane and pineapples. The final year for growing commercial pineapples on Maui was 2010, while the island's sugarcane industry folded in 2016.

HI 30 reaches the southern shore of the Isthmus of Maui at Mā'alaea. From here, traffic toward Lahaina can be fast, leaving few opportunities for roadside geology drivers. But a few stops are worth exploring.

As the highway turns west toward Lahaina at McGregor Point, a scenic pullout with parking near milepost 8 provides an outstanding view of the Kīhei coastline of Maui and of Haleakalā's southwest rift zone where it reaches the sea at Pu'u 'Ōla'i. The youngest lava flow on Maui erupted just upslope from Pu'u 'Ōla'i a few centuries ago.

The Island of Lāna'i looms across the channel 15 miles to the west. A single volcano built this island, which began growing a few hundred thousand years after West Maui did. Lāna'i's high point, Lāna'ihale, rises about 3,370 feet. The asymmetry in the profile of the island relates to a gigantic submarine landslide that carried part of Lāna'i away long ago.

Somewhat closer to the south is Kaho'olawe, also a single volcano but even younger. Between Kaho'olawe and Maui is the tiny tuff cone of Molokini. It grew during a single big eruption about 150,000 years ago, along the submarine extension of Haleakalā's southwest rift zone.

The roadcut opposite the parking area at McGregor scenic pullout exposes pale-gray trachyte that erupted during late-stage Honolua volcanic activity. The trachyte shows excellent flow banding, created as the flow sheared internally while oozing downslope. Small vesicles form zones, each several inches wide, that separate denser layers within the lava. Other roadcuts along this section of highway expose basalt flows, buried reddish soils, and basalt rubble. Safety nets meant to catch falling rocks cover many faces.

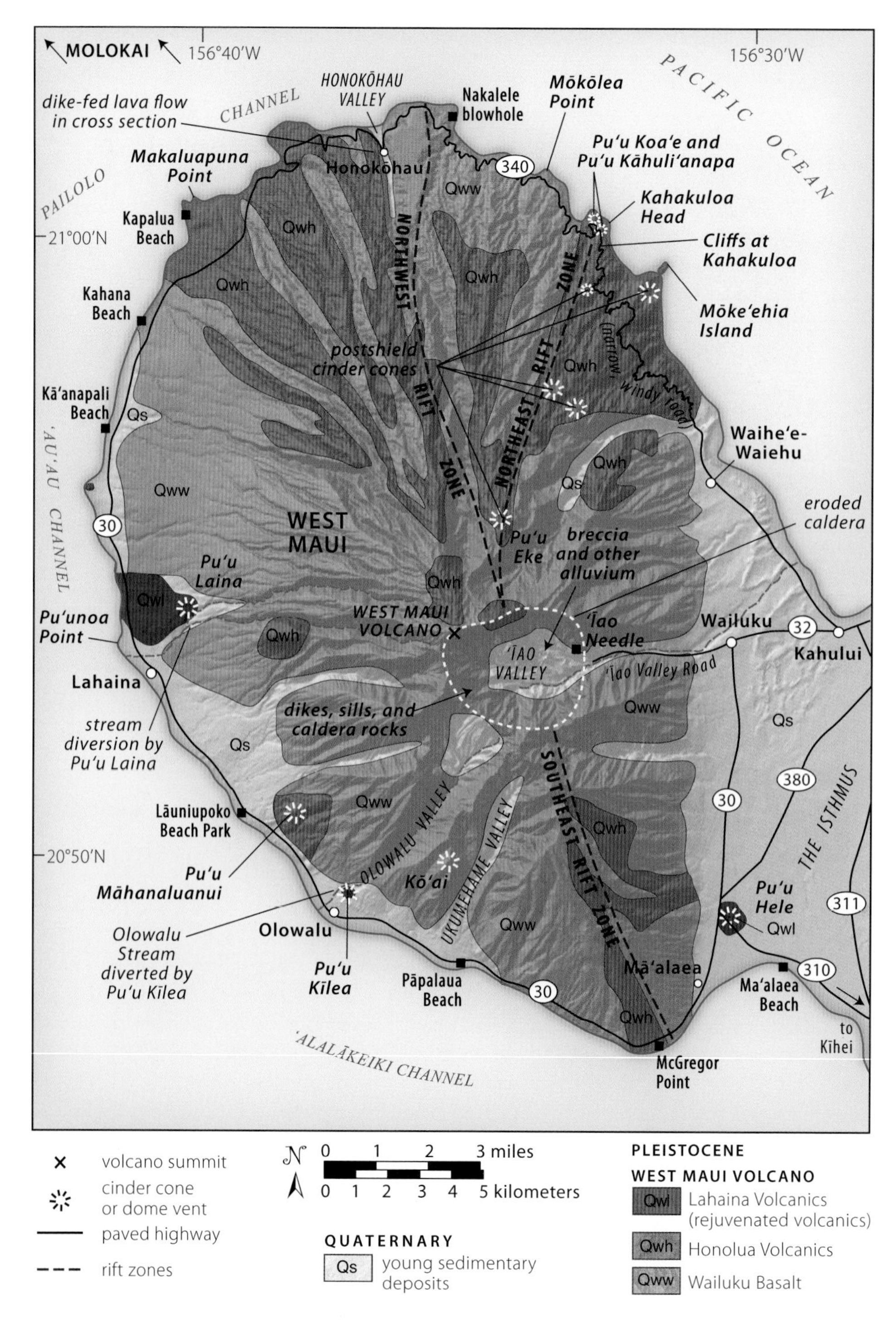

Major geologic features of West Maui.

Flow-banded trachyte across the highway from McGregor Point.

Around mileposts 12.5 and 13, look for pullouts or side roads to the right that allow you to peer up Ukumehame Valley, just inland. The head of the valley is eroded into the caldera of West Maui, where the rocks include a complex network of dikes associated with the volcano's main magma chamber. Ukumehame Stream carried abundant moonstones from this area to the coast. You can find a few in the sand at nearby Pāpalaua Beach, which consists mainly of black basalt pebbles and red cinder mixed with pieces of white coral.

From the highway, you cannot see the caldera's dense network of dikes because they are too far away. But you can see Kō'ai, a vent rising as a pale-pinkish-brown cliff in the left (western) side of Ukumehame valley. This lava dome of trachyte is nearly 600 feet high, part of the Honolua Volcanics. Erosion has exposed its interior together with the huge dike that fed it, farther along the valley wall. If you look closely with binoculars, you can see that the lava flows of older Wailuku Basalt surrounding the dike are bent upward and broken, presumably by the force of the intrusion. Other Wailuku flows exposed in the eastern side of Ukumehame Valley dip seaward, as expected.

Between mileposts 14 and 14.5, look for Luawai Street, a right turn leading a couple of miles upslope to a residential development at the mouth of Olowalu Valley. Driving a short distance in, you will see the quarried interior of an old alluvial fan terrace extending from the mouth of the valley. The rounded, oxidized boulders and cobbles of lava indicate powerful stream action and flooding followed by burial for thousands of years. Widespread fan development occurred across this landscape following the end of Honolua volcanism. Sedimentary deposits such as this accumulated up to a couple hundred feet thick.

Kō'ai lava dome forms the massive cliffy outcrop at the center of this landscape. Erosion has exposed a prominent trachytic dike that fed the dome in the lower slope to the right.

The mouth of Olowalu Valley with remnants of ancient oxidized alluvial fan deposits in the foreground.

Looking up Olowalu Valley on a clear enough day, you can see a long, jagged ridge. This feature marks the approximate boundary of West Maui's caldera, and the divide between Olowalu Valley and 'Īao Valley on the other side. As at the head of neighboring Ukumehame Valley, the ridge line is stitched with countless dikes.

A water well nearby yields 92-degree water, showing that hot rock still underlies this landscape. The ground heat probably relates to the rejuvenated volcanism that built the four Lahaina cinder cones, including small Pu'u Kīlea, which lies a few hundred feet northwest of Luawai Road on the alluvial fan. As Pu'u Kīlea grew, it diverted Olowalu Stream, forcing it to cut a new channel across its fan. The three other cones from the Lahaina Volcanics are a few miles farther northwest.

Near milepost 16 on HI 30, a gravel pit exposes the interior of a large Honolua cinder cone on the right side of the highway. Just past this location, look upslope to see Pu'u Māhanaluanui, a dome of pale Honolua trachyte. The seaward flank of the dome split during the final phase of the extrusion, releasing a small flow of viscous lava. The flow is thick because trachyte lava is too viscous to run into a thin sheet. To get a closer look at the dome you can drive a mile or so up Kai Hele Ku Road, a residential street directly across the highway from the entrance to Lāuniupoko Beach Park.

Pu'u Māhanaluanui, with a thick trachyte lava flow bursting from its base.

As you approach Lahaina town from the south, look up ahead on the slope above town to see another cinder cone that grew during the greatest, and westernmost, of West Maui's rejuvenated-stage eruptions—Pu'u Laina. A much bigger cone just upslope at the foot of the ridge is an older Honolua vent, Pa'upa'u.

Lahaina residents get their water from runoff, rain catchment supplied by stream diversion ditches, and wells. One well near town produces 82-degree water. Like the warm-water well near Olowalu, it is probably heated by rocks still warm from rejuvenated volcanism.

Offshore, a fringing coral reef extends about 1 mile north and south of the marina at Lahaina. It shelters the quiet water near shore, which is a good place for snorkeling and surfing.

Pu'unoa Point, at the north edge of Lahaina, exposes Lahaina Volcanics lava flows. Some are ordinary alkalic basalts rich in little green crystals of olivine. Others are nephelinite, a rare kind of basalt so extremely rich in sodium that it contains a mineral called nepheline in place of the usual plagioclase feldspar. Unless viewed under a microscope, nephelinite looks like most other kinds of basalt.

HI 30 AND HI 340
Lahaina—Wailuku via Windward West Maui
31 miles

HI 30 north of Lahaina is a driving adventure, beginning as a fast-moving wide highway, then changing by stages to a tortuously winding, single-lane mountain road that is occasionally closed due to landslides, though the route is paved all the way through (watch for falling rocks). HI 30 becomes HI 340 past Honokōhau Valley. Road conditions improve again approaching Wailuku from the north, but about 8 miles of HI 340 requires great care driving. To compensate, scenery and roadside geology are memorable.

From Lahaina north, development keeps HI 30 traffic crowded with no significant geology to view. But at Kapalua, last of the big resorts, is an amazing landscape to explore. Turn left on Office Road and follow it to its end, a T-junction. Look for public parking nearby to access the coastal walk. The trail leads south several hundred yards to Makaluapuna, a prominent point composed of Honolua alkalic lava flows and related breccias. Salty sea sprays have corroded the lava, some of which shows near-vertical layering, into pockmarked surfaces, knobs, and spires locally called "dragon's teeth." Durable salt crystals grow in microscopic openings within rock surfaces when films of spray evaporate under the hot sun. As the crystals grow they are strong enough to shatter the surrounding rock. The resulting pocked pattern is called tafoni (ta-PHONE-ee), a term from French or Sicilian. The pock holes at Kapalua are mostly small, no more than an inch or two across. Elsewhere, individual pockets formed this way may be large enough to fit an entire person!

The coastal walk here also provides excellent views of well-formed wave-cut benches along the shore. Mussels and snails have bored into the rock at tide level in dense concentrations. Between points, white calcareous sands from nearby reefs form pocket beaches excellent for surfing and swimming. Ten miles across Pailolo Channel, East Molokai rises from the sea. Though older than West Maui, the shield volcano of East Molokai still retains much of its original shape, seen from this direction.

Tafoni at Kapalua forms when salt crystals grow and expand in pores, shattering the rock.

HI 30 becomes HI 340 beyond the village of Honokōhau. From milepost 33 to milepost 37, numerous roadcuts reveal Honolua lavas and breccias, including pale-gray trachytes as well as darker alkali basalts. Both ʻaʻā and pāhoehoe flows stand out in cross section. Countless smaller roadcuts are exposed along the rest of HI 340, with trachyte becoming more abundant as you travel south. Trachyte is a rare lava type in Hawaiʻi, but not on West Maui, where it is easy to find.

The road skirts numerous pocket beaches and rocky bays, though vegetation obscures vistas in most places. Between mileposts 37 and 38, look for roadcuts with rich red laterites and saprolites, deeply weathered volcanic layers on their way to becoming soils. These soils or partial soils are older than the lavas exposed to the west along the road; probably they are early Honolua or even Wailuku in age.

Strongly weathered, nearly horizontal Wailuku pāhoehoe flows crop out in roadcuts near the north cape of West Maui. A material that has largely gone to soil but which still preserves relict structure of the original parent rock is called a saprolite. Excellent examples of saprolite occur in many places along the windward coast of West Maui.

At milepost 38.5, wide pullouts on the ocean side allow access to a spectacular blowhole via a downslope scramble of several hundred yards over weathered lava outcrops. The blowhole penetrates a wave-cut bench and can explode like a geyser to heights of more than 70 feet! Beware of approaching too closely; "eruptions" are sudden and can be violently powerful. The rock here features numerous tafoni patches and contains conspicuous chunky crystals of plagioclase feldspar and larger olivines that have oxidized to reddish brown in many rocks.

The upper end of Honokōhau Valley and one or two younger Honolua cones can be seen if you look upslope when the area is cloud free. Honokōhau Valley is cut by West Maui's longest stream, stretching about 15 miles. Honokōhau Stream tumbles over 1,120-foot-tall Honokōhau Falls near the head of the valley. Unfortunately, access to it is nearly impossible in this very rugged landscape.

A blowhole erupts near Mōkōlea Point on West Maui.

Lateritic soils, weathered well beyond the stage of saprolite development, form prominent banks visible on slopes just northwest of the blowhole.

Closer to the highway, note the thick lateritic beds exposed by erosion on the slope to the west. While the geology of the leeward coast of West Maui illustrates the growth of new land through alluvial fans and rejuvenated volcanism, here on the windward side many features are actively eroding.

Near milepost 40, look for the ʻOhai Trail pullout on the left. ʻOhai Trail is a pretty, 1.2-mile loop through native coastal shrubland. Directly upslope is a bluff of saprolite and more mature lateritic soil. From the headlands along the trail, you'll see a prominent rock tower to the southeast—630-foot-tall Puʻu Koaʻe, a trachyte dome. A series of domes and related trachyte flows crops out alongshore from Puʻu Koaʻe for several miles toward Wailuku.

On a day with few clouds, the wide, flat summit of 4,500-foot-high Puʻu Eke dominates the inland horizon. The summit, a dwarf cloud forest that often receives over 400 inches of rain a year, is a bog choked with acid-loving, water-tolerant plants. The summit is one of the wettest places on Earth. Geologist Harold Stearns camped here in the 1930s and discovered that the bog partly drains through enormous crevices cut into the rim of Puʻu Eke's circular tableland.

A pullout at the sign indicating the start of the "Narrow and Winding Road" provides a closer view of Puʻu Koaʻe—the closest you'll have if you decide to turn around here. The treacherous road hugs numerous cuts exposing Honolua alkali basalts and trachytes. Ascending from the tiny valley-floor village of Kahakuloa, the road skirts the inland base of Puʻu Koaʻe, passing through a zone of explosion debris and breccia enclosing the dome and the solid lava inside. The main mass of the dome displays rough columnar cooling joints easily visible from the road. These are

Two large trachyte domes, Puʻu Koaʻe on the left and smaller Puʻu Kāhuliʻanapa on the right, rise above Kahakuloa Bay.

larger and more widely spaced than the cooling joints you typically find in thinner basalt flows. Although Pu‘u Koa‘e rises starkly out of the sea, it formed well inland several hundred thousand years ago. Tectonic subsidence has lowered its base (coincidentally) to sea level since then. The domes of Pu‘u Koa‘e and Pu‘u Kāhuli‘anapa appear to mark the northern end of the northeast rift zone of West Maui.

South of Pu‘u Koa‘e, the road continues to hug West Maui's numerous gulches and ridges. Near the Cliffs at Kahakuloa development about 5 miles down the road and farther along, tall roadcuts reveal spectacularly weathered trachyte and benmoreite flows. Oxidizing soil solutions percolating downward, mostly along joints, have created stained red bands in the weathered lava called liesegang bands. If you catch a glimpse shoreward in places, you may see large white patches of exposed trachyte. HI 340 returns to Wailuku near the mouth of ‘Īao Valley.

Crudely developed, giant columnar joints in the flank of Pu‘u Koa‘e are visible from the roadside near Kahakuloa.

Liesegang banding in weathered alkalic lava, exposed along HI 340 near the Cliffs at Kahakuloa.

SOUTH MAUI (KĪHEI/WAILEA/MAKENA)
MA'ALAEA—LA PEROUSE MONUMENT

17 miles

North Kīhei Road (HI 310) runs along the south coast of the Isthmus of Maui. Engineers built North Kīhei Road along the crest of a 1-mile-long spit separating Keālia Pond from Mā'alaea Beach. The pond, beach, and spit all formed in recent geologic times. During the latest ice age, sea level was about 300 feet lower and the shore lay at least 5 miles from here. A large stream, the Waikapu, flowed through the area. Then, between 11,000 and 10,000 years ago, sea level suddenly rose and restabilized at its present position, turning Waikapu Stream's valley into a shallow bay. Fresh beach sands developed a spit all the way across the bay, enclosing Keālia Pond, where you'll find flocks of many Hawaiian marsh and shore birds.

Continue straight down the coast along North Kīhei Road, which becomes South Kīhei Road as you enter the Kīhei resort area. The route passes a series of fine-grained, calcareous sand beaches and small rocky points for the next 10 miles. The

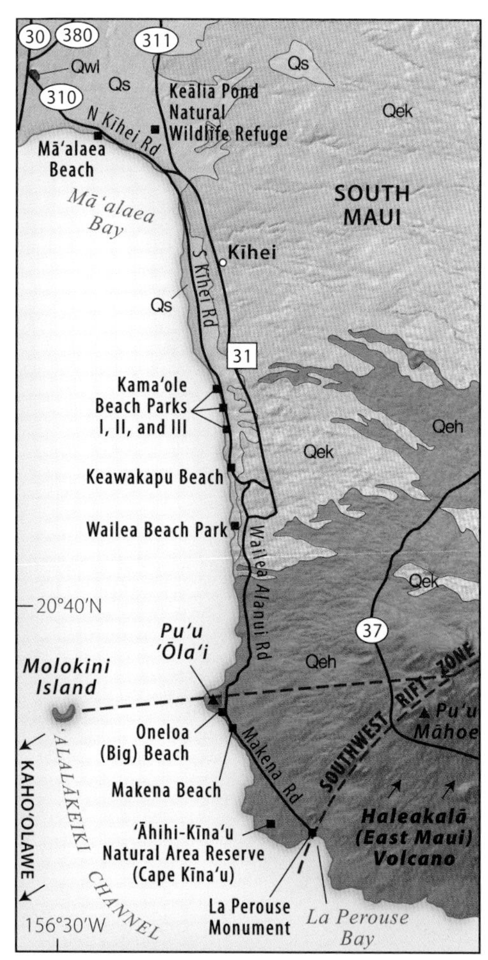

Geology of South Maui from Mā'alaea Bay to La Perouse Bay.

lava making up these points erupted from Haleakalā's southwest rift zone between about 700,000 and 350,000 years ago. The cooling lava trapped steam bubbles and other volcanic gases, forming numerous vesicles. Many outcrops also contain small apple-green crystals of olivine.

A continuous row of tourist establishments lines the coast, but county parks preserve many of the beaches for public use. This shoreline has suffered severe erosion in recent decades. Waves have driven beaches as much as 300 feet inland since the mid-1950s. Most erosion comes during fierce winter storms, called Kona storms because they blow in from the southwest, or Kona, direction. Efforts are underway to slow or stop the erosion, including the construction of seawalls in places.

Three sections of Kama'ole Beach Park between mileposts 4 and 7 separate South Kīhei Road from the ocean. The rocky points between Kama'ole Beach Parks II and III are eroded remnants of black pāhoehoe lava containing many lava tubes.

South of Keawakapu Beach, South Kīhei Road becomes Wailea Alanui Road. At Wailea Beach Park, waves have eroded rubble and cinder from beneath a thin basalt flow, undercutting it to make a scalloped beach with undertow furrows in the sand.

Follow Wailea Alanui Road farther south down the coast to the Makena Park parking areas on the ocean side of the road. These give access to Oneloa ("Big") Beach. The first parking area encountered lies at the foot of Pu'u 'Ōla'i ("Earthquake Hill"), a prominent cinder cone at the coast where Haleakalā's wide southwest rift zone meets the sea. Pu'u 'Ōla'i grew to 350 feet tall during an eruption sometime in the past 100,000 years.

Strolling at the northwest end of Oneloa Beach, you see the interior of Pu'u 'Ōla'i well exposed by storm-wave erosion. The cone is an amalgamation of red oxidized cinders that form massive walls. Deposits like this typically form from pulsating

Dipping layers of cinders in the wall of Pu'u 'Ōla'i at the northwest end of Oneloa Beach, just south of Kīhei.

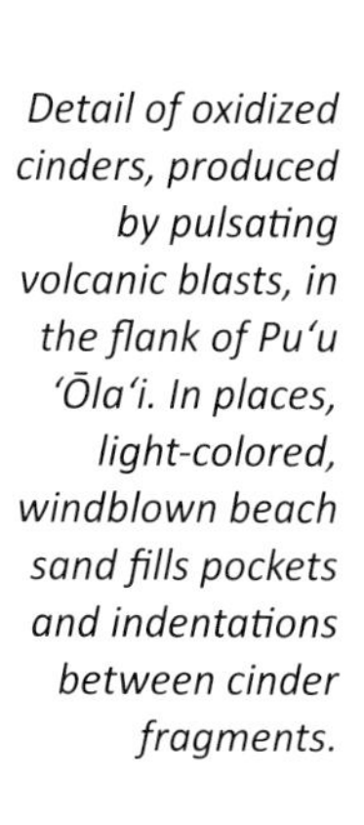

Detail of oxidized cinders, produced by pulsating volcanic blasts, in the flank of Pu'u 'Ōla'i. In places, light-colored, windblown beach sand fills pockets and indentations between cinder fragments.

bursts of gas that throw vast showers of lava clots out of a vent. This type of volcanic behavior is called Strombolian eruption (after a famous Mediterranean volcano that characteristically erupts this way nearly all the time). Almost all the cones you see on Haleakalā, including inside the summit basin, resulted from Strombolian activity.

Late during the Pu'u 'Ōla'i eruption, as gases waned, a lava flow emerged from the vent, forming the point of land extending seaward of the cone. The source fissure of the eruption also stretched this way. Some geologists think that the dike that supplied magma to the fissure is visible in cut-away section at the point. Crashing surf makes exploration treacherous here, however. No doubt, owing to tectonic subsidence, Pu'u 'Ōla'i originally formed well inland and not at the coast.

The crescent shape of Molokini Island 3 miles offshore is the eroded top of a cinder cone that was almost totally submerged when sea level rose at the end of the latest ice age. The sea now floods most of the cone's crater. Thin layers of dark-brown to black cinders crop out all around the rim of the cone. Once used for military target practice, Molokini is now a Marine Life Conservation District Seabird Sanctuary and a popular destination for snorkeling and sightseeing boat trips from Maui. You can see many kinds of birds and fish, as well as the rocks. The underwater scene features coral heads in a rainbow of colors, along with yellow butterfly fish, stripped Moorish idols, and dark-gray-and-orange surgeonfish. The long, low profile of Kaho'olawe, the uninhabited island next to Molokini, rises from the ocean about 7 miles away.

South of Pu'u 'Ōla'i a short distance, the coast road enters 'Āhihi-Kīna'u Natural Area Reserve. It crosses a field of clinkery 'a'ā lava largely devoid of vegetation. You can trace the black river of basalt to its source, Pu'u Māhoe, a split cinder cone low on the flank of the volcano. Myriad older southwest rift zone cones are scattered farther upslope.

This flow is the youngest erupted at Haleakalā. For many years, geologists thought that it erupted around the year 1790 because a map made by a surveyor from the French La Perouse expedition seemed to indicate that a bay previously mapped a

Young basalt flow from Pu'u Māhoe in 'Āhihi-Kīna'u Natural Area Reserve. Older cones of the Haleakalā southwest rift zone, part of the Hāna Volcanics, are scattered upslope.

few years before had been filled with lava. Anecdotes from the descendants of native Hawaiians living nearby seemed to support this. But more recent radiocarbon dates suggest an older age, perhaps as early as the fifteenth century. The road continues only 1 mile or so farther past the young 'a'ā flow, ending at a turnaround at the monument commemorating La Perouse's landing on the island.

HI 36, HI 360, AND HI 31
(HĀNA AND PI'ILANI HIGHWAYS)
Kahului–Hāna–Kīpahulu Visitor Center—Kaupō—'Ulupalakua

Almost 100 miles

You can drive all the way around Haleakalā in one long day, but about 5 miles of this route along the southern coast, between Kīpahulu and Kaupō, are so narrow and rough that many drivers prefer to turn around at Kīpahulu (about 10 miles past Hāna) and return home from there. Nevertheless, the route is passable for ordinary passenger vehicles, and we describe the complete circuit below.

Along the first 7 miles heading east from Kahului, HI 36 crosses the alluvial plain of Maui's Isthmus. Just past the Maui County Country Club golf course is the entrance to Baldwin Beach Park. This beautiful beige sand beach is well worth walking, though generally rough for swimmers and beginning surfers. The beach is part of an active sediment transfer process. In the offshore reefs, waves and coral-eating fish create the sand that longshore currents wash to the beach. Then, trade winds blow the beach sand inland, where it catches against obstacles in the landscape such as stones and trees to grow dunes. Vegetation eventually fixes older dunes in place via roots, leaf litter, and other deadfall. A walk to the west end of the beach shows how dunes have grown there for thousands of years. They extend beneath the neighboring golf course, as shown by the lumpiness of the fairways. Early Hawaiians used parts of this dune field for burials.

In January 1938, about 25 miles north of Baldwin Beach and 65 miles below the surface, the Earth generated a magnitude 6 to 7 earthquake—the strongest to strike Maui in recorded history. The shock jiggled seismographs as far away as New York. Fortunately, damage was slight, and no tsunami followed. The cause remains

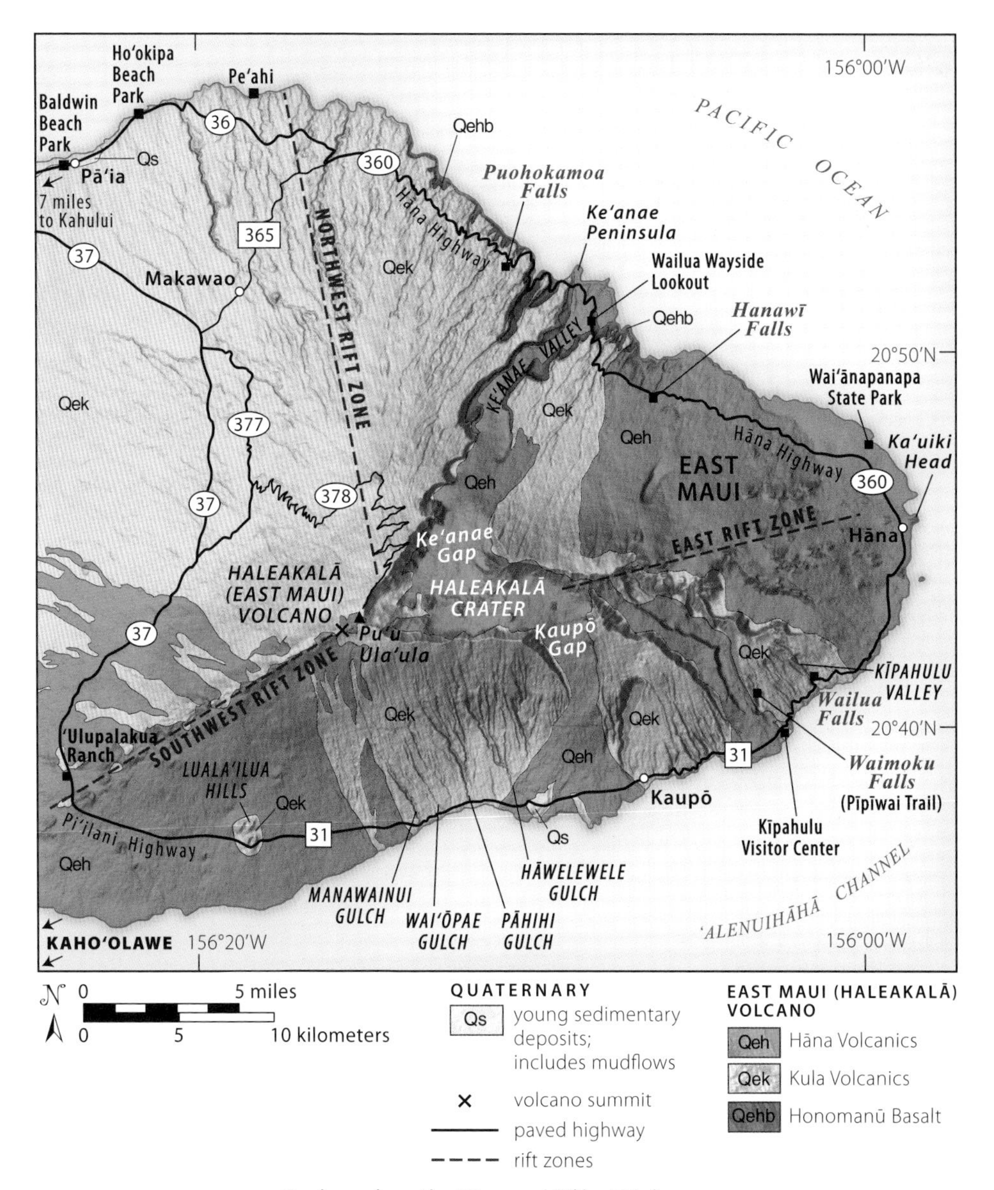

Geology along the Hāna and Pi'ilani Highways.

a mystery. Perhaps it relates to the slow sinking of the island. Though it is well north of the Hawaiian hot spot now, Maui is still settling into the ocean at a rate of about an inch every decade.

Looking inland from the Baldwin Beach parking area, note the low, vegetated cinder cones along the horizon nearshore. These Kula-age cones are part of Haleakalā's extinct (or probably extinct) northwest rift zone.

About 1.5 miles east of Pā'ia, watch for the Ho'okipa Lookout turnout, the entrance to Ho'okipa Beach Park, a popular surfing ground. No dunes have developed here because of the cliffiness of the shoreline. But the exposed cliff face at the east end of the beach below the bluff parking area exposes several noteworthy features. A mudflow deposit of rounded boulders of vesicular lava embedded in hardened ash and mudstone crops out in the middle cliff. The mudflow carrying this mess came from an interior valley. Mudflows can transport big boulders like these because mud is much denser than clear water and exerts a correspondingly greater buoyant effect. The mudstone overlies a bright-red ancient ("fossil") lateritic soil layer at the base of the cliff, indicating the existence of a surface that was stable for a long time, perhaps several thousand years, before the mudflow buried it. A bed of Kula volcanic cinder and an 'a'ā flow interior also crop out here.

HI 36 meets HI 365, the road to Makawao, east of milepost 16, where HI 36 becomes HI 360—the Hāna Highway. Mileposts reset to zero here. HI 360 crosses countless forested gulches with cascades and waterfalls eroded into the slope of the volcanic shield. The exposed rocks are mostly in the Kula Volcanics. Older Honomanū volcanic rocks that erupted in the shield-building stage of activity lie in deeper gulch bottoms and in the lower walls of larger gulches and at the base of sea cliffs. Rock exposures along the highway are weathered and sparse owing to dense vegetation. Just north of milepost 10, however, the road passes Puohokamoa Falls. The resistant ledge of a well-exposed 'a'ā flow core forms the lip of the falls.

Ke'anae Peninsula to Hāna

Near milepost 13, look for the first good views of the Ke'anae Peninsula, up ahead to the east. This half-mile-long coastal platform extends from the mouth of Ke'anae Valley. The road descends into Ke'anae Valley just east of milepost 16. Beneath the lushly vegetated rims lie old Kula lavas. The middle valley walls are even older Honomanū Basalt, the lavas that built up the main shield of Haleakalā. Honamanū flows also form the floor of Ke'anae Valley but have been mostly buried by younger sediments and Hāna lavas originating from Haleakalā's summit basin 10 miles to the south and 8,000 feet higher. On at least four occasions, very fluid Hāna lava flows followed Ke'anae Valley to the coast, in places leaving only thin veneers of basalt plastered against valley walls. The youngest flow, 10,000 years ago, built most of today's Ke'anae Peninsula.

The side road to the peninsula, well worth taking, leads to a county park at the point, providing breathtaking views of sea cliffs up to a few hundred feet high extending for miles along the coast. The crashing surf has carved the lava and breccia at the shore into jagged spires and knobs—a fantastically rough landscape. Look upslope to see 2-mile-wide Ke'anae Gap on the horizon, one of two openings in the rim of Haleakalā's summit basin, and the head of Ke'anae Valley through which Hāna lavas have repeatedly poured. The high point on the left side of the gap, Hanakauhi, is almost 9,000 feet high.

Wave-eroded Hāna lava at the tip of the Ke'anae Peninsula.

Just before milepost 19, a bit over 1 mile east of the Ke'anae Peninsula, Wailua Valley State Wayside Lookout provides a panorama upslope into the lower part of Ke'anae Valley with its strips of level floor enclosed by steep walls. Wailua Valley is basically a topographic extension of Ke'anae Valley. Valleys like this form as the pace of eruptions slows down on aging Hawaiian volcanoes. Long periods of erosion are interrupted by episodes of lava eruption, building up valley floors. In contrast, young, fast-growing Kīlauea Volcano on the Big Island has no valleys at all!

Watch for roadside waterfalls just south of mileposts 19 and 20.5. Near milepost 22, Pua'aka'a State Wayside highlights additional falls and plunge pools at the bases of resistant lava ledges. The large sizes of the plunge pools are an indication of the strong flooding that occasionally pours down this watershed, and most others on Haleakalā's wet windward side.

The highway crosses Hanawī Stream near milepost 24, with another waterfall upslope. Hanawī Stream is almost the farthest east that sugarcane and pineapple plantation engineers constructed a stream-water diversion ditch for irrigating Maui's dry Isthmus—25 miles away. The diverted water simply flowed under gravity through concrete or pipe channelways, twisting and turning to maintain a steady, gentle grade as it crossed the rugged flank of Haleakalā. Unfortunately, the Hanawī diversion structure isn't visible from the road. Nevertheless, modest streams such as this were certainly vital to Maui's twentieth-century agricultural economy—an important part of Hawaiian history.

The landscape begins to level out around mileposts 29 to 30. The Hāna Highway is approaching Haleakalā's east rift zone. Numerous cinder cones, green with forest cover, begin to appear up ahead. At least a half-dozen cones have formed in the east rift zone within the past thousand years.

At milepost 32, a side road leads to Wai'ānapanapa State Park. A short walk from the parking area at road's end leads to a small black sand beach. Unlike the black

Looking upslope from Wailua Wayside Lookout at valley floors covered by young Hāna lavas that poured from Haleakalā's summit area.

Wai'ānapanapa shore of dark basalt. This coastline also features a prominent lava cave with pools, natural arch, and other features of historic and geologic interest.

sand beaches that form where molten lava explosively enters the sea, the sand here comes from wave erosion of basalt in the sea cliffs, as shown by the abundance of rounded beach pebbles and cobbles. The continuous process will probably keep the beach supplied with black sand for a long time. Look a few hundred yards along the coast to the right (southeast) to see an arch that waves have eroded in the basalt.

A short pathway from the beach leads to a large lava cave. Parts of the roof have collapsed and form open skylights. Pools of beautiful clear water inside are popular swimming holes. They contain brackish water, which at this low elevation lies close to the surface and seeps through the floor of the cave.

Entering Hāna town, you can drive along Uakea Street to the harbor in a cove largely sheltered by a 380-foot-tall cinder cone, Ka'uiki Head. In part, Hāna's historical reason for being was the presence of an anchorage here, rare on this coast, and an abundance of land suitable for sugarcane and other agriculture thanks to the rich, weathered ash deposits and related soils in the vicinity of the east rift zone.

Continuing along Uakea Street to where it ends just past the Hāna Community Center, a short but treacherous footpath leads downslope to Kaihalulu Beach, along the south side of Ka'uiki Head. This unusual red sand beach formed from erosion of the adjacent cinder cone. The oxidation, shown by the red color, resulted from a long period of heating as hot gases escaping the freshly formed cone filtered through the cinder and mixed with oxygen in the air. The heat allowed the oxygen to combine with iron in the cinder clots, turning them red. The color of bricks forms in a similar way. Unoxidized cinder is black, just like other forms of basalt.

A side street leading upslope from the highway through Hāna leads to a lookout at the 540-foot-tall summit of Pu'u o Kahaula, an excellent place to view east rift zone features and the coastal landscape below.

Kīpahulu and Kaupō

Milepost numbers begin to decrease west from Hāna, and HI 360 becomes HI 31 near Kīpahulu. The road narrows as the surrounding landscape once again steepens.

Between mileposts 45 and 44, 95-foot-tall Wailua Falls tumbles into a large plunge pool next to the road. Between mileposts 43 and 42, HI 31 enters the Kīpahulu section of Haleakalā National Park.

The Pīpīwai and Kōloa Trails begin at the parking lot of the Kīpahulu Visitor Center. The short Kōloa Trail Loop provides access to an overlook of large plunge pools carved into columnar-jointed 'a'ā flows near the mouth of 'Ohe'o Stream. The more ambitious 3.7-mile (round-trip) Pīpīwai Trail leads upslope with an elevation gain of 800 feet to Waimoku Falls. The face of the falls exposes multiple, stacked Hāna 'a'ā flows, with thinly layered, platy pāhoehoe at the top. The trail passes through spectacular non-native bamboo forest and flood-sculpted streamside landscapes to get there.

West of Kīpahulu, HI 31 deteriorates significantly for about 5 miles, hugging the shoreline between mountain and sea. Plenty of roadside outcrop includes lava flow cores, breccias, and volcanic conglomerates in the older Kula Volcanics. But all eyes tend to fix on the winding, narrow road. The countryside becomes drier and more open along the way.

Road conditions improve past the settlement at Kaupō. Look upslope to see the mile-wide flat notch of Kaupō Gap, the southeastern entrance to Haleakalā's summit basin. The ramping slope below the gap widens seaward, with few stream cuts, into

Along the Pīpīwai Trail, numerous powerful floods sculpted linked pools into the ʻaʻā. These features reflect the tremendous power of floods that can course down this drainage out of Kīpahulu Valley upslope.

Kaupō Gap and Kaupō Valley, viewed from the yard at St. James Church historical site. The church rests on a peninsula built largely by lava, debris flows, and other sediment derived from Haleakalā Crater through Kaupō Gap.

Kaupō Valley. One of the best places to contemplate these features is at St. James Church, a lonely but well-kept historical site almost 1 mile past Kaupō village.

The steep mountainside, abundant loose volcanic ash and cinder from recent Hāna eruptions, and a climate that produces occasional heavy rains provide ideal conditions for mudflows in Kaupō Valley. Rain mixes with volcanic ash and yields dense mud, easily capable of moving large boulders. Hāna lava flows (originating from within Haleakalā's summit basin) and mudflows fill most of the valley floor all the way to the sea, where they have built a broad coastal shelf. One prehistoric mudflow is nearly 300 feet thick near the lower end of the valley.

Kaupō to 'Ulupalakua

Between Kaupō and 'Ulupalakua Ranch, HI 31 crosses the steep southern flank of Haleakalā between the volcano's east and southwest rift zones. The cliffy rim of Haleakalā's summit basin forms the crestline above, and like a dam, it has prevented the youngest Hāna lavas from spreading down this side of the mountain. Erosion has carved enormous gulches into the mountainside as a result, best seen near mileposts 30 and 29. The first crossed is Hāwelewele, followed by Pāhihi, Wai'ōpae, and Manawainui Gulches.

Look for excellent cross sections of 'a'ā and pāhoehoe flows, baked underlying ancient soil beds, breccias, and columnar jointing exposed in gulch walls and well

Erosion along Wai'ōpai Gulch upslope from the highway exposes a columnar-jointed mass of younger Hāna lava that flowed into and mostly filled an older valley tens of thousands of years ago ("paleo-Wai'ōpai Gulch"). The lava originated from a vent high on the flanks of Haleakalā. The horizontally layered reddish rock beds exposed on the right (east) wall of the Gulch are older Kula lava flows.

preserved by the dry climate. In places, you may be able to make out angular unconformities, where one set of younger flows truncates the layering of older flows beneath. A good example can be seen inland from HI 31 at the Wai'ōpae Gulch crossing. A small shed (at the time of this writing) marks the site. Here, a 75,000-year-old Hāna lava flow with long cooling columns unconformably overlies several-hundred-thousand-year-old Kula lavas. The Hāna flow poured down and partly filled an older version of Wai'ōpae Gulch that developed during a preceding period of erosion.

Just past Wai'ōpae Gulch, watch for a large natural arch in the lava at the shore. On a clear day, you may be able to see the profile of the Big Island to the south across 'Alenuihāhā Channel, about 30 miles away.

Continuing past Manawainui Gulch, the road begins climbing toward the southwest rift zone. At nearly 2,000 feet elevation, it skirts four small cinder cones—the Luala'ilua Hills—which lie oddly off the trend of the rift zone, 4 miles upslope from here. Finely layered ash layers, some containing blocks of older lava blasted out by the volcanic explosions, crop out on slopes to the right of the road. There are also several circular collapse structures in olivine basalt pāhoehoe flows that ponded at the base of the cone closest to the road.

Approaching 'Ulupalakua Ranch—the crest of the southwest rift zone—HI 31 crosses the Hanamanioa Flow, a lightly vegetated, youthful-appearing 'a'ā flow. Radiocarbon dating indicates that it erupted nearly 1,000 years ago, during a period

KAHO'OLAWE

The distance from 'Ulupalakua Ranch, where HI 37 turns northeast into the Maui Upcountry, to the island of Kaho'olawe, is around 10 miles. The island rises 1,500 feet above sea level and measures 11 by 6.5 miles. It is built mainly of tholeiitic basalt of a shield volcano, with a thin cap and caldera fill of alkalic lavas that erupted during postshield volcanic activity around 1 million years ago. They erupted mostly from a rift zone that trends southwest from a caldera about 4 miles wide at the eastern end of the island.

Like all other Hawaiian Islands, Kaho'olawe has shed big pieces of itself in the form of huge submarine landslides. A large part of the island that once extended from the caldera toward Maui is gone. Lavas erupted from Haleakalā have built up against the old avalanche scar and cover much of the seafloor debris from this event. Eruptions from a modest rejuvenated stage have left small patches of lava in the caldera.

Kaho'olawe's upper slopes—about 25 percent of the island's landmass—are covered with a thick mantle of red lateritic soil. The soil is not to be found below about 800 feet. It probably was washed away by a giant tsunami in the wake of one of the more recent cataclysmic landslides. The 'Alika-2 debris avalanche that dropped into the ocean from the west coast of the Big Island about 127,000 years ago is the likely culprit. What soil remains is continually blowing and eroding away because past overgrazing and military bombing practice have destroyed the plant cover. Local geologists call the center of the island the Dust Bowl. Efforts to restore a drought-resistant plant cover and protect remaining archaeological remains are underway with indigenous Hawaiian stewardship. Kaho'olawe, lacking much freshwater because it lies in the rain shadow of Haleakalā, was never heavily populated by Polynesian settlers, unlike neighboring Maui and the Big Island. Nevertheless, it is rich in prehistoric artifacts, with more than five hundred recorded archaeological sites.

of time when Haleakalā was more active than it is now. Looking downslope from here to the southwest point of Maui (Cape Kīna'u), even younger-looking 'a'ā spreads out along the shore, the youngest known flow from Haleakalā. It erupted sometime between the mid-fifteenth and nineteenth centuries. Also look for tiny Molokini, a low, Haleakalā-related cone poking above the sea to the right of Kaho'olawe Island. From 'Ulupalakua the drive back to Kahului along HI 37 is direct, providing beautiful Upcountry views of the Kīhei coast and West Maui, across the Isthmus.

HI 378 (HALEAKALĀ CRATER ROAD)
HI 377—Haleakalā National Park

21 miles to summit

HI 378, the Haleakalā Crater Road, begins at the intersection with HI 377—the Haleakalā Highway—about 13 miles southeast of Kahului in the Maui Upcountry. HI 378 is a winding, well-paved, well-graded ascent of almost 7,000 feet to Haleakalā's summit at the southwest end of the summit basin, or "crater." Mileposts on HI 378 begin at zero at the junction.

All the rocks along the road are Kula basalts erupted from Haleakalā between 900,000 and 150,000 years ago. Most are alkalic basalt, rich in sodium and generously speckled with glassy grains of green olivine. Cross sections in roadcuts, beginning around milepost 3, show mostly 'a'ā flow cores and associated oxidized breccias. The flows are thin, characteristic of lava pouring down steep slopes. Little has been lost to erosion on this flank of the mountain since their eruption.

Just uphill from milepost 9, watch for a large gravel quarry that has opened the interior of a cinder cone, Pu'u Nianiau. You can see layers of oxidized, glassy cinders that blew out of the vent and settled on the flanks of the growing cone, layer upon layer. Hot steam and escaping gases filtering through the cinders as the eruption ended altered much of it to a yellowish-brown mineral-like substance called palagonite (pa-LAG-oh-nite).

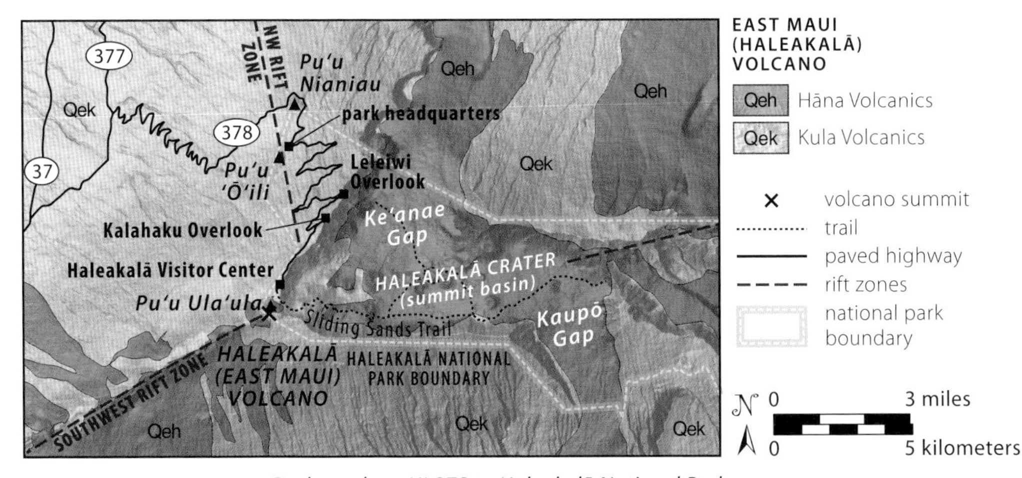

Geology along HI 378 to Haleakalā National Park.

Palagonitized and oxidized cinder layers in a quarry on the flank of Pu'u Nianiau.

The road enters the extinct northwest rift zone of the volcano where it crosses the boundary of Haleakalā National Park near milepost 10. On a clear day you can see cinder cones along the line of the rift zone all the way to the coast, 9 miles east of Kahului. The Park Headquarters Visitor Center offers excellent exhibits and information about hiking trails in and around the volcano. The large cinder cone next to the road as it switchbacks just past the Visitor Center is Pu'u 'Ō'ili, part of a half-mile-wide vent complex that marks one of the latest eruptions in the northwest rift zone.

The road winds for another 9 miles to the rim of Haleakalā's breathtaking summit basin.

Leleiwi Overlook

A walk of a few hundred feet from a small parking area next to the road takes you to Leleiwi Overlook, the best location to view up close the Ke'anae Gap opening into the summit basin. Multiple flows of Hāna lava, erupted as cinder cones grew and partly filled the basin, have poured through Ke'anae Gap, including two within the past 1,000 years. The longest flows have reached the sea, building Ke'anae Peninsula along the Hāna Highway, 9 miles to the northeast. The latest eruption within the basin took place only 800 years ago.

From the Leleiwi parking area, look oceanward to see West Maui, with the island of Molokai beyond. A clear day allows you to make out the profile of East Molokai's volcanic shield, with a wide, flat notch showing the location of its eroded caldera near the extinct volcano's summit. The island of Lāna'i is a bit south of west, and Kaho'olawe is to the southwest. Tiny Molokini punctuates the strait between Maui and Kaho'olawe.

Walk down the road about 100 feet from the overlook and look to the left of the highway to find an outcrop of Kula basalt containing numerous glassy green olivine crystals about the size of peas and black pyroxene crystals up to twice that size. They are concentrated in the upper several feet of the flow. This mineral composition is typical of ankaramite, a rare variety of alkalic basalt. The lower part of the flow contains few large crystals and looks more like ordinary alkalic basalt. Both upper and lower parts appear to be separate lobes of the same flow. What accounts for the striking change in composition? Geologist Gordon Macdonald suggested that the eruption tapped first the upper part of the magma chamber, then the lower part, where the large crystals had settled. So now we see the lower part of the magma chamber represented in the upper part of the flow.

Ankaramite lava flow exposure near Leleiwi parking area. The change in crystal load going from top to bottom tells a tale!

Kalahaku Overlook

You can see tilted slabs of compressional ridges in a broken lava flow around the Kalahaku Overlook parking area. Compressional ridges develop when part of a flow slows, causing the advancing lava from behind to buckle. The medium-gray rock is alkalic basalt, part of the Kula Volcanics.

Each of the successive views into Haleakalā's summit basin is different; the panoramas become more breathtaking with increasing elevation. Look northeast from the overlook to see the large gap cut by Ke'anae Stream as it eroded the western half of the basin. Look east to see Kaupō Gap, 7 miles away, which Kaupō Stream

Looking across the head of Ke'anae Gap from Kalahaku Overlook at the northwest corner of Haleakalā's summit basin. Projection rightward of steeply dipping Honomanū shield flows in the mountain face across the gap illustrate how much taller Haleakalā's summit stood before erosion began to transform the volcanic landscape. The darkest, youngest lava flows filling the gap below are 800 to 1,000 years old.

eroded through the eastern half. Long after these streams cut their big, steep-walled valleys, renewed volcanic activity of the Hāna stage filled the valley floors with lava and cinder to form the present basin floor.

The rough alignment of cinder cones across the basin continues the row of Hāna vents from the southwest rift zone across the summit into the east rift zone. The cinder deposits owe their spectacular red colors to oxidation by steam and volcanic gases during and shortly after the eruptions. The enclosing green walls are composed of Kula lavas.

No one knows exactly when the current spasm of postshield eruptions began in the summit basin, but there have been a lot of them. A radiocarbon-age date on charcoal from a buried soil bed suggests that as many as twenty eruptions have taken place in the past 2,500 years.

The basin floor lies 2,000 to 2,500 feet below Kalahaku Overlook. The highest of a dozen cinder cones rises about 600 feet. But these statistics do not convey an accurate sense of scale as vividly as an occasional glimpse of hikers on the trails far below.

In the distance to the south on a clear day you may spot the towering summits of Mauna Kea and Mauna Loa on the Big Island. Before Haleakalā sank deep into the Earth's mantle, it was probably as high as they are, more than 13,000 feet.

Haleakalā Visitor Center and Sliding Sands

Haleakalā Visitor Center, a short distance downslope from the volcano's summit, offers an encompassing view of the basin. The road beyond leads about a half mile to a lookout shelter at the actual summit, atop Pu'u Ula'ula, a Hāna cinder cone. The summit elevation is a bit over 10,000 feet, Hawaii's third-tallest mountain. The parking area largely fills the shallow crater next to the shelter. Pu'u Ula'ula also marks the upper end of Haleakalā's southwest rift zone.

Escaping steam and gases power the explosive eruptions of cinder cones. The cones expand as magma rises to the surface and then blows from the vent, coughing up scraps of molten lava that cool as they fall through the air. The larger scraps fall close around the vent and build up a cinder cone; the smaller particles drift on the wind as a cloud of ash. The last few explosions leave a summit crater shaped like a cup.

Imagine a cinder cone eruption from a safe perspective from the rim: Clouds of black ash rapidly churn from a steaming gray cone only a few miles away, rising like a dark tower into the sky above the basin floor. Incandescent bombs arc through the air from the ash clouds and fall as glowing globs of molten lava, tumbling down the slopes of the growing cone. An echoing roar reverberates from mountainside to mountainside, punctuated by booming sounds like cannon blasts. The ground shudders.

From the summit, you can peer down the rugged spine of the southwest rift zone to the ocean, 15 miles away. Kaho'olawe stands across the channel, together with Molokini, a largely submerged cone on the underwater continuation of the rift zone.

Science City, a research complex, perches on a nearby cone downslope from the summit lookout. Precise instrumental measurements, made by shining lasers from Science City off reflectors left by Apollo astronauts on the Moon, showed for the first time that the Pacific Plate moves northwest at about 4 inches a year, carrying the Hawaiian Islands along for the ride. Facilities here also include lunar and solar observatories, a satellite tracking station, and a radio repeater station.

The only way to see the interior scenery of Haleakalā's summit basin is on foot, most conveniently on the Sliding Sands (Keonehe'ehe'e) Trail, which descends from the parking area of Haleakalā Visitor Center to the basin floor. The vertical descent all the way to the bottom is 2,800 feet, and the round-trip is 8 miles. Of course, you don't have to go all the way; many people prefer to turn around where the slope becomes much gentler after the first 2.5 miles. Morning is best for any outing, before clouds fill the basin.

The volcanic landscape of barren, oxidized cinder seen along the trail seems as surrealistic as would the surface of another world. It is easy to see how Sliding Sands got its name. This landscape formed where explosive Hāna eruptions broke out, building a cindery ramp on a cliffy, preexisting slope. Much of the debris composing the upper section of the trail came from Pu'u Ula'ula cone as it grew.

Lava bombs as much as 2 feet across litter the terrain. Escaping gases blew the bombs out of the vent as blobs of pasty magma full of steam and gas. Most of them are rounded, even streamlined. The broken ones show dense rims around highly vesicular cores. The dense outer rind chilled before the expanding gases inside could create vesicles like those in the cores.

To the northeast as you descend, look midway up the basin walls for an enormous dike of light-gray alkalic basalt, part of the magma plumbing system that existed during Kula times. It was exposed by erosion from Ke‘anae Stream as the pace of volcanic activity slowed down. The dike is one of many intrusions concentrated near Haleakalā's summit. It can't be more than a few hundred thousand years old.

Keonehe'ehe'e—the Sliding Sands, with Haleakalā's summit basin stretching more than 7 miles beyond.

HAWAI'I ISLAND

Hawai'i, the youngest and largest Hawaiian island, includes as much land as all the others combined, hence its nickname: the Big Island. It stretches 95 miles from north to south and 80 miles from east to west—bigger than some eastern states. Five large volcanoes coalesce to form the Big Island. A sixth older one, once part of the island, has sunk into the sea just off 'Upolu Point, the northwest cape. A seventh, a very young volcano (Kama'ehuakanaloa, formerly called Lō'ihi) is growing underwater off the south coast. Its mass could eventually add new land to the island but not for tens of thousands of years. By that time much of the northern end of the Big Island will have subsided into the ocean, so it is unlikely that the island's size will increase much.

The two north-to-south volcanic trends in the Hawaiian Islands, first noted by James Dana in the 1840s, show up best in the alignment of the Big Island's volcanoes. The Loa Trend includes Māhukona, Hualālai, Mauna Loa, and Kama'ehuakanaloa, while the Kea Trend includes Kohala, Mauna Kea, and Kīlauea. The ages of these volcanoes appear to alternate from one alignment to the other, though volcanoes have remained active simultaneously along both trends for hundreds of thousands of years.

Listed oldest to youngest, the sequence of formation of the Big Island's volcanoes is Māhukona→ Kohala→Mauna Kea→Hualālai→Mauna Loa→Kīlauea→and Kama'ehuakanaloa. This pattern suggests that a new unnamed volcano could start growing someday a few tens of miles to the southeast of Kīlauea on the deep ocean floor. Presuming geologists are still around then, they won't be surprised if this happens!

Eruptions began to build the Big Island from the seabed a little over 1 million years ago, and the land first appeared above the waves within about 250,000 years. Lava flows continue to erupt frequently in the southern half of the island, adding hundreds of acres of new land. To the north the volcanoes are extinct or dying. Of all Hawaiian Islands, the Big Island shows the greatest diversity of rocks and landscapes, and it gives us a glimpse into the ancient roots of the older islands.

MĀHUKONA VOLCANO

Ancient Māhukona, now submerged off the northwest cape, once formed a small, gently sloping elliptical island about 1 million years ago. It rose 800 feet above sea level and was the origin of the Big Island. Like all young Hawaiian volcanoes, it developed a summit caldera and a set of rift zones. A remnant of an old west rift zone can still be seen in the underwater topography. Younger lavas from neighboring Kohala have long since buried Māhukona's east rift zone.

Well-formed coral reefs developed around Māhukona's shoreline. As the volcano kept sinking, though, it eventually dragged these reefs too deep underwater for their corals to continue thriving. The drowned reefs appear as sediment-covered steps on the flanks and are submerged as much as 3,800 feet. Māhukona's summit finally vanished beneath the waves between 435,000 and 360,000 years ago.

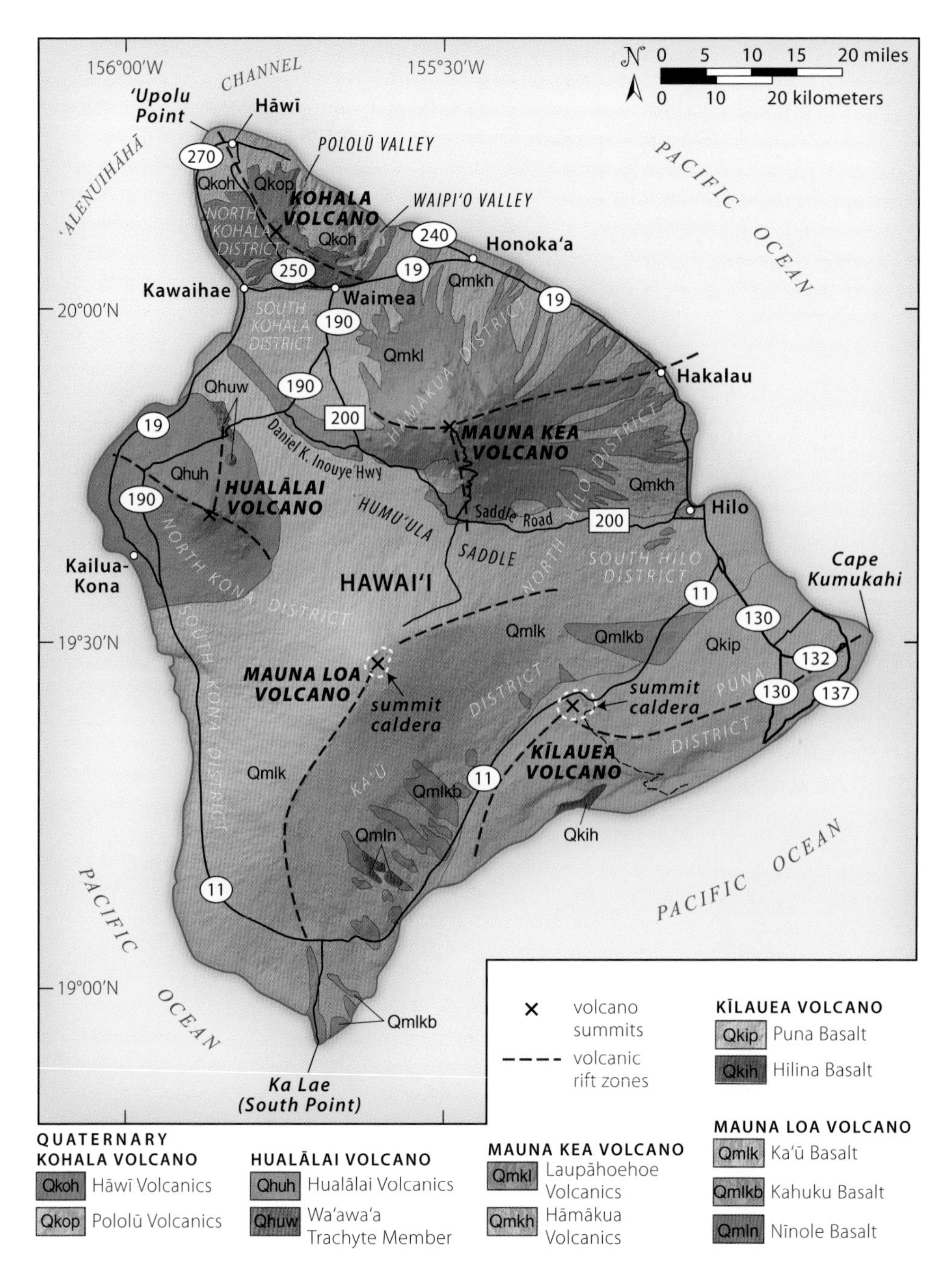

Geologic map of Hawai'i Island.

GEOLOGIC DEVELOPMENT OF HAWAI'I ISLAND

about 800,000 years ago

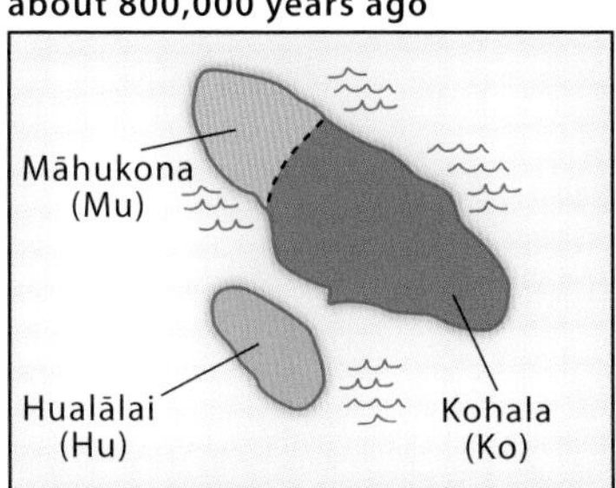

about 700,000 years ago

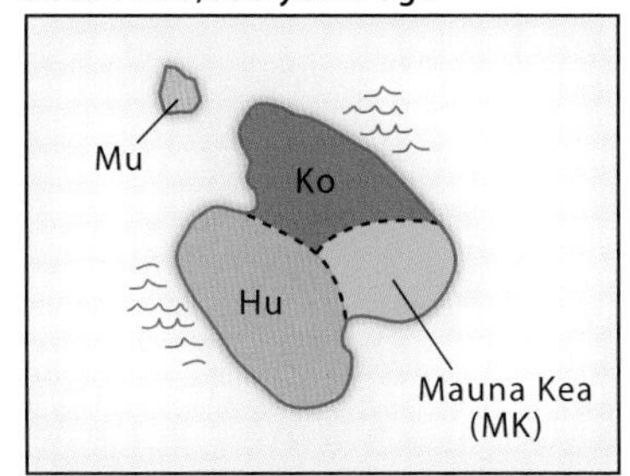

about 500,000 years ago

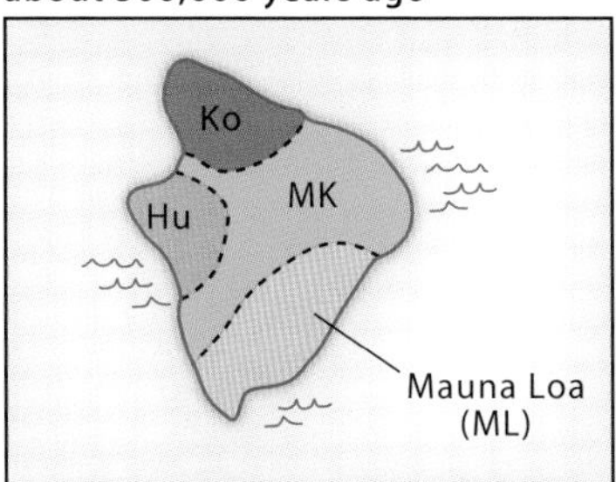

today

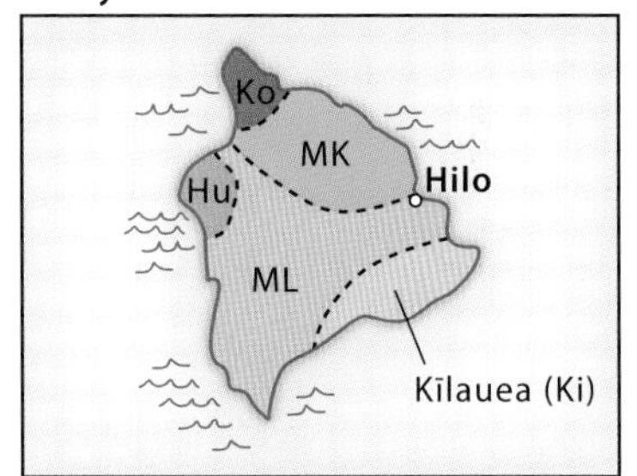

distant future

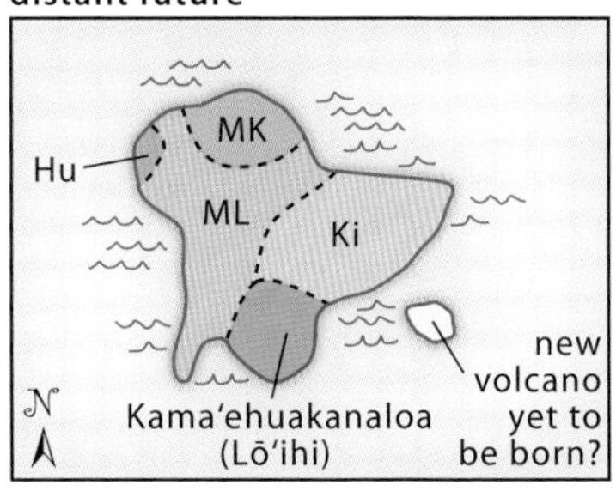

KOHALA VOLCANO

Even before Māhukona drowned, Kohala was building new land to the east. Kohala remained very active as a young shield volcano until around 350,000 to 250,000 years ago, producing vast amounts of pāhoehoe lava and forming two rift zones that extend from a summit caldera, much like Māhukona. Kohala's northwest rift zone stretches about 30 miles toward the island of Maui, while the other may be traced to the southeast over 80 miles, mostly underwater. This is the Hilo Ridge, largest known rift zone in the Hawaiian Islands, though much of it is overlain by younger Mauna Kea Volcano.

The usually cloudy summit of Kohala, Kaumu o Kaleiho'ohie, just north of the town of Waimea, tops out at 5,480 feet. Kohala once made up most, if not all, of the Big Island. Now it covers only about 6 percent of it, owing mostly to slow but steady subsidence. How much taller was Kohala before the rate of tectonic subsidence began outpacing the rate of adding new land? Average subsidence of the northeast

Kohala Volcano viewed from the south with Mauna Kea cinder cones in foreground.

Big Island is roughly one inch per decade. Radiometric dating suggests that this rate has held steady for at least 600,000 years. If you take into consideration that Kohala's eruptions tapered off significantly a couple hundred thousand years ago, the volcano probably rose more than 10,000 feet above sea level during its primacy.

Toward the close of Kohala's shield-building stage, a few hundred thousand years ago, its unsupported northeast side began collapsing into the Hawaiian Deep. We see evidence for at least three giant landslides. The first two were slumps: slow-moving landslides that produced a steplike seafloor topography. Each may have been active for many tens of thousands of years. The older Laupāhoehoe slump broke off from Kohala's northeast rift zone. The younger and smaller Pololū slump rumpled the ocean floor north of the volcano's caldera as far as 30 miles offshore—almost to the Island of Maui. Both slumps cover a submarine area about the size of the state of Delaware.

A younger, more violent event—the Kohala debris avalanche—quickly ripped away part of the volcano's northeast flank, spreading countless fragments of crust over older Pololū slump deposits and even farther underwater to the east. This debris avalanche doubtlessly caused a gigantic tsunami. The shoreline cliffs of the Kohala Coast between Waipi'o and Pololū Valleys, as much as 2,000 feet tall, are the scars left in Kohala's side from this last major failure. Some geologists believe that the two big valleys are simply eroded, upslope extensions of the avalanche margins. Radiometric dating of corals growing across the debris field indicates that the avalanche took place sometime between 370,000 and 350,000 years ago.

During late-shield-building time, an 8-mile-long, fault-bounded trough, or graben (German for "graves"), formed across Kohala's summit. The summit graben lies upslope from the Kohala debris avalanche scar and is parallel to it. Possibly it formed during the avalanche as the summit region stretched in response to the sudden disappearance of the mountainside to the northeast. The graben is about 2 miles wide and originally was no more than a few hundred feet deep. Younger volcanic rocks and sediments now largely fill it.

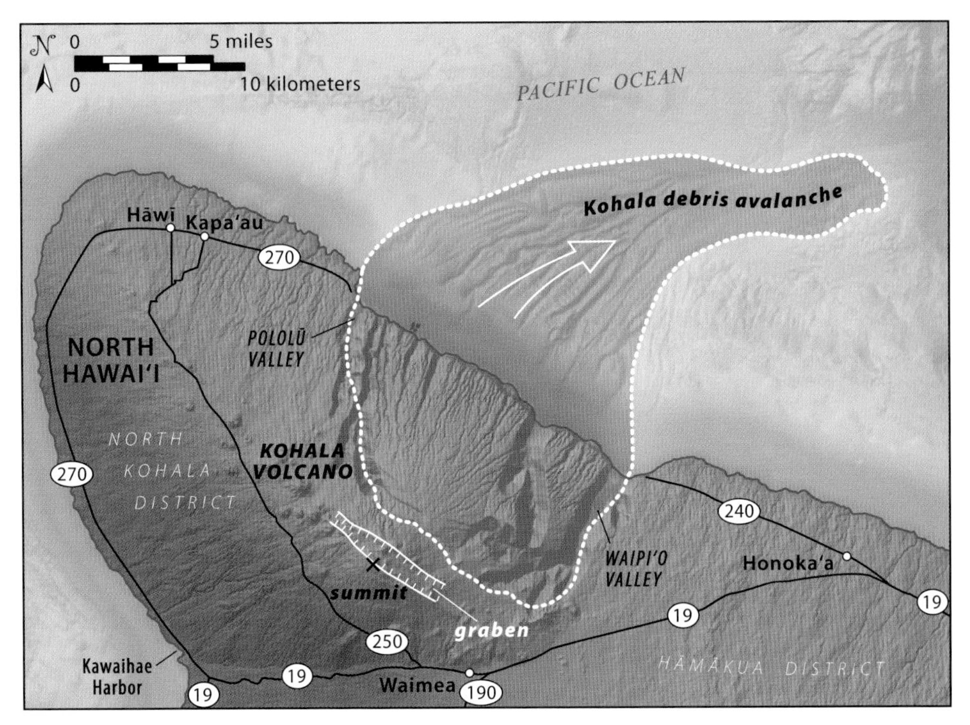

Major landslides have torn the northeastern flank of Kohala Volcano, most recently the Kohala debris avalanche between 370,000 to 350,000 years ago.

The older tholeiitic lavas that constitute Kohala's shield are the Pololū Volcanics. The alkalic cinder cones and 'a'ā lava flows that thinly veneer most of Kohala are part of the Hāwī Volcanics. These postshield stage rocks erupted mostly between 250,000 and 120,000 years ago, completely burying the old shield caldera and much of the summit graben. Some Hāwī activity may have lingered as recently as 60,000 years ago.

During Hāwī volcanism the trade winds blew huge volumes of volcanic ash downwind from active vents, spreading it across a wide area. The ash weathered into soil many feet thick that contains a zone of impermeable clay, called hardpan, just below the surface. A hardpan layer covers most of the summit graben. Heavy rain falling in the graben neither drains nor infiltrates, so water accumulates close to the surface to form a boggy landscape, often wet from cloud mist. This environment supports one of the world's most unusual ecosystems—a Hawaiian cloud forest. Dwarf trees rise from beds of thick sphagnum moss, a landscape impossible to cross on foot. Where infiltrating water manages to percolate through the hardpan, erosion has created funnel-shaped holes in the forest floor, some with dark pits at their deep bottoms. The water may emerge through the walls of nearby valleys as springs, producing filmy waterfalls that seem to appear out of nowhere.

Lavas in the Pololū Volcanics are the oldest exposed at the surface on the Big Island. Their internal magnetic fields, frozen into the rocks when they solidified, are aligned parallel to the Earth's present magnetic field. This means none could have erupted before 770,000 years ago when the Earth's magnetic field assumed its current orientation. Most Pololū samples dated so far range from 450,000 to 320,000 years old.

MAUNA KEA VOLCANO

Mauna Kea (or Maunakea) was the next volcano to add new land to the Big Island. Mauna Kea means "White Mountain," or "Mountain of Wakea," the Sky Father. It is the tallest mountain in Hawai'i at 13,803 feet. This measurement, from the National Geodetic Survey in 2016, will change with time, as will all elevations on this youthful island. Measured from its deepest base on the seafloor, Mauna Kea is even taller than Mt. Everest. With generally clear, dry air and a thin, clean atmosphere at the top, it is no wonder that scientists from around the world operate some of the world's most sophisticated telescopes here, providing humanity with many important discoveries about the universe. Mauna Kea is also sacred in the Hawaiian tradition, a bridge to the cosmos and home of the snow goddess, Poli'ahu, one of Pele's antagonists.

You might infer from Mauna Kea's great size that this volcano was more active than its neighbor, Kohala. But this conclusion is probably wrong. Mauna Kea largely grew up on the flank of Kohala's east rift zone, so its elevation is considerably boosted from the buried mass of Kohala. Mauna Kea is still one of the world's most impressive mountains, but fans of Mauna Kea should recognize that it stands on the shoulders of earlier giants.

Like Kohala, Mauna Kea is a smoothly arching tholeiitic shield capped ruggedly with at least three hundred postshield alkalic cinder cones and associated lava flows. Mauna Kea presumably once had a summit caldera, and a loose concentration of

Snowcapped Mauna Kea and Humu'ula Saddle area with Mauna Loa lava flows in the foreground.

cinder cones in three bands radiating from the summit also seems to suggest the existence of rift zones during the shield stage.

Mauna Kea's oldest exposed lavas are the somewhat alkalic Hāmākua Volcanics. Along HI 19, on the northwest coast, the many flows cropping out in numerous roadcuts and valley walls are all Hāmākua vintage. They erupted mostly between 250,000 and 65,000 years ago, primarily as pāhoehoe flows. Older Hāmākua tholeiitic shield-stage lavas have been found 1,000 feet down in the drill holes.

The alkalic cap, which forms the steep upper slopes of Mauna Kea, consists of the younger Laupāhoehoe Volcanics. The gentle, lower shield profile of the mountain suggests that if the Laupāhoehoe Volcanics hadn't erupted, Mauna Kea would stand only about 7,000 feet tall today.

Eruption of Laupāhoehoe cinders and lavas is almost certainly not finished. A set of young cinder cones and flows, the Pu'u Loaloa volcanic field, extends southward from Mauna Kea's summit to Humu'ula Saddle and last erupted just 4,500 years ago. Minor earthquake activity continues to show that Mauna Kea is restless inside, though this could largely result from the steady sinking of the mountain into the mantle underneath. Mauna Kea is a good example of a dormant volcano: one with no recorded history of eruptions but with fresh enough evidence of volcanic activity to indicate that it could one day spring back to life. No doubt this remote possibility is a concern for those astronomers working on top.

HUALĀLAI VOLCANO

Buttressed by the western flank of Mauna Kea and southwestern flank of Kohala, Hualālai Volcano began forming perhaps as early as 900,000 years ago and built rapidly above sea level over the next few hundred thousand years. The flank support of nearby Mauna Kea guided development of two rift zones radiating from Hualālai's summit. The 40-mile-long northwestern rift zone extends underwater most of the way as the Kīholo submarine ridge. It enters the ocean near Kona International Airport, just north of the town of Kailua-Kona. The generally inaccessible southeastern rift zone slopes away from the summit much less steeply than the northwest rift zone, implying that it originally could have been longer. Younger Mauna Loa lavas mostly bury it. Only about 9 miles of it can be followed today. Like Kohala, Hualālai experienced a major submarine landslide in its flank as its shield grew. This feature is the 40-by-15-mile North Kona slump.

Hualālai has entered its ongoing postshield stage of alkalic activity. Around one hundred cinder cones dot the rift zones above sea level. A large trachyte flow and a 450-foot-tall, 1-mile-wide pumice cone, Pu'u Wa'awa'a, erupted from the north flank of the volcano around 100,000 years ago.

Ongoing postshield activity at Hualālai has certainly been more vigorous than at Mauna Kea, which might appear to be more active because it is much taller. However, only a small fraction of Mauna Kea's surface area is covered with alkali volcanic rocks less than 5,000 years old. This compares to about 80 percent of Hualālai's surface. Some two hundred eruptions have occurred at Hualālai during the past 10,000 years, an average rate of one fresh outbreak every fifty years. Though some geologists argue about the dates, the latest eruption appears to have taken place in 1800–1801. Two large alkali basalt flows burst from northwest rift zone vents at different elevations, overrunning at least one community and adding new land at the

Hualālai Volcano viewed from the northeast. The jagged profile of Hualālai results from numerous postshield alkalic cones erupted within the past 130,000 years.

coast. These flows brought up large concentrations of diverse xenoliths from deeper magma reservoirs within the volcano.

Strong, shallow earthquake activity in 1929 suggested that Hualālai was getting ready to erupt again. Though weeks of damaging shaking occurred, the mountain kept its magma pent-up inside. Now thousands of people live on the steep flanks of the northwest rift zone; this eruption potential is no minor concern. On the other hand, Hualālai's eruptions appear to bunch together in time, and only three have taken place in the past 1,000 years. Hopefully this dangerous volcano will remain in a state of slumber.

MAUNA LOA

Mauna Loa ("Long Mountain"), at nearly 13,700 feet high, is the world's largest *active* volcano, though larger extinct volcanoes are known on Earth's ocean floor. Like Mauna Kea, Mauna Loa has grown on the flanks of preexisting older neighbors. Still in its shield stage of development, Mauna Loa is likely to surpass Mauna Kea in elevation in the foreseeable geologic future, perhaps even exceeding 14,000 or 15,000 feet. Mauna Loa's summit has gained 500 to 900 feet of elevation in the past 12,000 to 10,000 years alone. About 95 percent of the surface of Mauna Loa—about half the Big Island—is covered with lava flows less than 4,500 years old.

Mauna Loa probably began erupting on the seafloor around 700,000 years ago and over the next 300,000 years built well above sea level. A summit caldera formed with an inner pit, Moku'āweoweo, measuring 1.5 by 3 miles across and 800 feet deep.

Two rift zones extend from Moku'āweoweo, their orientations determined by the buttressing effects of neighboring Hualālai and Mauna Kea. The 25-mile-long northeast rift zone is entirely landlocked, with only the upper half of it erupting in recent geologic times. Perhaps the rapid growth of Mauna Loa's younger southeastern neighbor, Kīlauea, has impinged upon the lower northeast rift zone, shutting down its activity. The southwest rift zone, in contrast, stretches 40 miles to the island's south

Mauna Loa, the world's largest active volcano, is also the classic example of a gently sloping shield volcano. The mountain is so broad that the summit, about 16 miles away, is not visible; most of the ridge is the northeast rift zone. The pāhoehoe flows in the foreground came from a vent high on the slopes in 1935.

Aerial view of Mauna Loa's summit caldera, Moku'āweoweo. Cones that formed during the 1940 and 1949 eruptions of Mauna Loa are visible in the background, surrounded by mostly younger lava flows. Fissures from 1984 extend through the caldera and past Lua Poholo, the pit crater in the lower part of the image. —Photo courtesy Hawai'i Civil Air Patrol

cape, then continues another 20 miles underwater. Future eruptions could break out anywhere along its length, even in the deep.

Mauna Loa's eruptions aren't narrowly restricted to the caldera and rift zones. Fissure vents also radiate from the summit across the northwestern flank, from Kealakekua Bay in the west to Hilo in the east. The masses of Hualālai and Mauna Kea strengthen this side of Mauna Loa so well that it is impossible for a third coherent rift zone to develop here, but magma still manages to intrude the interior of the volcano throughout this region.

Mauna Loa has erupted more than thirty times in the past 180 years. Almost 40 percent of the eruptions have occurred solely at the summit, while 25 percent have broken out in the southwest rift zone, 30 percent in the upper northeast rift zone, and only 6 percent on the northwest flank. Eruptions typically start with a days-long summit phase, then spread into one of the rift zones. The rate of lava eruption is far greater than on neighboring Kīlauea. In fact, very few volcanoes in the world routinely produce lava as fast as Mauna Loa does. Its southwest rift zone vents erupt on average around 14,000 cubic feet of lava each second—enough lava to fill an Olympic-size swimming pool every four seconds! On the other hand, Mauna Loa has rarely shown continuous lava lake activity like Kīlauea—at least not historically. Mauna Loa's eruptions last for months at most, while at Kīlauea individual eruptions may linger for decades.

Rubble from three or four debris avalanches off Mauna Loa's steep western flank lies strewn across the ocean floor as far as 60 miles offshore. They broke the flank of Mauna Loa along 50 miles of coastline—most of the current South Kona District—and bit into the Big Island's interior as far as ten miles. Enormous landslide scarps, cliffs up to several thousand feet tall, must have formed, but the volcano has since mostly rebuilt itself across these scars. The most recent avalanche, the 'Alika-2 slide, took place 127,000 years ago and is the youngest giant slide to occur in Hawai'i. The slide displaced around 120 cubic miles of Mauna Loa's western flank. The sea wave it produced may have washed coral blocks well ashore as far away as Molokai, 130 miles to the northwest. By comparison, the destructive 1980 Mt. St. Helens eruption and collapse moved about 1 cubic mile of material.

Mauna Loa's western flank is unusually steep and prone to giant avalanching for several reasons. Bending of the oceanic crust under Mauna Loa's great weight has caused the seafloor to slope as much as 10 degrees toward the volcano's western base. As Kīlauea grows to the east, the stress it produces may cause Mauna Loa's southwest rift zone to shift westward. Intrusion of new dikes along the shifting rift zone could also cause the mountainside to bulge in an unstable way. On the other hand, some geologists argue that the geologically recent jump in position of the southwest rift zone could simply be a passive response of the volcano's magma plumbing system to the big debris avalanches. Cause and effect can be hard to ascertain.

No part of the Big Island is more threatened by lava flows than Mauna Loa's southwest flank. The steep mountainside together with the rapid eruption rate of lava from vents directly upslope mean that lava flows can reach the coast within a few hours, even from vents close to the summit. There may be little time for residents to pack up for a move during the next eruption—they may just have to "get up and go." A single highway available for evacuees is also a concern. Fortunately, potentially threatened areas are currently lightly populated.

While Mauna Kea is tall enough to preserve a record of ice age glaciations in its summit deposits, no such deposits may be found atop almost equally tall Mauna Loa. Young lava flows have covered any trace of past ice fields or glaciers there.

KĪLAUEA VOLCANO

Kīlauea ("Rising Smoke Cloud") began forming a couple hundred thousand years ago, a mere shoulder on the flank of Mauna Loa. Around 100,000 years ago the volcano started adding new land to the Big Island, building Kīlauea's summit to its present 4,000-foot elevation. The young shield formed a caldera and two rift zones, each rift zone strikingly different.

Kīlauea's caldera, Kaluaopele or "the Pit of Pele," measures 2 by 3 miles, with outer walls averaging about 450 feet tall. An inner crater, Halema'uma'u—the focal point of summit eruptions—has historically ranged from under 200 feet to more than 1,000 feet deep. Lava lakes, overflows, and subsidence frequently reshape Halema'uma'u, changing its appearance dramatically. At times the caldera disappeared entirely under the mass of growing lava shields. Then it redeveloped in collapse episodes—one after the next—over centuries.

Aerial view of Kīlauea caldera (Kaluaopele) in 2019, looking to the south. South Sulphur Bank (yellowish cliff) and Halema'uma'u pit are center, and the tan patches in the lower right are the Steam Vents and North Sulphur Bank. —Photo by C. Parcheta, US Geological Survey

Kīlauea's east rift zone stretches 80 miles, with only about 40 percent of it above sea level. Eruption of lava from Pu'u'ō'ō, a continuously active vent in the upper east rift zone, added over 1 square mile of new land to the island between 1986 and 2018. The 2018 flank eruption in the lower Puna District added another 1.5 square miles. Simultaneously, where the east rift zone enters the ocean at Cape Kumukahi (East Cape), land subsidence is much faster than observed elsewhere on the Big Island—up to 7 inches per decade. This subsidence is due to the weight from the rapid addition of flows to this shoreline. Lava less than 1,000 years old covers 90 percent of the land surface of Kīlauea, mostly from summit overflows and east rift zone eruptions.

Kīlauea's southwest rift zone stretches a little over 20 miles, almost entirely above sea level. The frequency and typical volumes of eruptions here are less than in the east rift zone, and the landscape is more visibly broken by faults and gaping ground cracks. Mauna Loa is slowly spreading seaward, and Kīlauea's upper southwest rift zone and western flank ride piggyback atop it.

The amount of slumping in Kīlauea's south flank is immense. Two separate masses of mountainside are presently undergoing collapse. The largest of these landslides measures 20 by 60 miles and involves most of the southern flank of the east

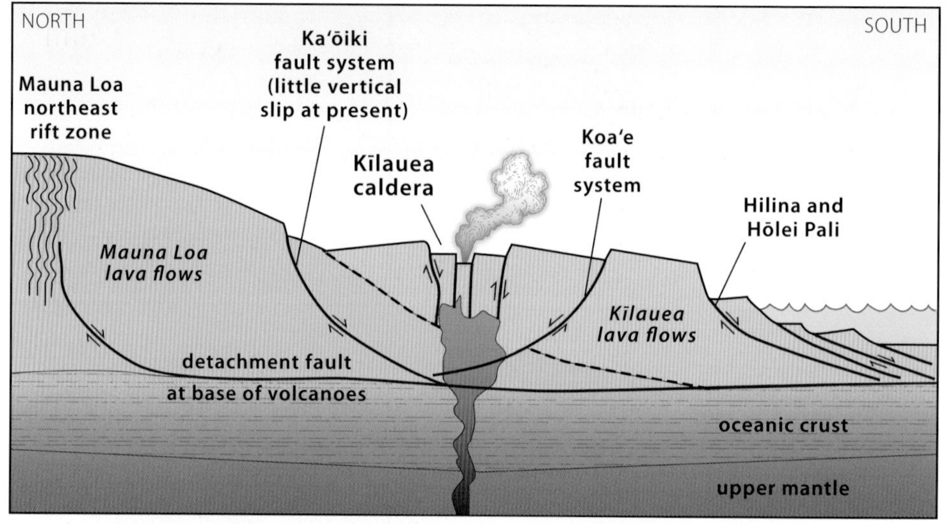

The interior of Kīlauea is sliced by many faults that facilitate the slow sliding of the volcano into the deep sea.

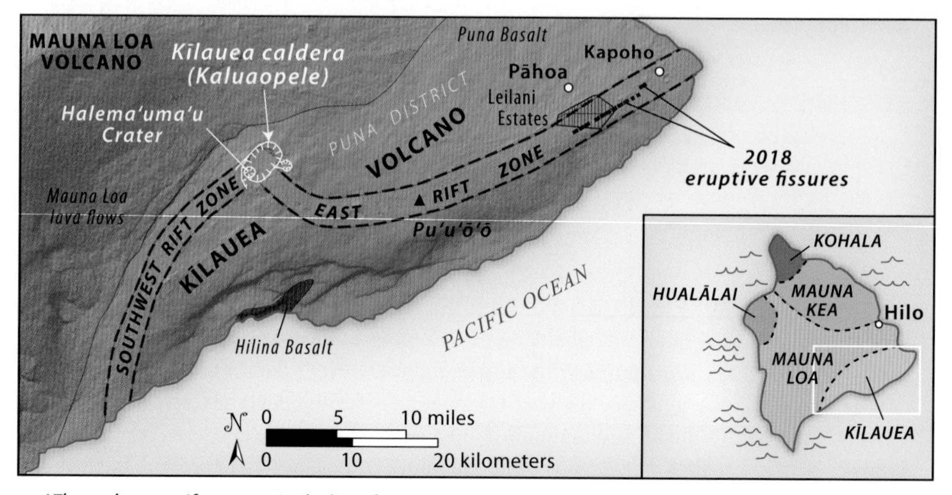

Kīlauea's two rift zones, including features related to the 2018 eruption in the lower Puna District near the eastern cape of the island. The east rift zone is the longer and more active of the two.

approximately 100,000 years ago

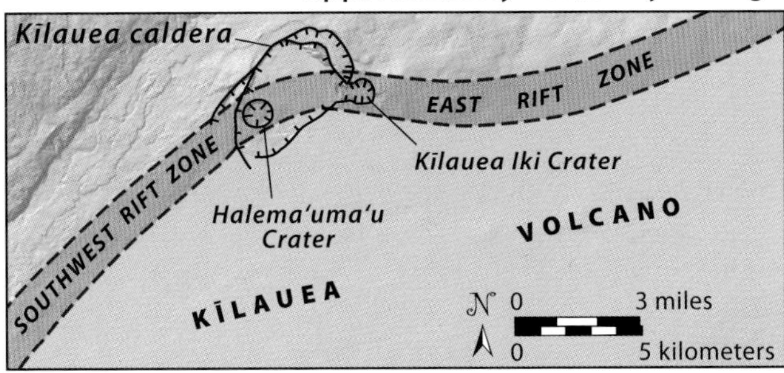

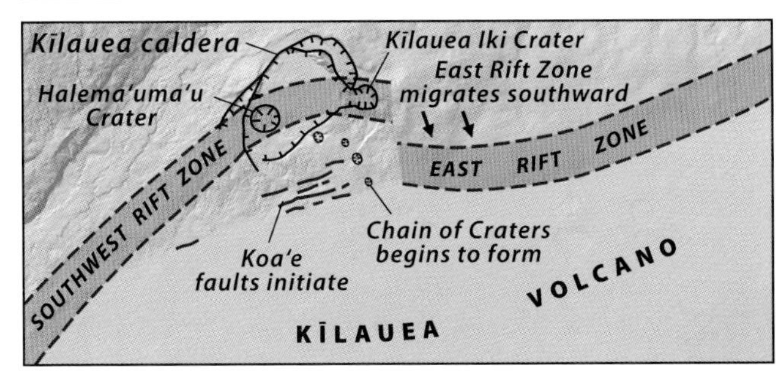

As the east rift zone of Kīlauea has shifted southward in response to slow collapse of the volcano's seaward flank, the Chain of Craters and Koa'e fault system have developed in its wake.

today

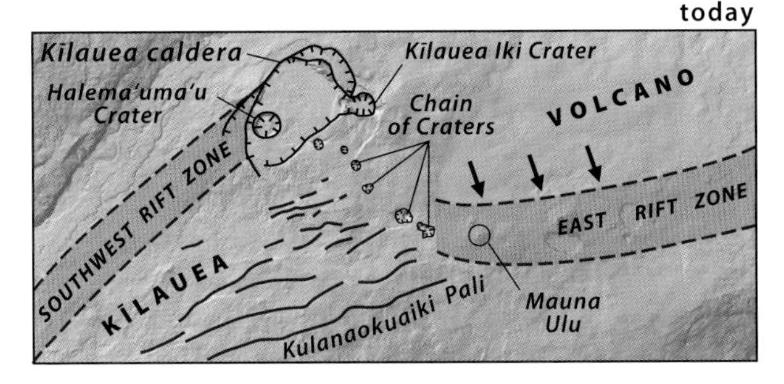

rift zone. Pali, or cliffs, are the trace of giant slump scarps, in places nearly 1,000 feet high. They trend parallel to the south coast of the island. Over a period of tens of thousands of years, the east rift zone has shifted southward 4 miles in response to the collapsing south flank, almost detaching from the summit magma reservoir. Simultaneously a great array of ground cracks and faults, the Koa'e fault system, has developed across the upper south flank of Kīlauea to accommodate the shifting magma intrusions and southward slumping.

A belt of collapse pits, called the Chain of Craters, has formed within the opening gap between the summit and the upper east rift zone. These pits appear where shallow magma bodies accumulate, then are tapped away to supply east rift zone vents that open farther downslope. The overlying land sinks passively into drained void spaces. In other words, none of the craters are volcanic vents. Except where flows have poured in, providing them with flat lava floors, their bottoms simply consist of rocky rubble. They are like the pits above old collapsed mine tunnels.

KAMA'EHUAKANALOA VOLCANO (LŌ'IHI SEAMOUNT)

Kama'ehuakanaloa Volcano, a seamount or undersea mountain, began forming around 300,000 years ago and now rises 2 miles above the submerged base of Kīlauea. A set of pit craters marks the lonely, rounded summit, which sits 3,000 feet underwater. A 9-mile-long north rift zone and 12-mile-long south rift zone slope away gently to either side of the caldera, with rugged, steep flanks hosting frequent landslides. The volcano has the profile of a long, narrow ridge; in fact, the original name, Lō'ihi, translates to "long." Observed undersea earthquake activity and submersible exploration suggest that Kama'ehuakanaloa last erupted in 1996. It will someday be as active as Mauna Loa and Kīlauea are today, but plate motion must first carry it closer to the center of the Hawaiian hot spot.

KA'Ū, KONA, AND KOHALA REGIONS

HI 11 (HAWAI'I BELT ROAD)

Kailua-Kona—South Point Road

53 miles

Before heading south on HI 11, you can find several pocket beaches of white sand near Kailua-Kona if you head south along the shore on Ali'i Drive. The largest and most popular beaches are Magic Sands and Kahalu'u Beach. Beachside outcrops of lava, ranging from 10,400 to 7,000 years old, are from Hualālai Volcano. The sand is ground-up calcium carbonate derived from patchy offshore reefs. No large fringing reef exists here as they do around older islands. The protected water at Kahalu'u is a particular delight for snorkelers. Look for schools of convict tangs, surgeonfish, wrasses, and the occasional honu (sea turtle). Don't touch or step on the coral!

Right across Kaleiopapa Street from the boat marina at Keauhou Bay, at the south end of Ali'i Drive, look for a low sea cliff in the forest. A footpath skirts the cliff base with informational signs highlighting the area's history as a center for Hawaiian royalty into the nineteenth century. Mo'ikeha Cave crops out near the western end of the cliff, flooded by seeping groundwater. The "cave" is not a lava tube but simply a cliffy, rubble-filled overhang. Many of the pāhoehoe lobes that are exposed as smooth, saclike masses throughout the cliff face have vesicles concentrated in the center where gases accumulated as the margins cooled. Some small caves or pockets are also visible where lava drained from small lava tubes. Further evidence of pāhoehoe may be seen in the ropy structure in rotated plates of old flow crust in a few places.

For the most part the cliff seems to expose the interior of a single, big, complex lava flow. Careful inspection shows that it is not entirely pāhoehoe. Patches showing the telltale structures of 'a'ā also occur. The flow was evidently undergoing a transition between the two lava types as it erupted around 10,000 years ago. The source of the flow was a now-buried vent in the southeast rift zone of Hualālai, 15 miles away. A prominent baked zone marks the bottom of this big flow near the base of the cliff. Beneath it lie thin 'a'ā flows with fine clinker, the sort of stretched lavas that develop where 'a'ā flows down steep slopes.

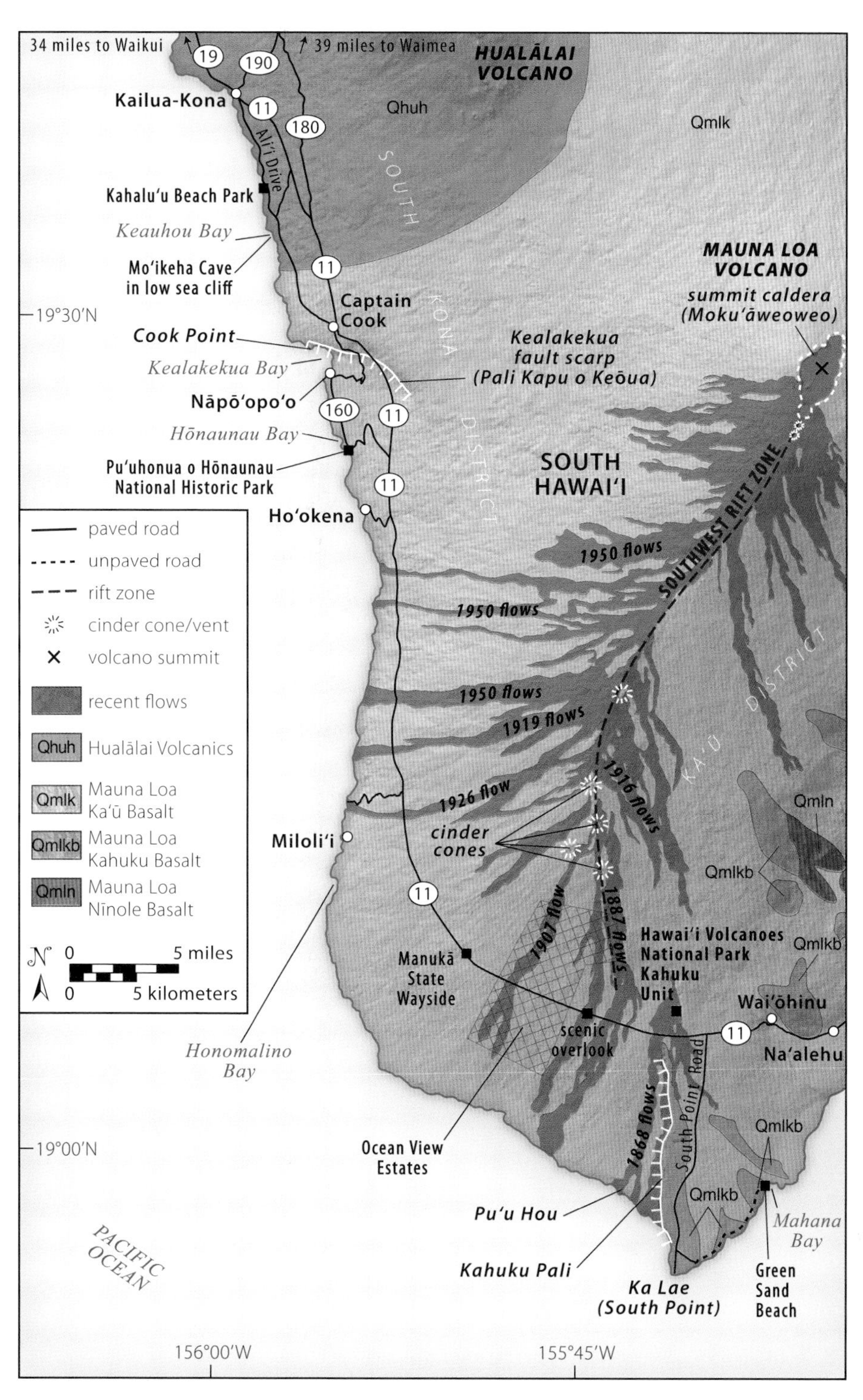

Important South Kona geologic features.

Cliff exposure of lava flows from Hualālai volcano in Keauhou Bay by Mo'ikeha Cave. A red horizon at the bottom of the stack of flows is an old weathered surface that was baked as the lava flowed over it. Close-up shows cross section of an individual pāhoehoe lobe with vesicles concentrated toward the center. The lobe is about 3 feet long.

Kailua-Kona to Captain Cook

HI 11 begins on Hualālai Volcano at Kailua-Kona, then traverses the western and southern flank of Mauna Loa. During the past 150 years, Mauna Loa southwest rift zone eruptions have sent massive lava flows across this road six times—1868, 1887, 1907, 1919, 1926, and 1950. These flows destroyed villages, burned forests and ranch land, blocked escape routes, and severed communications.

The road begins on a region of Hualālai Volcano mostly covered in 12,000- to 2,000-year-old flows. At around mile 114 the road passes a ridge of 33,000-year-old Mauna Loa lava. This upland is Hawai'i's most famous coffee region, home of true Kona coffee. The most productive zone for growing coffee lies between 700 and 2,000 feet above sea level in a belt 20 miles long and 2 miles wide. Soils in most places here are deep, porous, and well drained, though not excessively leached given their young volcanic age. Before Kīlauea's continuous eruptions began in 1983, South Kona rainfall averaged about 68 inches annually—nearly perfect conditions for coffee production. Clouds of volcanic gas drifting to this side of the island lowered that average to only around 49 inches until late 2018 when the air once again cleared. The combination of cleaner air and increased rainfall quickly boosted prospects for local coffee growers.

At stops along the road into Captain Cook and between mileposts 107 and 106, you may catch glimpses of Pali Kapu o Keōua, a towering sea cliff fringing Kealakekua Bay. This pali is the headwall of a slump block or giant landslide from the unstable western flank of Mauna Loa. It isn't clear if this pali is related to the much shorter sea cliffs at Ho'okena Beach to the south.

HI 160 TO KEALAKEKUA AND HŌNAUNAU BAYS

At the stoplight in Captain Cook, turn onto Nāpō'opo'o Road (HI 160) to explore Kealakekua and Hōnaunau Bays. At the seaside parking area in Nāpō'opo'o, you can look across Kealakekua Bay to Pali Kapu o Keōua, the 600-foot-high sea cliff left by either the great North Kona slump or a related landslide. Occasional slippage and earthquakes continue along the mostly buried scarps of the slump in a zone of weakness extending inland. Within the past few hundred thousand years, at least four great slumps and mega-landslides have torn the southwest flank of Mauna Loa across 40 miles, from here all the way to South Point. Kealakekua Bay is at the northern end of this region of weakness known as the Kealakekua-Kahuku fault system. Seaward swelling of Mauna Loa's frequently active southwest rift zone has largely triggered this instability.

The most recent mega-landslide, the 'Alika-2, originated mostly from an area about 10 to 20 miles south of Kealakekua Bay, though some geologists also attribute Pali Kapu o Keōua to this event. In any case, as much as 600 cubic miles of mountainside broke free during that collapse, spreading as far as 60 miles out to sea underwater. No doubt local shorelines were pounded by a tremendous resulting tsunami. Evidence for such in North Kohala, 40 miles to the north, indicates that the waves swept to hundreds of feet above sea level as far as 4 miles inland. In March 2011 a tsunami related to the far more distant Great Tohoku Earthquake in Japan did considerable damage all along the Kona coastline, sweeping away one seaside house in Nāpō'opo'o and damaging resorts in nearby Kailua.

The fault plane that defines Pali Kapu o Keōua extends underwater, too, and served as a pathway for lava that erupted within the bay in 1877. Three vents opened at water depths between 300 and 3,000 feet, as close as 1 mile west of the road's end. The February 28, 1877, *Hawaiian Gazette* reported:

> *In the afternoon of the 24th, three boats from [a] steamer visited the scene of the eruption, cruised directly over the most active part, where the water was in a state of peculiar activity, boiling and appearing as if passing over rapids . . . . While the boats were in position, blocks of lava, two feet square, came up from below, frequently striking and jarring the boats . . . . Nearly all the pieces on reaching the surface were red hot, emitting steam and gas, [and they were] strongly sulphurous.*

The magma feeding the eruption infiltrated from beneath the summit of Mauna Loa, about 25 miles east of Nāpō'opo'o. This is the one and only definitely witnessed submarine eruption in Hawai'i.

Cook Point, directly across Kealakekua Bay from Nāpō'opo'o, is composed of 2,000-year-old Mauna Loa pāhoehoe. The 27-foot-high obelisk on its shore commemorates Captain James Cook, who was killed at this site in 1779. Great Britain owns the obelisk. The waters

View northwest across Kealakekua Bay from Nāpō'opo'o cobble beach. The scarp of Pali Kapu o Keōua looms above the bay. A fault extending inland from this site remains active.

around the monument are home to one of the most beautiful coral reefs on the island, highly popular with divers and snorkelers.

South from Nāpō'opo'o, HI 160 takes you about 4 miles to Pu'uhonua o Hōnaunau. The empty, overgrown plain here was densely populated when Captain Cook visited this area in 1778. In 1782 it was also an important battleground—the site of Kamehameha's first great victory to unify the Hawaiian Islands. Noncombatants fled south to shelter within the walled enclosure of Hōnaunau. The plain owes its existence to multiple lava flows that spread at sea level to build new land after the latest collapse of the western flank of Mauna Loa along the North Kona slump. Had that collapse and subsequent eruptions not taken place, the slope of the volcano would continue to slant steeply into the ocean as it does elsewhere along this coast.

The lava flows at Pu'uhonua o Hōnaunau National Historical Park are from Mauna Loa and range in age from 1,000 to 750 years. Take forty-five minutes or so to walk through the park. In addition to showcasing ancient Hawaiian history and culture, the loop trail leads to a broad coastal lava platform. Separate pāhoehoe flows form the points that enclose Hōnaunau Bay. A concentration of smooth boulders on the platform closest to the near-shore palm grove and heiau may be the product of strong storm surges or possibly of tsunamis that loosened chunks of lava from the edge of the platform and washed them inland. Some rocks show pits and scallops made by salt corrosion from sea spray.

In addition to a wide area of level ground and a place to land canoes, Hōnaunau attracted Hawaiians because of springs that were used to form pools for aquaculture and water supply. Fresh groundwater from inland comes to the surface here because it floats on top of a deeper saltwater aquifer continuously recharged by the ocean.

Ho'okena Beach to Ocean View

At about milepost 101.5, the turnoff to Ho'okena heads 2 miles down a narrow, winding road that drops 850 feet to the coast. At the end of the road is Ho'okena Beach Park, a lovely black-and-white sand beach sitting at the base of large cliffs of older lava flows. The scalloped cliffs appear to be the headscarps of landslides, though large normal faults cut through this flank of Mauna Loa, too. The faults may or may not be related to individual landslide events. Look south from the beach to see the dark swaths of 1950 lava dramatically highlighted against the green vegetation. Behind the beach you can see where prehistoric lava flows poured over and plastered against the older cliffs approximately 2,000 years ago. These are most likely the source of the pāhoehoe lava and sand at the shoreline. Camping by permit is allowed here and facilities are available.

Fault scarps form the steep cliffs above the black-and-white sand of Ho'okena Beach.

The next section of highway passes through three strands of 1950 lava. The flow strands just north of milepost 98 and at milepost 97 are narrow and overgrown. The third one at milepost 92.5 is covered in lichen and sparse 'ōhi'a trees. The 1950 eruption was one of Mauna Loa's largest historic eruptions. On June 1, 1950, a long fissure split open from 9,000 to 10,000 feet elevation in the southwest rift zone and began erupting lava. The roar of lava fountains could be heard from HI 11, 15 miles away. As the fissure continued opening farther down the rift zone, floods of lava streamed downslope. Within only two hours, the first of these flows crossed the highway and inundated the village of Pāhoehoe. All villagers reached safety, some escaping with only the clothes on their backs. Soon afterward the flow entered the ocean, creating a towering steam cloud. From vent to sea, this massive 'a'ā flow traveled 13 miles in only about three hours. Two additional flows south of the first one reached the ocean in about 14 and 18 hours, respectively. Before ending three weeks later, the 1950 eruption destroyed nearly two dozen structures and cut HI 11 in three places, burying more than 1 mile of the road.

Between mileposts 91 and 90 is another young-looking lava flow in the Kīpāhoehoe Natural Area Reserve. On September 29, 1919, a vent high on Mauna Loa's southwest rift zone erupted fountains of lava up to 400 feet high and sent a river of lava down the volcano's forested slopes. Within 20 hours an ʻaʻā flow several hundred yards wide crossed the precursor to the current highway and buried the small village of ʻĀlika.

Just south of milepost 89 is the turnoff for Miloliʻi Road, a narrow, winding road that descends 1,700 feet to the coast over 5 miles of mostly older lava. Thin-bedded, finely granulated ʻaʻā flows drape the steepest slopes, which also provide excellent views of the coast. Dark 1926 lava is visible to the south, the 1919 flow to the north, and much development of new housing in between. The last part of the road crosses over the 1926 lava and through the native Hawaiian community to Miloliʻi Beach Park, with parking and a shelter. A twenty-minute hike along a marked trail through thick vegetation, past several ancient sites and over lava flows, takes you to Honomalino Bay, which hosts one of the larger beaches in South Kona. This beautiful, out-of-the-way, black-and-white sand beach is fringed with coconut palms. You may see an active blowhole on the south end of the beach at high tide when waves are forced into a small cave opening. There are no facilities here; please stay on the trail and off private property.

At milepost 87 you will pass over a narrow ʻaʻā flow erupted in 1926. The 1926 eruption began on April 10 at the summit of Mauna Loa, but fissures soon migrated down the volcano's southwest rift zone. By April 14, three main vents were sending huge ʻaʻā flows downslope. Two days later, the main flow—advancing at 7 feet per minute—crossed the road. The flow thickened and widened as it rushed down a steepening slope toward the sea. On April 18, the 40-foot-high, 1,500-foot-wide ʻaʻā flow plowed through the Hawaiian fishing village of Hoʻōpūloa. People were evacuated by boat to nearby Miloliʻi.

Near milepost 81, watch for Manukā State Wayside. A 2.1-mile trail loop crosses heavily forested old lava only a few hundred years old, circling by archaeological sites and a large lava tube skylight. South of here the road continues into mostly overgrown lavas that are 3,000 to 750 years old. Around milepost 78, HI 11 crosses a branch of January 1907 lava, erupted from the southwest rift zone about 8 miles upslope. Two main branches of 1907 lava cut through Ocean View Estates, and many homes are now built atop them, some within the southwest rift zone itself. Most homes, however, are located on older prehistoric flows, including many lava tubes and channels. One cave in the residential area, Kula Kai Caverns, offers visitor tours and is noted for its archaeological significance, complex passageways, and bright mineral deposits that line passage walls.

Watch for the scenic pullout on the south side of the road at milepost 75. The thick lava seen here is also from the 1907 eruption. The cut across the road from the pullout shows a classic cross section of an ʻaʻā flow: a sandwich-like structure of crumbly clinker on top and bottom with a dense, smooth interior that was once the molten core of the flow.

Downslope from the pullout, you can see South Point (Ka Lae) and the massive fault scarps that extend inland from the point, a testament to the instability of Mauna Loa's western flank. These scarps, up to 900 feet tall, would be even more spectacular and continuous but for the fact that countless southwest rift zone eruptions have

Aerial photo showing Kahuku Pali, the prominent fault cliff extending inland from South Point. The upslope part of the Kahuku Pali is called Paliomāmalu, the coastal part is called Paliokūlani. The tree-filled circular depressions atop the pali are ancient pit craters, one partly cut open when Paliomāmalu formed. The 1868 Mauna Loa lava (dark, young-looking flow) lapped at the base of the pali and flowed to the ocean in a matter of hours.
—Photo by J. Griggs, US Geological Survey

since built new land across them to extend Mauna Loa's flank smoothly seaward. North of the scarps along the shore, look for giant Pu'u Hou and other littoral cones (or half-cones) formed along the coast where lava flows poured into the sea. In this arid environment, the gradual color change and growth of plant life reveal a complex landscape of overlapping lava flows of various ages.

HI 11 crosses another dark, fresh-looking 'a'ā flow east of the scenic pullout, then passes through an older kīpuka around milepost 74.5.

West of and around milepost 73 the road crosses several black, sparsely vegetated 'a'ā flows of the 1887 eruption. This eruption followed a typical pattern for Mauna Loa. A short-lived breakout at the summit was followed a few days later by lava gushing from fissures on the volcano's southwest rift zone. The lava took only about a day to travel the 15 miles from vents all the way to the sea. The eruption lasted a little over two weeks and was accompanied by frequent and sometimes strong earthquakes. A well-formed lava channel is nicely exposed just east of milepost 74.

HI 11 crosses lava flows of the 1868 eruption from mileposts 72 to 70.5. Upslope, the view reveals the cones and young lava flows of Mauna Loa's southwest rift zone, along with the many homes built practically to its edge. Much of this area is designated as Lava Zone 1, land with the highest volcanic risk on the island.

KAHUKU UNIT OF HAWAI'I VOLCANOES NATIONAL PARK

The entrance to the 116,000-acre Kahuku Unit, added to Hawai'i Volcanoes National Park in 2003, is a north turn near milepost 70.5 on HI 11. Hours are limited, so check ahead, and roads are largely unpaved but mostly drivable by car. A short drive takes you to the main parking area and visitor center to find information and maps showing miles of well-maintained local hiking trails. This region brims with fascinating stories: terrifying volcanic activity, Hawaiian and military history, paniolo—Hawaiian cowboys—and cattle ranching, and the conservation and restoration of endemic forests.

One short trail near the visitor center leads to a grassy cinder cone called Pu'u o Lokuana, with a longer option to walk through the 1868 lava flows on the west side of the road. A quarry at the top of Pu'u o Lokuana reveals layers of oxidized red cinder—the classic sign of an eruptive vent. The original cone was much taller before mining and was used as a radar tower in World War II.

A short drive north of the visitor center brings you to the lower parking area for the Palms Trail, a 2.6-mile loop that begins by crossing a narrow strip of the 1868 lava flow, including its eruptive fissure and large spatter rampart. (See the sidebar on the 1868 earthquake in the road guide HI 11: Hilo—Wai'ōhinu for a description of this harrowing eruption.) Even just a few minutes exploring this flow is rewarding. As you cross the lava, notice the abundance of bright-green olivine crystals. Crystals weathered from the flow have formed pockets of rich green sand, much like the sand making up the beach at Pu'u Mahana near South Point. The 1868 lava, like others in Mauna Loa's lower southwest rift zone, may have drained from a deep level in Mauna Loa's magma reservoir, where abundant grains of dense olivine had settled and accumulated over a long period of time. As you near the 1868 spatter rampart, note the shelly, fragile appearance of the pāhoehoe, typical of the gas-filled lava discharged in the immediate vicinity of active vents. As the gas is released, the lava becomes denser as it flows farther downslope.

Beyond the 1868 flow, Palms Trail leads through historic ranchland and past quarries in 10,000- to 5,000-year-old cinder and spatter cones, now largely grassy and forested. To the east of the lower Palms Trail parking area is another ancient volcanic feature, a grassy ridge that is an exceptionally thick, 5,000- to 3,000-year-old lava flow erupted from a vent just a short distance upslope.

The Pit Crater Trail begins about 3.5 miles up Kahuku Road from the entrance. This strenuous hike, about 5 miles round-trip, climbs 1,165 feet of elevation to a remarkable ancient pit crater. Within its sheer walls, pristine Hawaiian rainforest was protected from grazing animals. The Pali o Ka'eo Trail at the upper end of the road also provides sweeping views of the Ka'ū coastline far below.

South Point (Ka Lae)

Watch for the turnoff to South Point from HI 11 near milepost 69.5. This 10-mile-long road drops in elevation nearly 2,000 feet to sea level. The road begins in cool upland forest and quickly descends to open, hot, windswept grasslands. Downslope 2.6 miles, it crosses a narrow tongue of 'a'ā erupted in 1868. Once closer to the ocean, look behind you to see the tremendous fault scarp cliff called Paliomāmalu. In places it rises over 600 feet. On the shore west of the cliff you can see young-looking lava flows and shoreline cones produced by eruptions in the southwest rift zone of Mauna Loa, most recently in 1868 and 1926.

Paliomāmalu, and its seaward extension, Paliokūlani, are part of the Kealakekua-Kahuku fault system, a series of large normal faults resulting from piecemeal collapse in the western flank of Mauna Loa. The fault system probably began forming a few hundred thousand years ago, and large submarine landslides have broken the western, downthrown block from time to time. Movement along these faults probably caused Mauna Loa's southwest rift zone to shift west to its present position west of South Point.

The road forks as you approach Ka Lae. The route to the right takes you to the point. Near the parking area, look for a large skylight in a lava tube (or sea cave) hollowed out by wave erosion. On the walk to the point, watch for examples of sea salt corrosion holes and lines in the lava. Ka Lae is the southernmost point of land in the United States. It has been used by people in a variety of ways, from worship to satellite tracking and generating electricity. Because the point is the closest potential landfall to the rest of Polynesia, some archaeologists think that Hawaiians landed on South Point when they first arrived, probably sometime in the mid-thirteenth century. The rough seas here demand respect; the offshore current is called Halaea after a chief who was swept to his death.

The paved fork to the left (Lio Lani Road) ends at a parking area—the starting point for the trek to Pu'u Mahana (also known as Puāuāomahana or Pu'uomahana) and Green Sand Beach. The walk is about 2.5 miles each way along the shore and can be hot and windy. The landscape consists mostly of deeply weathered, yellow-ochre Pāhala Ash, rutted by erosion and heavy vehicle traffic in many places. The origin of Pāhala Ash is unclear because it is so badly weathered. It probably consists of multiple thick ash layers from Kīlauea, deposited over a period of thousands of years. NASA has studied this soil as an Earth analog of what might be found on the planet Mars. It is easy to imagine why.

Mahana Bay is rapidly eroding into the loose tuff layers of Pu'u Mahana tuff cone, leaving behind dense olivine crystals to form Green Sand Beach.

Salt corrosion of pāhoehoe near the southwest end of Mahana Bay. As the lava groundmass dissolves away, the less soluble olivine crystals drop out and collect to form pockets of green sand. A quarter coin is shown for scale.

Green Sand Beach consists largely of fragmented olivine crystals eroded from the flanks of Pu'u Mahana, an ancient tuff cone. You may be tempted to think that the original crater of the cone lies in the bay just off the beach, but the inclination of tuff layers indicates that it must have existed a short distance inland toward the east. Most of the cone, likely more than 50,000 years old, has simply eroded away. Younger lava flows poured around the remainder of the cone to form the points enclosing Mahana Bay, a natural shoreline sand trap. Outcrops close to shore include great concentrations of olivine crystals, typical of lava erupting from low on Mauna Loa's southwest rift zone. Salt corrosion has formed pockets in the lava, and the grains of insoluble olivine minerals have collected in and around these pockets to create additional green sand. Enjoy one of the world's best examples of an extremely rare green sand beach.

HI 19 (QUEEN KA'AHUMANU HIGHWAY)
KAILUA-KONA—KAWAIHAE
33 miles

With the island's busiest airport, dry weather, and sandy beaches, Kailua-Kona is heavily visited and rapidly growing. HI 19, the Queen Ka'ahumanu Highway, follows the coastal route between Kailua and Kawaihae. It crosses young Hualālai and Mauna Loa lava flows, provides access to some of the state's finest beach parks, and ends at the foot of Kohala, oldest of the island's five volcanoes.

Just north of Kailua, a set of foot trails through Kaloko-Honokōhau National Historical Park leads across various Hualālai pāhoehoe flows mostly 10,000 to 5,000 years old. As the lava entered the sea, it must have spread out rapidly to build the remarkably level coastal plain. Maliu Point, at the south end of the beach close to the boat harbor, is mostly composed of a large tumulus, an uplifted ridge of fractured lava. The lava here erupted only about 2,000 years ago. Remains of an ancient Hawaiian temple, Pu'uoina Heaiu, sprawl nearby.

North from the Ellison Onizuka Kona International Airport at Keāhole, HI 19 crosses lava from the Hu'ehu'e Flow of 1800–1801, the most recent eruption of Hualālai. The flow added new land to about 15 miles of shoreline, burying many fishing villages and filling a bay in the process. Reverend William Ellis compiled an account of the eruption from interviews with witnesses:

> *Stone walls, trees, and houses, all gave way [before the flow], even large masses of hard, ancient lava, when surrounded by the fiery stream, soon split into small fragments, and, falling into the burning mass, appeared to melt again, as borne by*

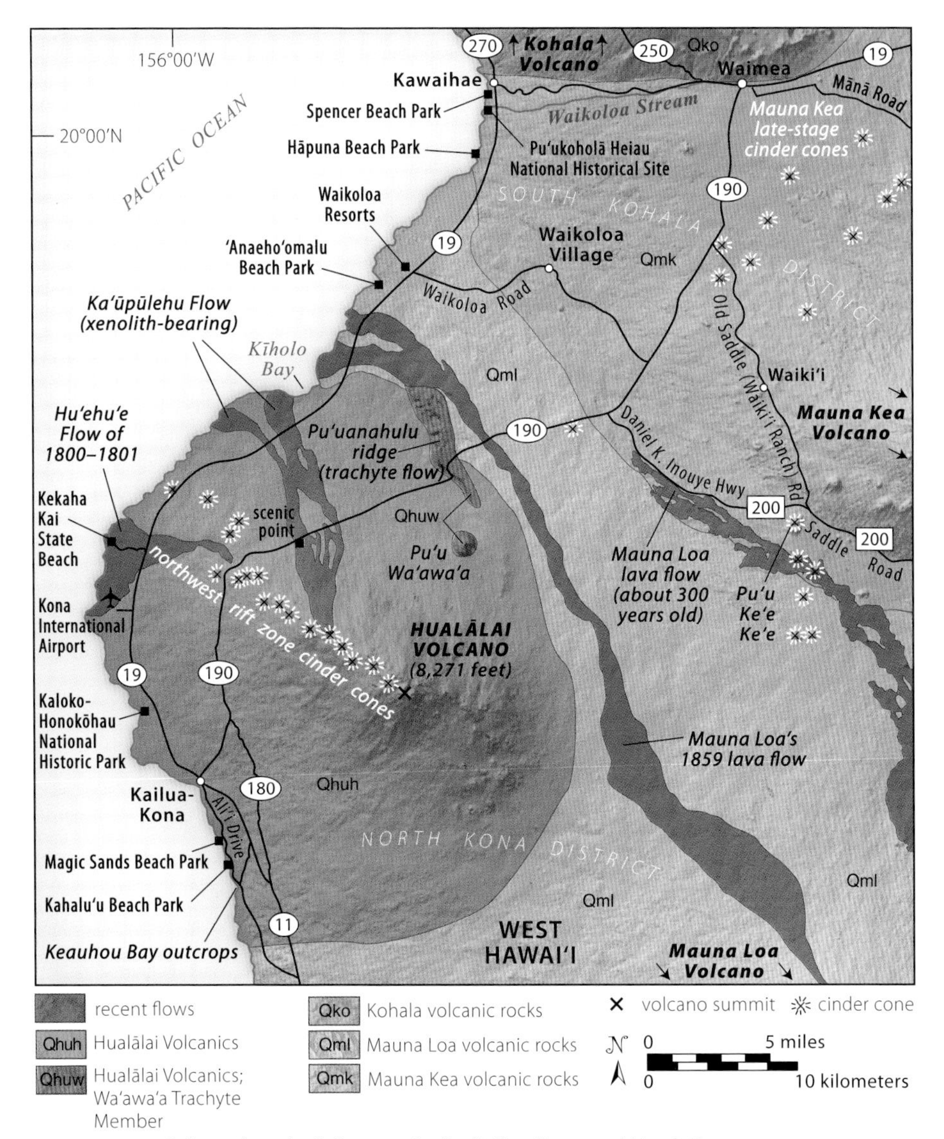

Points of geologic interest in the Kailua-Kona and North Kona area.

it down the mountain's side. Offerings were presented [by the natives], and many hogs thrown alive into the stream to appease the anger of the gods, by whom they supposed it was directed, and to stay its devastating course. All seemed unavailing, until one day the king Kamehameha, went attended by a large retinue of chiefs and priests, and, as the most valuable offering he could make, cut off part of his own hair, which was always considered sacred, and threw it into the torrent. A day or two after, the lava ceased to flow. The gods, it was thought, were satisfied. (Brigham 1909, 14)

Like most young lava flows on Hualālai, this lava is alkalic basalt that erupted during the postshield stage of volcanic activity. Many cinder cones upslope from here mark the northwest rift zone of Hualālai.

North of the Kekeha Kai Beach turnoff, a rough road leading to beautiful beaches, HI 19 heads across lava that is at least 3,000 years old. Between mileposts 89 and 87, it crosses the coastal part of Hualālai's northwest rift zone. Ancient Ku'ili cone is visible near the coast, while other cones dot the ridge that extends upslope to Hualālai's typically cloud-cast summit. From mileposts 84.5 to 83, the road crosses another set of lava flows attributed to or slightly preceding the 1800–1801 eruption and originating high on the rift zone. Known as the Ka'ūpūlehu Flow, it contains xenoliths and is discussed in more depth in the road guide for HI 190: Kailua—Waimea.

KEKAHA KAI STATE BEACH

About 2.5 miles north of the airport and after milepost 91 is the turnoff to Kekaha Kai State Beach. The 1.7-mile access road is rough, unpaved, and not recommended for vehicles with low clearance. The road stays on 1800–1801 lava, which varies from 'a'ā or slabby pāhoehoe near the highway to classic inflated pāhoehoe closer to the shore. This flow carries many small xenoliths. If you look around the upper parking area near the road's end where the insides of thick pāhoehoe lobes have been exposed, you may find olivine-rich fragments, generally under 2 inches long, embedded in the lava. Some fascinating lava features can also be seen in the flow surface, including the edge of an older coastal shelf with fragile driblets of lava flowing over it—sometimes called lava draperies—and a wide, flat area where two large streams of lava flowed around the shelf and coalesced before crusting solid. Other signs of inflated pāhoehoe appear here as well, including huge up-tilted slabs and tumuli.

Several white sand beaches are accessible at Kekaha Kai. Right below the lower parking area is Kekaha Kai State Beach, fringed with palms and spring-fed brackish-water ponds. Just inshore from the north end of the beach, you might discover several small, circular hornitos—rootless vents formed where highly fluid spatter burst under great pressure from the top of the flow. In places nearby the liquid core of the inflated flow drained away abruptly, causing much of the overlying solid crust to collapse. Much of the flow around you is hollow; short pillars support higher areas of arching crust that has not yet collapsed. Small lava tubes, channel troughs, and irregular void spaces abound. Similar structures have been identified during recent explorations of the deep ocean floor where highly inflated sheet pāhoehoe flows suddenly drained. At this point along the shore, the lava poured into and filled several Hawaiian fishponds. Perhaps abrupt drainage into these ponds accounts for development of the pillars, tubes, and other unusual void spaces in this part of the flow.

The calcareous white sand at Kekaha Kai State Beach overlies 1800–1801 pāhoehoe flows. Cinder cones dot the ridge of Hualālai's northwest rift zone on the horizon.

At Kekeha Kai Beach, a lava flow of 1800–1801 from Hualālai Volcano shows draining of an inflated pāhoehoe flow. Small pillars support intact sections of the flow, though much of the surface crust dropped several feet.

A short walk north from the parking area will take you to Mahai'ula Beach. Endangered Hawaiian monk seals may occasionally be found lounging on the sand. Federal law requires that you keep your distance; please do not disturb these beautiful, rare animals. Another half mile beyond Mahai'ula, over an older lava flow, is secluded Makalawena Beach—one of the prettiest in Hawai'i.

A scenic pullout another mile or so north offers a sweeping view of Kīholo Bay, where Hualālai lavas come into contact with younger Mauna Loa flows. The darker pāhoehoe on the north side of the bay is from Mauna Loa's 1859 eruption. This lava is the longest historic flow in Hawai'i, having originated 32 miles upslope from a radial fissure vent a short distance northwest of Mauna Loa's summit. Early-erupted 'a'ā reached the ocean in just eight days, moving as fast as 14 miles an hour downhill—far faster than a person can run. Eventually, the flow transformed partly to pāhoehoe, spread along the coast, and built the delta on the far side of the bay that is visible today.

Look inland from the overlook to see Hualālai's 100,000-year-old Pu'u Wa'awa'a trachyte cone and associated ridge-forming lava flows. These features are the products of the most energetic known Hawaiian eruption, comparable to the most powerful eruptions of the past century worldwide.

Around milepost 79 the road crosses onto the 1859 Mauna Loa pāhoehoe, then 'a'ā from the same eruption within the next half mile. North of the 1859 flow, roadside lavas are thousands of years old but appear surprisingly fresh due to the arid climate and lack of vegetation.

HI 19 meets Waikoloa Beach Drive about 25 miles north of Kailua. Turn west toward the shore to reach 'Anaeho'omalu Bay and a palm-shaded beach. Look for a beach access sign. The white sand comes almost entirely from offshore coral patches. Two ancient Hawaiian fishponds are just inshore from the beach. The rugged 'a'ā defining the south side of the bay, which is full of caves and inlets, erupted from Mauna Loa. On a clear day, you can look inland to see the summits of four of the

The 5,000- to 3,000-year-old 'a'ā lava flows from Mauna Loa's north flank underlie much of the Waikoloa resort coast.

island's volcanoes all at once: Hualālai, Mauna Loa, Mauna Kea, and Kohala parade from right to left.

Around milepost 73, about 3 miles north of Waikoloa Beach Road, the highway passes onto hummocky and sparsely vegetated Mauna Kea lava flows. The hummocky terrain is typical of land formed in hawaiite, a common type of alkalic lava that erupts during postshield volcanism.

Approximately 7 miles north of Waikoloa is a turnoff to Hāpuna Beach. This popular, 0.35-mile-long strand is made of tan calcareous sediment washed in from nearby reefs. The beach has a gradual submarine slope, ideal for swimming and bodyboarding. Ledges at the north end of Hāpuna Beach are composed of a rare variety of basaltic lava called ankaramite. This flow erupted from Mauna Kea during postshield activity tens of thousands of years ago. Look for blocky crystals of black pyroxene and glassy yellowish-green olivine sparkling in the lava. Snorkeling can be very good around adjoining rocky headlands, at least on calm days. On clear days, the crest of Haleakalā is visible, 10,000 feet high and 50 miles to the northwest on Maui, across the rough ʻAlenuihāhā Channel.

HI 19 continues north to a T-junction. From the intersection, turn left to continue north on HI 270 or turn right to follow HI 19 as it climbs to Waimea, following the contact between Mauna Kea and Kohala lavas. Waikoloa Stream, running just south of HI 19, may occasionally carry runoff from the wetter Kohala summit area in stark contrast to the surrounding arid landscape.

HI 190 (MAMALAHOA HIGHWAY, HAWAI'I BELT ROAD)

Kailua-Kona—Waimea

39 miles

See map on page 207.

HI 190, part of the Hawaiʻi Belt Road that circumnavigates the island, rounds the northwest rift zone of Hualālai Volcano about 9 miles north of Kailua town just north of the Kalaoa subdivision. A scenic lookout on the west of the road between mileposts 28 and 27 perches on the steeply sloping Kaʻūpūlehu Flow, which probably erupted around 1800. Take care crossing to this pullout; traffic is fast on this curvy section of road. The view of the coastal platform below is rewarding, showing that the flow spread across a wide area and traveled all the way to the sea.

The source vent of the Kaʻūpūlehu Flow lies about 4 miles upslope from the scenic pullout, at the 5,200-foot elevation in the northwest rift zone. In 1800–1801 a vent at only 1,600 feet produced the big Huʻehuʻe Flow on which the Kona International Airport is built. Two other young, though much smaller, lava flows burst from the rift zone between these principal vents, though their timing is not well constrained. All these flows were once thought to be part of the same eruption of 1800–1801, but recent work suggests that Kaʻūpūlehu may be older, though only by a few decades at most.

The Kaʻūpūlehu Flow is known for its fluid—and presumably fast-moving—lava, high rate of discharge, great depth of origin, and abundance of xenoliths. A few hundred feet north of the lookout, just before reaching a prominent channel in

Roadcut through circa 1800 pāhoehoe along HI 190 at the scenic lookout near milepost 28 shows stacks of thin flows with pockets where lava drained away on the steep slopes.

the flow, look for a spectacular display of xenoliths embedded in an outcrop along the right highway shoulder. The shoulder is narrow, so again be mindful of traffic. Dozens of angular xenolith chunks, ranging up to several inches in diameter, include dunite, gabbro, and peridotite. Xenolith clusters like this may form like the flood bars of river rock found in streambeds: carried downslope by the fluid stream—whether water or lava—and accumulating in pockets where the fluid slows down. These xenoliths are much denser than the lava and suggest that the lava had to be moving very fast to carry them from deep in the crust where they form. *Please do not sample or deface the exposure here.* It is an important teaching resource because similar localities are much less accessible.

With some careful scrambling, you can ascend the lava channel just past the xenolith locality. Spectacular draperies of lava splashed up against channel walls in places to form stalactitic, almost gothic walls and overhangs. It is easy to imagine lava flowing to the coast extremely fast based on exposures like this. Recent earthquakes have partly broken down these walls and exposed multiple thin layers of overflows that built the channel levees. Widely scattered, embedded xenoliths may be discovered here, too.

Around milepost 25 you'll see your first good view of Pu'u Wa'awa'a, a forest-cloaked pyroclastic cone about 2 miles upslope of the highway. The ridge lying just beyond is Pu'uanahulu, a stack of immense trachyte lava flows associated with this cone.

A turnout on the right side of the road between mileposts 22 and 21 leads through an automated gate to the Pu'u Wa'awa'a trailhead. Bear left to reach the shelter at the trailhead. This 8-mile round-trip walk to the top of the cone climbs 2,000 feet in elevation. From there you can enjoy sweeping views of Hawai'i's northwest coast, Pu'uanahulu ridge, and the rugged summit region of Hualālai, studded with alkalic cinder cones.

The Pu'u Wa'awa'a eruption took place around 110,000 years ago, one of a series of similar eruptions of Hualālai that may have continued for about 20,000 years. This eruption occurred close in time to the collapse of the southwestern flank of neighboring Mauna Loa—the giant 'Alika-2 landslide. Perhaps there is a correlation

Xenolith bed in the Ka'ūpūlehu Flow near milepost 28 shows variety of xenolith types: dunite (all green olivine), peridotite (olivine plus black pyroxene), and gabbro (white plagioclase plus olivine and pyroxene). A 4.75-inch-wide field book for scale.

Delicate draperies of lava along this flow channel crossed by HI 190 attest to the great speed and fluidity of the circa 1800 Ka'ūpūlehu lava. Xenoliths are scattered in the lower channel walls.

Erosional gullies in this unique trachyte pumice cone on Hualālai volcano inspired its name: Pu'u Wa'awa'a or "many-furrowed hill."

with these eruptions: a sudden change in crustal stress inside Hualālai caused by the breakdown of its volcanic neighbor. In any event, younger basalt flows have buried almost all the trachyte from other eruptions except Pu'u Wa'awa'a.

Trachyte is a product of the slow crystallization and changing chemical composition of cooling alkali-rich magma. This magmatic "sludge" can erupt very explosively, producing huge amounts of pyroclastic material. As steam and gases escape, however, the remaining trachyte oozes out as thick, pasty flows. For the most part, then, Pu'u Wa'awa'a is slightly older than Pu'uanahulu, though both originated from the same host magma.

The cone is composed of countless layers of pumice and ash mixed in with dark fragments of obsidian—a major source of glass for early Hawaiians. The dense glass suggests that Pu'u Wa'awa'a began as a volcanic dome of highly pasty lava, much like the one that filled the crater of Mt. St. Helens following its 1980 eruption. Explosions tore the dome apart, however, blowing frothy bits of magma foam—pumice—and fine ash high into the atmosphere. All this material fell back to the land surface, accumulated as countless layers blanketing the landscape, and buried any remains of the original dome. Some layers indicate that occasional pyroclastic surges and flows raced down the side of the cone while it grew. The same deadly phenomena destroyed Pompeii during the AD 79 Vesuvius eruption. University of Hawai'i geologist Tom Shea has identified at least one deposit that was hot enough to weld pyroclastic fragments together. You may see much of this layering exposed in an abandoned pumice quarry along the first part of the trail where the road ends at the foot of the cone. In addition to pumice and obsidian fragments, the eruption blasted out blocks and lapilli of older basaltic lava, stripped from the vent walls as the frothing magma violently escaped.

The volume of eruption products from the Pu'u Wa'awa'a vent was tremendous—the equivalent of thirty-five years' worth of Kīlauea eruptions at average eruption

View of south wall of Pu'u Wa'awa'a quarry, showing fine layering of pumice, ash, and other blocky material forming the steep flank of the trachyte cone. At the top, 3,000- to 1,500-year-old Hualālai basalt flowed over and baked the underlying, weathered pyroclastic bed to produce the red, oxidized contact.

Detail of typical pyroclastic deposits in the flank of Wa'awa'a cone. Layers contain abundant angular fragments of light pumice, glassy obsidian, and older basaltic lava torn from the wall of the vent during the eruption. Sorting describes the degree to which average fragment sizes stay the same or show variation within a particular layer. Some Wa'awa'a layers are well sorted (show little variation in particle size from bottom to top), indicating that they accumulated from blasts that maintained the same energy level throughout their duration (right). Others are graded, becoming coarser upward (left). This could indicate that the blasts that produced these fragments grew stronger before reaching a crescendo then ending quickly.

rates for that volcano. It probably took several years for Pu'u Wa'awa'a to grow to its full height. If you hike all the way to the top, you'll notice that the cone is horseshoe shaped. A persistent wind from the southeast could have been blowing to pile cone-building debris to one side, or perhaps lava issuing from the vent as the magma lost gas pressure simply pulled the flank of the cone off in that direction in the final phase of the eruption. Gradual erosion in the millennia since the eruption has created a set of drainages in the slopes of Pu'u Wa'awa'a.

As you return to the trailhead, look northeast at the profile of Pu'uanahulu, the trachyte lava that issued from Pu'u Wa'awa'a. Notice the rolling topography along the horizon of the ridge where lava rumpled into irregular folds as it hardened toward the end of the eruption. Drilling demonstrated that two separate trachyte flows, one stacked atop the other with an intervening pumice-rich layer and a combined thickness of as much as 600 feet, compose the interior of Pu'uanahulu. The intervening pumice shows that the eruption switched back to highly explosive behavior for a short while before it gave up the last of its magma as flowing lava.

HI 190 crosses the crest of Pu'uanahulu at the Pu'u Lani subdivision. The golf course at milepost 20, built near the end of the trachyte flow, features huge lava compressional ridges, providing rolling fairways. Just northeast of the golf course on the south (inland) side of the highway, look for a large outcrop of fragmented light-gray rock, the best roadside exposure of trachyte lava on the island.

Massive Pu'uanahulu trachyte lava flows form a ridge. The roadcut where HI 190 crosses the far (northeast) edge of the ridge shows the light color of trachyte under the weathered surface.

At milepost 19, HI 190 crosses onto the 1859 flow erupted from Mauna Loa. The source vent opened on the northwest flank at an elevation of nearly 11,000 feet, releasing lava for a period of 10 months—one of Mauna Loa's longest recorded eruptions. The flow reached the ocean 35 miles away in just the first eight days, destroying a Hawaiian village at Wainanali'i.

Between mileposts 16 and 15, the highway skirts a Hualālai cinder cone to the southeast surrounded by 5,000- to 3,000-year-old Mauna Loa lavas. The cone may be as much as 10,000 years old. About 1.5 miles past the cone is the intersection with HI 200, the Saddle Road. For the next 10 miles north of the HI 200 intersection, HI 190 crosses numerous hummocky Mauna Kea alkalic lava flows carpeted with weathered ash and windblown glacial silt and several postshield cinder cones as it approaches the lowland between Mauna Kea and Kohala. Look north to see the town of Waimea and Kohala's eroded volcanic shield. Like Mauna Kea, Kohala is crowned with cinder cones that grew during the terminal phase of volcanism.

HI 270 ('AKONI PULE HIGHWAY)
KAWAIHAE—HĀWĪ—POLOLŪ VALLEY LOOKOUT

27 miles

From the T-intersection with HI 19, HI 270 leads north through Kawaihae to Hāwī through a dry and windy landscape. Rainfall in this coastal region is the lowest on the island, normally less than 10 inches per year, technically classifying this area as true desert.

A short distance from the intersection, look for a sign marking a paved half-mile drive downslope to Samuel M. Spencer Beach Park and Pu'ukoholā Heiau National Historical Site. The beach at Spencer Park is a wide stretch of gently sloping calcareous sand, much like the ones at Hāpuna and nearby Mauna Kea Beach Resort. Harbor works and an offshore reef provide calmer waters than at other North Kona beaches. Swimming and snorkeling here are excellent, though it is often quite windy. Pu'ukoholā Heiau, the remains of a large sacrificial temple to the war god Kūkā'ilimoku ("Kū, the Eater of Land"), overlooks Spencer Park. It is said that between 1790 and 1791 Kamehameha had this heiau built of rocks *hand carried* nearly 20 miles from Pololū Valley. Following the destruction of rival chief Keōua's army by pyroclastic blasts during the powerful 1790 Kīlauea eruption, Kamehameha sacrificed Keōua—his own relative—at this site.

North from Kawaihae, HI 270 crosses calcareous sand and coral dredged from harbor construction and passes onto shield-stage lava flows of the Pololū Volcanics of Kohala Volcano. At a few hundred thousand years old, these are some of the oldest outcrops on the island. From mileposts 5.5 to 7, the highway crosses younger, postshield flows of Hāwī Volcanics, mostly ranging in age from 250,000 to 120,000 years. Many of these rocks show white calcareous staining produced by high evaporation rates in the arid microclimate.

Around milepost 8 a small cone of Pololū age is visible next to the coast as you pass through a short section of older flows and back onto Hāwī flows. After milepost 10 the highway stays on older, heavily weathered Pololū lavas, except for one

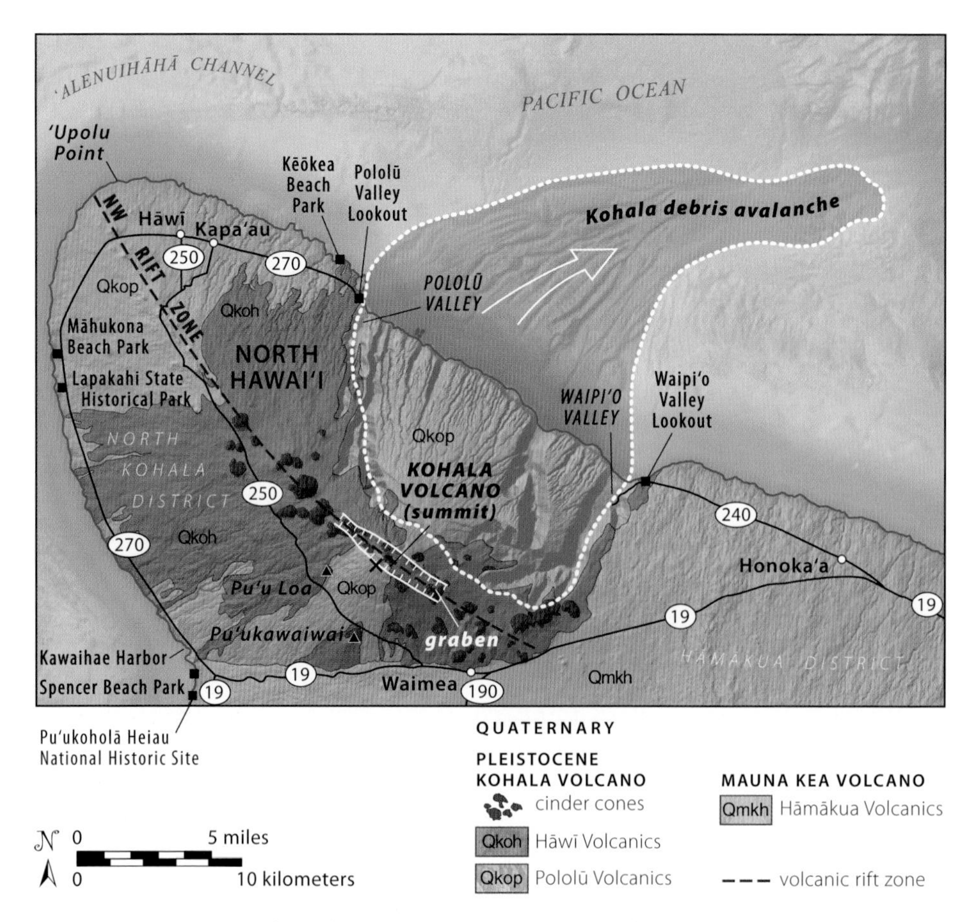

Geology along HI 270 and HI 250 in the Kohala region.

younger flow around milepost 12. In any event, it is difficult to distinguish individual flow boundaries in this greatly weathered, brushy landscape.

About 11 miles north of Kawaihae, Lapakahi State Historical Park—inhabited by Hawaiians until the mid-nineteenth century—is an archaeological center with many partially restored structures. Koai'e Cove, within the historical park, has many sea caves and abundant marine life. Most of these caves are lava tubes.

About 1 mile north is the entrance to Māhukona Beach Park, an abandoned port for the Kohala Sugar Company. World War II and the 1946 tsunami put the port out of business, but the remains of the sugar mill, railroad, and lighthouse can still be found. The harbor is now home to some colorful marine life and is a decent spot for snorkelers.

North of the Māhukona turnoff, HI 270 parallels the Kohala coast and gradually turns east toward the town of Hāwī. You cross the northwest rift zone of Kohala around mileposts 18 to 19. Note the subdued ridge speckled with cinder cones on both sides of the road. One reason the southwestern side of Kohala never suffered giant landslides or slumps is due to the large undersea shield volcano, Māhukona.

It lies 30 miles offshore, providing flank support to its younger neighbor. The Big Island originally extended many miles west of Kohala's present shore, but slow tectonic subsidence has steadily drowned it over many years.

The change in vegetation between Kohala's dry southwestern slope and wet northeastern side is striking. So is the geologic contrast. While the southwestern slopes of Kohala are barely eroded, deep stream gullies, valleys, and collapsed mountainsides distinguish the opposing northeastern half. Lava flows are deeply weathered on the northwest, with thick, highly oxidized soil covers and abundant spheroidal corestones. Northeast Kohala is the oldest, most eroded landscape on the island.

About 6 miles east of Hāwī town, around milepost 28, look for the turnoff to Kēōkea Beach Park. Follow the narrow, winding road for about 1 mile to the coast at cliffy Kēōkea Bay. The small jetty is composed of boulders of alkalic basalts, including ankaramites full of large crystals of olivine and pyroxene. The sea cliffs to the right of the jetty expose two deeply weathered Pololū-age flows of Kohala Volcano. You can see remnant pāhoehoe lobes in the flow at the base of the cliff, though the rock is very soft. The overlying flow has weathered mostly to reddish soil with a denser, still somewhat stony core.

HI 270 ends at the Pololū Valley Lookout. Parking is along the road and can be tight, but the coastal views are spectacular. Pololū Valley is the northwestern end of an angular indent in the coastline that stretches 15 miles to Waipi'o Valley on Kohala's northeastern flank. The enormous sea cliffs that dominate the view looking east from Pololū Lookout represent the eroded headwall of the giant Kohala debris avalanche that occurred between 370,000 and 350,000 years ago.

Pololū Valley developed from erosion over thousands of years along the shattered edge of the gigantic avalanche. Streams beheaded by the avalanche initially tumbled

The sea cliff at Kēōkea Beach Park exposes heavily weathered Pololū-age lava flows of Kohala Volcano. Inset shows remnant pāhoehoe lobes.

over the fresh scarp thousands of feet into the ocean. Gradual headward erosion exercised by the waterfalls slowly notched the scarp with gulches and smaller valleys, which stretch miles inland. The heads of valleys do not reach the summit of Kohala but instead end in huge amphitheaters with nearly vertical walls through which countless springs emerge. Sea stacks looming above the ocean just east of Pololū Valley imply that wave erosion and undercutting have shifted scarp-related sea cliffs at least a short distance inland. The sea stacks could also consist of the eroded remains of avalanche blocks that never fully submerged.

The trail from the parking area drops 490 feet over about a half mile to the black sand beach at the mouth of Pololū Valley. The bedrock valley walls extend downward in a canyon that was once hundreds of feet deeper. It flooded as the island slowly sank and sea level rapidly rose at the end of the most recent ice age. Streams poured sediment into the submerged part of the canyon, filling it to create the flat valley floor that you see today. These sedimentary deposits consist mainly of muds, sands, and gravels washed down from weathered lava flows during periods of heavy rain. Just inland from the beach, 50-foot-high sand dunes, now stabilized with ironwood trees, constrict the outflow of Pololū Stream and protect adjoining wetlands.

The valley was once inhabited, extensively cultivated to grow taro (kalo) along the freshwater stream, and hosted a quarry that some people believe was used to construct Pu'ukoholā Heiau 20 miles away near Kawaihae, although geochemistry reveals this is unlikely. Diversion of water to sugarcane plantations drove most inhabitants out of the valley, and then the devastating tsunami of April 1946 struck Pololū Valley especially hard. First, the water withdrew from the shore, exposing broad bedrock flats on a wave-cut bench. It surged back, rising over 50 feet as it swept across the beach at the mouth

Pololū Beach has both fine sand and a storm beach of rounded cobbles to boulders. The cobbles host a variety of rock types; look for large plagioclase crystals in some.

Inland view of Pololū Valley and Pololū Stream from behind the large coastal dunes. The dunes have acted as a natural dam to impound stream waters during the present high stand in sea level and create a large, estuarine pond that is gradually silting in.

of the valley. The steep front of the sand dunes is thought to have been scoured during the tsunami, depositing the sand in a matching layer found hundreds of feet inland. Archaeological evidence suggests the dunes themselves may once have been home to some of the earliest inhabitants of the island.

The walls of Pololū Valley are mostly shield-stage basalt flows from the older Pololū Volcanics. About 2 miles inland, however, alkalic lava flows of the Hāwī Volcanics erupted at least 120,000 years ago during the youngest stage of volcanic activity. Lava poured down headwater cliffs and onto the valley floor, indicating that the valley already existed. Rounded cobbles of the different Kohala lava flows and vent deposits, some crystal rich, can be found along the beach.

HI 250 (KOHALA MOUNTAIN ROAD)
Hāwī—Waimea

20 miles

See map on page 218.

From Hāwī, HI 250 climbs high up on the flank of Kohala Volcano, skirting the volcano's summit area. The first few miles of this road take you through heavily vegetated lava flows of the older, late-shield-stage Pololū Volcanics. Driving about 3 miles to around milepost 19, you reach the northwest rift zone, where grassy or tree-covered cinder cones mark ancient eruptive vents. The road follows the axis of the northwest rift zone for about 6 more miles and passes numerous cinder cones.

Just north of milepost 15 the highway crosses onto younger flows of the Hāwī Volcanics before veering off the rift zone. Additional cinder cones up to 500 feet high may still be glimpsed upslope near the summit. Minor erosional gullies are worn into rock by highland rainfall, inconspicuous in themselves yet strikingly different from the massive, headward-eating valleys of the windward (northeast) side.

HI 250 reaches a crest at 3,564 feet in an area that was heavily terraced for dryland agriculture by Hawaiians during the reign of Kamehameha in the late eighteenth century. The countless terraces can still be seen in satellite and aerial imagery intricately woven across this formerly heavily vegetated landscape, though they can't be distinguished from the roadside.

Just north of milepost 8 is a scenic overlook with a wide view of Hualālai Volcano and the North Kona coastline. The huge bulk of Mauna Loa rises above Hualālai to the east. You can see the full lengths of the 1859 flow from Mauna Loa, and the flows of circa 1800 from Hualālai, both of which reach the coast in North Kona.

A dome of an extremely unusual type of alkalic lava called benmoreite is exposed in a roadcut near the scenic overlook by milepost 8. Benmoreite is so rare that most geologists have never even heard of it. Its distinguishing features are its light-gray color and crystals of black amphibole. You rarely see amphibole in Hawai'i because few Hawaiian magmas contain its required ingredients. You can distinguish amphibole crystals from pyroxene by their shape and luster: amphibole crystals tend to be long instead of blocky and have glossy rather than dull surfaces. The benmoreite also contains crystals of pale plagioclase as well as scattered small crystals of olivine weathered to rusty specks. Benmoreite is a viscous lava, and flows and domes like this one tend to be very thick.

A bit less than 1 mile up the slope from the overlook is Pu'u Loa, one of the youngest cinder cones on Kohala. It also erupted benmoreite 120,000 years ago. Lava flows from it surround the older dome, and one narrow tongue even reaches the coast at Kawaihae.

The road passes back onto thick, older 'a'ā flows as it winds downslope toward the town of Waimea. Just past milepost 5.5 the quarried mound on the right (south) is Pu'ukawaiwai, a large hawaiite cinder cone complex of the Hāwī Volcanics composed of four aligned vents and thick piles of stratified tephra. The quarry wall nearest the road shows the profile of a crater that was buried as the original cinder cone grew. Activity shifted downslope to the three other vents as spatter and cinders filled the crater. Lava bombs can be seen in the quarry walls and scattered on the surface of the cone. Lava flows associated with this cone spread downslope in a broad fan until they are overlapped by younger Mauna Kea lavas. Pu'ukawaiwai derives its red color from an iron oxide stain that formed as hot gases, mainly steam, baked loose cinder inside the cone—a natural brick kiln.

As you continue down into Waimea town, note the many postshield-stage cinder cones on neighboring Mauna Kea to the south. For the most part these are part of the Laupāhoehoe Volcanics, younger than those you just passed on Kohala Mountain.

HĀMĀKUA AND HILO REGIONS

MĀNĀ ROAD
WAIMEA—HUMU'ULA SADDLE
43 miles (four-wheel-drive only)

The 43-mile-long Mānā Road is a mostly rough, unpaved track skirting the eastern flank of Mauna Kea. The route reaches 7,300 feet at its highest point after nearly 4,000 feet of elevation change between one end at a posted intersection with HI 19, 3 miles east of Waimea, and the other end on Mauna Kea Access Road near Humu'ula Saddle.

Mānā Road was established to support ranching operations beginning in the late nineteenth century. The Civilian Conservation Corps improved the road significantly in the 1930s to support ranching and forestry, though erosion in years since has undone much of this improvement. While easily passable on fair weather days, Mānā Road can be a slippery, muddy marathon under rainy conditions. Most rental companies do not allow people to drive this road due to both remoteness and often-fickle weather conditions. Four-wheel drive is essential for long stretches. It is scenically spectacular, however, and there are definitely geologic points of interest. We describe this route from north to south. (Set your odometer at the turnoff from HI 19 and be sure to close all gates behind you as you drive past.) A complete drive typically takes about five hours.

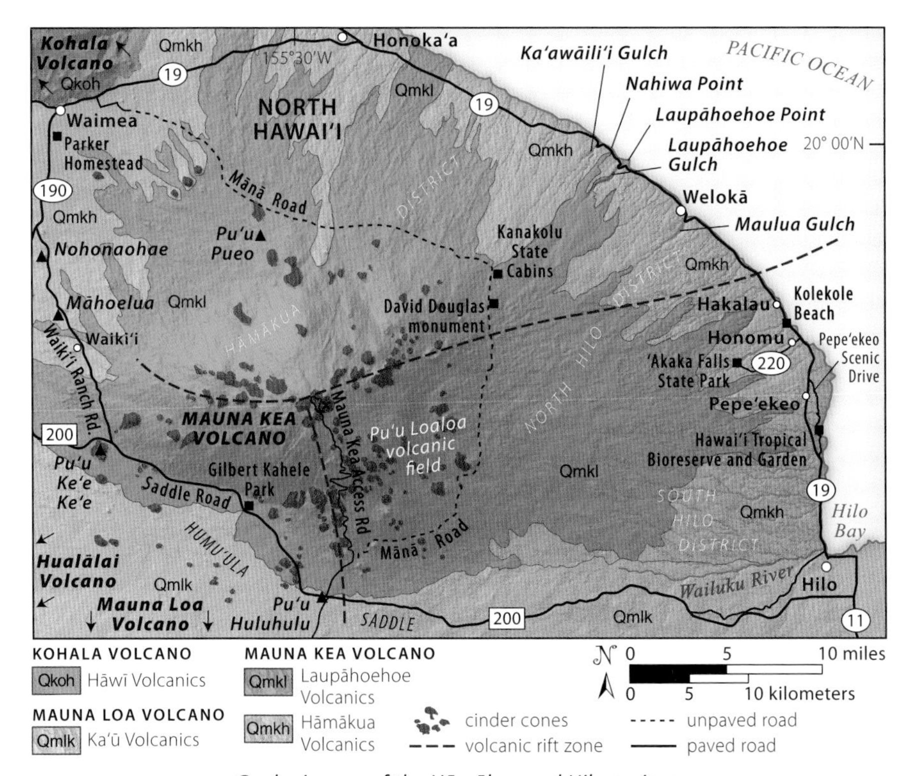

Geologic map of the Hāmākua and Hilo regions.

The first 4 miles of Mānā Road approach Mauna Kea's steep northern flank, with a scattering of older cinder cones of the Hāmākua Volcanics in the adjoining grassy lowlands. To the southwest on a clear day is a sweeping view of Hualālai volcano, 35 miles away. The dirt part begins at mile 2.7, though it remains reasonably passable without four-wheel drive for the first 20 miles or so.

At mile 3.3 large blocks on a hillock to the left of the road mark the site of the historic nineteenth-century Purdee home, now on private Parker Ranch land. At mile 4.3 grassy hummocks to the right are the end of hornblende-bearing benmoreite flow, an alkalic lava of the Laupāhoehoe Volcanics erupted from Mauna Kea. The road rises onto this big flow at mile 5.2. The lava it overlies is all Hāmākua Volcanics, as are the cinder cones closest to the road, mostly upslope.

At mile 6.1 Mānā Road passes the site of the original Parker homestead, from which developed the vast holdings of today's Parker Ranch, one of the largest cattle ranches in the United States. Some ruins and a family cemetery lie nearby. Large cement silos seen through the trees to the left of the road farther upslope were used to store corn grown in support of World War II troops training at the now-abandoned Camp Tarawa, just outside Waimea. The rusty remains of water tanks and catchments abound too.

From here until mile 11.5 the road continues crossing benmoreite lava, skirting the subdued cone Pu'u Pueo ("Owl Hill") just upslope at mile 10.4, one of several vents that erupted benmoreite in this area. Glance back from time to time to catch sweeping vistas of the Waimea saddle separating Mauna Kea from Kohala to the northwest.

At mile 11.5 Mānā Road leaves the benmoreite. Between miles 11.5 and 13.3, it crosses a Laupāhoehoe flow that originated at the prominent skyline cinder cone thousands of vertical feet upslope. The cabin next to the road at mile 13.3, which also marks where the road crosses onto older Hāmākua lava, is a livestock management line station still maintained by local cowboys.

Mānā Road returns to Laupāhoehoe lava after crossing a big gulch at mile 14.2. The grassland in this area is sprinkled with small, weather-beaten koa trees. It is worth noting that prior to the onset of ranching, this area was completely covered with koa-māmane-native strawberry forest. The grasses that presently carpet the landscape were introduced early in the last century—an utter ecological transformation.

At mile 21.8 road conditions become more challenging. After crossing another gulch near mile 23, you'll reach Keanakolu State Cabins, a former Civilian Conservation Corps camp that is now a wayside in a magnificent koa forest. It includes restrooms, a historic orchard, and several miles of well-maintained foot trails. Cabin rentals for overnight stays are publicly available through the Department of Parks and Recreation.

At mile 25 the road begins crossing a large Laupāhoehoe flow that erupted from a vent 3 miles upslope and traveled all the way to the coast 15 miles to the east, where it built rugged Laupāhoehoe Point. Look for the pullout to the David Douglas monument on the left side of the road. A short trail leading downslope takes you to the memorial obelisk, Kalaukauka ("Pit of the Doctor"), commemorating the famous horticulturalist who lost his life near this site in 1834.

Between miles 25 and 30.4, Mānā Road skirts the upslope edges of densely wooded stands, including the Hilo Forest Reserve and native Hakalau Forest National Wildlife Refuge. More open ranch country upslope provides views on clear days of Mauna Kea's ruddy mantle of cinder cones and pyroclastic deposits.

The flood-scoured streambed of a headwater tributary of the Wailuku River along Mānā Road on Mauna Kea.

Past mile 30.4, the road enters a landscape of cinder and ash beds invaded by New Zealand gorse, a thorny shrub. Travel by foot is treacherous. At mile 34.3 you reach a rock gully containing one of the headwater tributaries of the Wailuku River. At mile 35 another tributary runs next to a deeply quarried cinder cone. All along the Wailuku, smoothly faceted, pale-colored lava exposed in the channels indicates the occasional passage of fierce floods.

At mile 37.7 you cross onto one of the youngest lava flows erupted from Mauna Kea. It erupted about 4,500 years ago from Pu'u Kole cone, often hidden in mist upslope. For the next 5 miles, the road rounds the bases of young cinder cones and crosses slopes deeply mantled with their deposits. The largest of these cones is Pu'u Loaloa at mile 41.3. This area is under Hawaiian Homelands management, so special permission is needed from the Homelands office to explore off road. Ravines provide excellent exposures of cinder beds mantling this terrain in many places.

HI 19 (THE HĀMĀKUA COAST, HAWAI'I BELT ROAD)
WAIMEA—HILO
60 miles

The pastureland in the saddle between Mauna Kea and Kohala is one of the prettiest places in Hawai'i. Remarkable changes in plant cover accompany the change from dry country on the west side to wet on the east side. The forested east rift zone of Kohala lies along the horizon to the north, and Mauna Kea commands the horizon to the south. The scattered cinder cones on the lower flanks of Mauna Kea are older, erupted as part of the Hāmākua Volcanics, while those visible higher up are Laupāhoehoe Volcanics, erupted during the past 65,000 years. The smooth, pale-gray surface around the summit of Mauna Kea is composed of glacial deposits. On clear days you can see the summit astronomical observatories, 15 to 20 miles away.

WAIPI'O VALLEY

HI 240 (see map on page 218) heads east from HI 19 at Honoka'a, following the coast for about 8.5 miles to Waipi'o Valley Overlook. Looking west from Waipi'o Valley Overlook, you can see along 5.5 miles of shoreline. The sea cliffs here rise as much as 2,000 feet, formed when the Kohala debris avalanche slid into the sea. On the far side of the valley, wide switchbacks of the Muliwai Trail lead 9 miles across very rugged, eroded tableland to Waimanu, another large, wilderness valley not visible from here. As of 2022, access into Waipi'o Valley was at least temporarily limited to residents only.

Like Pololū Valley near Hāwī (see page 219 in the HI 270 road guide), Waipi'o Valley was once a much deeper canyon; it, too, is gradually sinking. Waipi'o Stream has partly filled the canyon with sediment, creating a broad valley floor, 1 mile wide at the shore. Waipi'o means "curved water" in Hawaiian, possibly referring to the gentle meandering of the stream. Waipi'o Valley was an important agricultural area along with Pololū Valley in early Hawai'i. As many as ten thousand people may once have lived here, though only a few dozen call the valley home today. Heiau (religious temples) and burial sites abound. The impact of diseases and immigration following contact with the outer world drastically transformed the population, and the 1946 tsunami wiped out most of the community existing before then. The wave towered nearly 40 feet as it swept up the valley about a half mile.

The flat floor of Waipi'o Valley is composed of sediments that filled the former much deeper canyon that existed here before sea levels rose at the end of the last ice age.

Rocks exposed in the walls of Waipi'o Valley are mostly pāhoehoe flows related to the Pololū Volcanics of Kohala Volcano. Those exposures in the lower walls show few if any phenocrysts. A few hundred feet below the rim, however, flows are somewhat alkalic and contain abundant white plagioclase crystals. A good place to look for them is in cuts along the road steeply descending below the overlook. The crystals occur in characteristic clumps, giving some rocks a salt-and-peppery appearance. Individual crystals as much as 1 inch long may be found. Many are weathered, soft and crumbly, owing to ongoing soil formation that is well expressed in outcrops. Such a concentration of large plagioclase crystals can accumulate as they float near the top of a magma chamber that is stagnant for a long time. Eruptions are too frequent during the main phase of shield growth for that to take place. It seems likely that the crystals grew and concentrated while eruptive activity was waning, toward the end of the shield-building stage.

After reaching the valley floor along the county road, you can take a hard-right turn along a dirt track leading to the beach. Here, cobbles of dull-gray tholeiitic lava are most common, but you may also discover scattered samples of plagioclase-rich rocks and picrites—basalts containing lots of green olivines. Waves crash and break quite close to shore here, showing that the seafloor slopes steeply away from the mouth of the valley. In fact, a submarine network of channels possibly related to Waipi'o Stream and its cascading

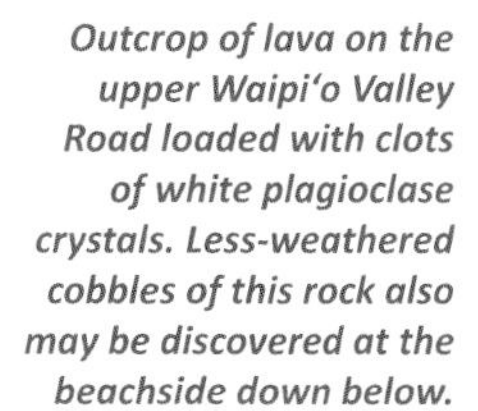

Outcrop of lava on the upper Waipi'o Valley Road loaded with clots of white plagioclase crystals. Less-weathered cobbles of this rock also may be discovered at the beachside down below.

Hi'ilawe Falls in upland Waipi'o Valley. The bulge in the wall to the left of the falls is a Mauna Kea lava flow that flowed into the valley after its formation.

underwater sediments continues another 7 miles or so northward. The sharp incline offshore no doubt contributed to making the 1946 tsunami here so large and destructive.

As the county road heads deeper into the valley, it follows a tributary valley carved out by Lālākea Stream, which pours into the valley via two narrow waterfalls, Hi'ilawe and Hakalaoa, with single drops as high as 1,200 feet. A lava flow from Mauna Kea, part of the Hāmākua Volcanics, spread down the steep slope east of the falls and blocked the stream until it cut a new channel. This flow is one of the few that entered Waipi'o Valley after it eroded. The lava is completely covered with vegetation but forms a prominent bulge in the eastern valley wall. The other post-erosional flows are somewhat older Hāwī Volcanics that erupted from the summit of Kohala.

Between Honoka'a and Hilo, HI 19 mainly crosses bedrock of older postshield Hāmākua basalt. Deposits of ash related to the Laupāhoehoe Volcanics, erupted from vents near the summit of Mauna Kea, widely carpet the older lavas. This aged volcanic terrain has weathered over tens of thousands of years into soils that nourish pasturelands, macadamia and coffee orchards, tree plantations, and other agriculture. Very little of the original island vegetation is visible along the highway.

Not quite 12 miles southeast of Honoka'a, HI 19 curves down through Ka'awāili'i Gulch. The large roadcut at the northern rim exposes a flow of Laupāhoehoe alkali basalt. Across the middle slopes of the gulch, the road slices past older postshield Hāmākua lavas, including flows of ankaramite rich in black pyroxene and green olivine. The lower slopes consist of older olivine-bearing tholeiitic basalt, perhaps erupted when Mauna Kea's shield was still growing rapidly.

About 1 mile after exiting Ka'awāili'i Gulch, just before reaching another large canyon, Laupāhoehoe Gulch, watch for the Laupāhoehoe Point Road on the seaward side of the highway—a short, recommended side trip.

About 4 miles southeast of Laupāhoehoe, HI 19 winds across the third large canyon encountered along this route—900-foot-deep Maulua Gulch. The oldest exposed rocks on Mauna Kea, the Hāmākua Volcanics, crop out on the floor of this gulch. They erupted 150,000 years ago, when Mauna Kea was still in its shield-building primacy.

A short distance south of Hakalau, the highway passes Kolekole Beach Park. A lovely shoreline side trip, it is at the bottom of another gulch, though less deep than the others mentioned above. The 1946 tsunami swept into Kolekole Gulch, with a crest 40 feet high. The huge waves undermined the bridge over the park, which was at that time part of the Hāmākua Coast Railway. A rocky black sand beach sits at the mouth of wide Kolekole Stream, and a big estuarine pool and small tributary waterfall lie behind the beach bar. Most other black sand beaches on the Big Island were formed when molten lava shattered into fine particles of glass as it entered the ocean. The supply of glassy black sand on those beaches is severely limited. The black sand at Kolekole Beach comes from slow stream erosion of basalt lava flows farther up the valley and from waves eroding rocks along the shore. These processes provide a continuing and reliable supply of black sand that will maintain Kolekole Beach indefinitely. The grains of black sand are nicely rounded and dull rather than sharp and glassy.

HI 220 branches from HI 19 through the historic plantation town of Honomū, then heads 4 miles upslope to 'Akaka Falls State Park. The drive ascends slopes of oxidized soil weathered from ash of the Laupāhoehoe Volcanics, formerly heavily cultivated with sugarcane. A short loop trail in the park leads to overlooks of 440-foot-high 'Akaka Falls and nearby Kahūnā Falls, almost as tall but largely hidden by vegetation. Both waterfalls launch across resistant ledges of Hāmākua

'Akaka Falls in 'Akaka Falls State Park spills over Hāmākua lava flows.

LAUPĀHOEHOE POINT

Narrow, paved Laupāhoehoe Point Road winds down the face of an ancient sea cliff to reach a county park, site of a tragedy on April 1, 1946, when a tsunami swept in, drowning twenty schoolchildren and four teachers. Ancient, weather-beaten lavas of the Hāmākua Volcanics crop out in roadcuts on the way down. The point grew when a 5,000-year-old alkali basalt flow entered the ocean through neighboring Laupāhoehoe Gulch, building a wide, flat-topped delta. The flow erupted from a cinder cone at the 9,500-foot elevation on the northeast flank of Mauna Kea, about 13 miles west of here. The lava also entered the sea to form Nahiwa Point a few miles to the north. Intensive wave erosion has sculpted the shore of the point into jagged pinnacles and spires, in places exploiting joints and fractures in the lava. On a clear day, you can look up the gulch from Laupāhoehoe Point and see all the way to the summit of Mauna Kea.

PEPE'EKEO SCENIC DRIVE

Between the 'Akaka Falls turnoff and Hilo is another byway worth exploring: the Pepe'ekeo Scenic Drive. It begins at Sugarmill Road with a left (seaward) turn off the highway between mileposts 11 and 10, just north of the town of Pepe'ekeo. Almost immediately after making this turn, turn right onto the Old Mamalahoa Highway. This predecessor of modern HI 19 was first built as a foot trail connecting villages in the nineteenth century. Highway signage provides guidance. Pepe'ekeo Scenic Drive winds 4 miles through tropical forest before reconnecting with HI 19, crossing seven streams with rapids and small waterfalls. The youngest flows exposed along the road are at least 100,000 years old. You can see how they have fared through centuries of weathering and erosion.

Pepe'ekeo Scenic Drive rounds Onomea Bay and passes Hawai'i Tropical Bioreserve and Garden. Waves 35 feet high raced into Onomea Bay during the 1946 tsunami, wiping out a fishing village. The cliffs around the bay expose thin flows of 'a'ā basalt, part of the Hāmākua Volcanics, and the layered, light-gray interior of an old tuff cone at the northwest point. Solid flow cores appear as pale layers separating the dark flow breccias.

Within the bioreserve a network of trails allows visitors to explore a couple of small gulches. In one such gulch, Onomea Falls features a series of cascades spilling over lava with well-developed cooling joints exposed in cross section. Along the shore lie exposures of a hardened debris flow consisting of broken lava rubble embedded in beige, sandy grit, now well cemented. This deposit, unusual for Hawai'i, is all that remains of an enormous flood of mud and debris that poured into the ocean from one or both gulches many thousands of years ago. You may also examine the deposit by taking a public access trail from the scenic drive to the shoreline, accessible just beyond the bioreserve.

lava in Kolekole Gulch. You can sometimes see scars from fresh debris avalanches on the over-steepened cliff faces above plunge pools. Floods typically follow winter storms and wash away the avalanche debris.

A network of little waterfalls resembles delicate shreds of lace on the vegetated cliffs. Water soaks into the porous lava flows high on the slopes, percolates through them, and issues as springs and seeps closer to the sea. Someday these deep gorges will evolve into large, amphitheater-headed valleys, much like the ones along the Kohala Coast.

The Hilo Area

Enjoy occasional views of Hilo Bay as you approach Hilo from the north on HI 19. The scale of the enormous Pana'ewa Flow from Mauna Loa that formed the far shore of Hilo Bay may be well appreciated from this distance. It erupted nearly 1,000 years ago. Just before entering downtown Hilo, the highway crosses the Wailuku River—Hawai'i's longest—via the Singing Bridge.

Hilo, population 45,000, is the largest city on the Big Island and the second-major harbor in the Hawaiian Islands after Honolulu—owing to a deep, 5-mile-wide, northward-facing bay. A nearly 1.5-mile-long breakwater protects the harbor from strong seasonal swells but does not prevent sometimes destructive tsunamis, originating primarily from North and South America, from impacting the shore behind it. The soccer fields and parks of the Hilo bayfront seem idyllic, but they can flood during periods of heavy rain and are highly vulnerable to tsunami damage. Hilo Bay acts as a funnel, focusing tsunami waves and increasing the run-up. The bayfront area was once densely populated, but Pacific-wide tsunamis in 1946 and 1960 destroyed most of buildings and caused many casualties. Building is now restricted in the tsunami zone.

Hilo owes its existence—and potential destruction—to Mauna Loa lava flows erupting from the northeast rift zone. These flows built a broad peninsula, 3 to 4 miles into the ocean, that defines the eastern side of Hilo Bay and culminates at Leleiwi Point. Much of the peninsula formed when the giant Pana‘ewa lava flow poured into the sea nearly 1,000 years ago, possibly associated with the formation of Moku‘āweoweo, the 1.5-by-2.5-mile-wide caldera atop Mauna Loa. Pana‘ewa lava crops out all along the coast from the hotel area on Banyan Drive east past Richardson Ocean Park. The important level ground for Hilo's airport, big box stores, racetrack, and many homes and gardens in the southern part of the city is Pana‘ewa lava. This flow is also heavily quarried for building and foundation stone.

Waiākea Pond is an estuary for shallow streams draining the more steeply sloping flank of Mauna Loa just west of the Pana‘ewa Flow. The main part of downtown Hilo, west of the pond, consists of lavas ranging in age from 8,000 years to the 1881 flow, which entered today's city limits, coming to rest within 1 mile of Waiākea Pond. Older, intensely weathered ash deposits probably originating from Mauna Kea also crop out, most noticeably in excavations. Some local tephra relates to the Hāla‘i Hills, a set of steep-sided explosive cones now largely dotted with homes, visible a short distance upslope from downtown. The Hāla‘i Hills represent a far-flung Mauna Loa vent of unknown age, but certainly thousands of years old.

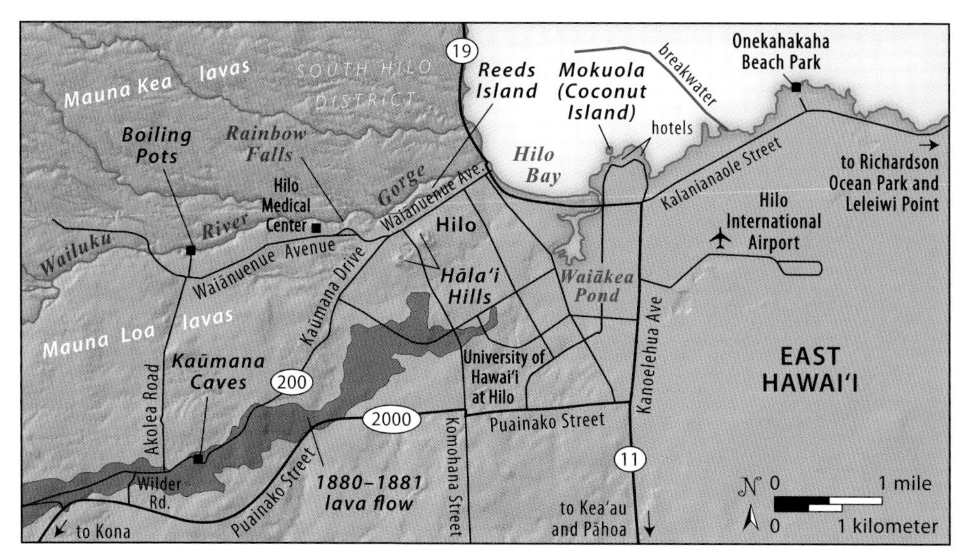

Points of geologic interest in the Hilo area.

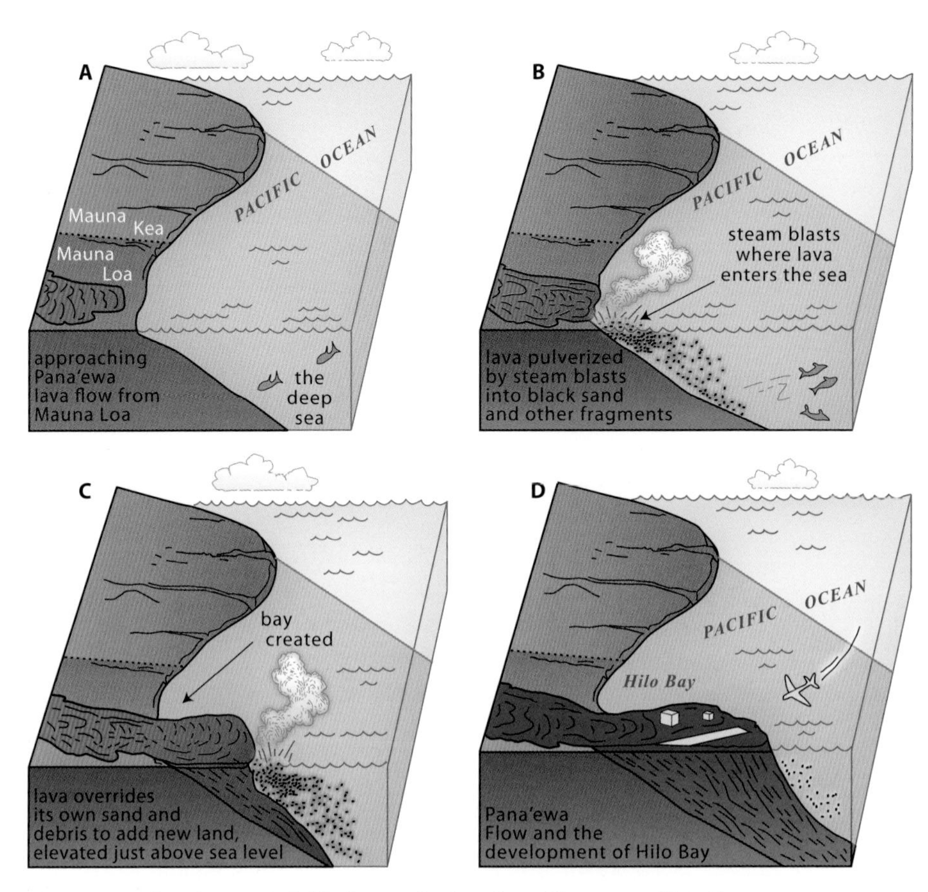

Development of Hilo Bay as the Pana'ewa Flow poured into the sea.

Outcrops of Pana'ewa basalt and freshwater springs are common along the eastern shores of Hilo. Lili'uokalani Gardens with Mokuola (Coconut Island) and Hilo Bay in background.

Wailuku River Gorge

The Wailuku River forms the northern border of Hilo where it follows a contact between Mauna Loa and the much older Hāmākua Volcanics of Mauna Kea. The river cut a spectacular gorge along this contact, including numerous waterfalls and plunge pools, breathtaking outcrops of columnar cooling joints, and paleochannels—some filled with younger lava, others still visible as depressions in forest-choked landscapes. As Mauna Loa lava flows continued to feed into the Hilo area, they displaced the Wailuku River farther northward onto the older, more rapidly eroding flank of Mauna Kea. As a net result, these flows have built up the landscape despite the rapid rate of erosion and downcutting taking place along the river. The contest between erosion and building of new land is well expressed in a number of places. Two outstanding examples are Reeds Island and Rainbow Falls.

Ka'iulani Street, a half-block uphill from the Hilo Library, leads across a historic wooden bridge that spans a jungle-filled gorge, called the Waikapu or Old Wailuku Channel, into Reeds Island, a pretty neighborhood atop a gently sloping terrace. This high terrace is not a true island but rather a slice of land wedged between the Waikapu Channel and the modern river gorge. The terrace measures from 700 to 1,500 feet wide and nearly 1 mile long.

About 10,500 years ago the Wailuku River poured through an ancient gorge, roughly bisecting Reeds Island lengthwise. Neither the Waikapu nor modern Wailuku Channels existed at that time. But then fresh lava from Mauna Loa, the 'Ānuenue Flow, poured down the ancient gorge and filled it nearly to its rim. The river was displaced to one side of the flow and eventually carved a new course along its southern margin—the Waikapu Channel (see figure on next page).

Over the next 7,000 years, the new drainageway deepened 35 to 80 feet and became 175 to 325 feet wide, comparable to parts of the modern river gorge. Meanwhile, the 'Ānuenue Flow also blocked and diverted a major Mauna Kea tributary entering from the northwest—Kioho'ole Stream, which began eroding another channel along the northern edge of the flow, roughly parallel to the Waikapu Channel a short distance to the south. Streams on both sides now bracketed the 'Ānuenue Flow.

There, things may have lingered with further erosion and downcutting up to the present day, but around 3,200 to 3,100 years ago, another Mauna Loa flow, the Punahoa pāhoehoe, suddenly dammed the Wailuku River at the upper end of Reeds Island. For a while the river managed to continue flowing across this obstacle, developing at least three alternate, shallow branches to maintain its course through Waikapu Channel. These feeder waterways were unable to keep a stable, long-term connection with the river upstream, however. The Wailuku overtopped the low divide now separating it from nearby Kioho'ole Stream, possibly during a period of flooding shortly following entry of the Punahoa Flow. As a result, the river abandoned the Waikapu Channel and combined its waters with Kioho'ole Stream to scour out the modern 100-foot-deep Wailuku gorge.

The Reeds Island story shows how the contact between Mauna Loa and Mauna Kea can shift substantially northward in a short period of time—in this case, 750 to 1,500 feet in a just few thousand years.

To view Waikapu Channel, walk onto the old wooden bridge where Ka'iulani Street crosses onto Reeds Island. You can see into the modern river gorge by walking

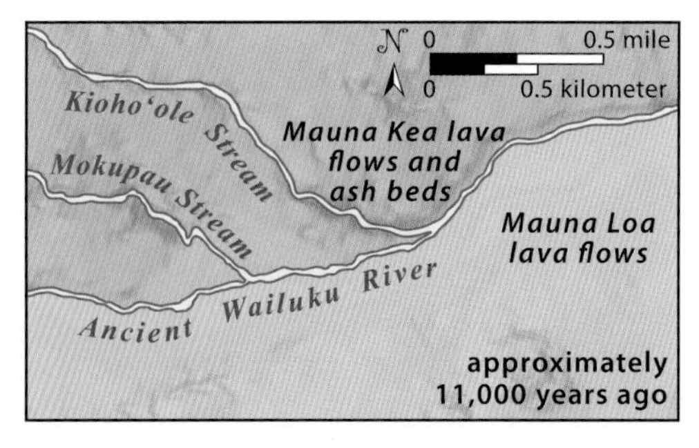

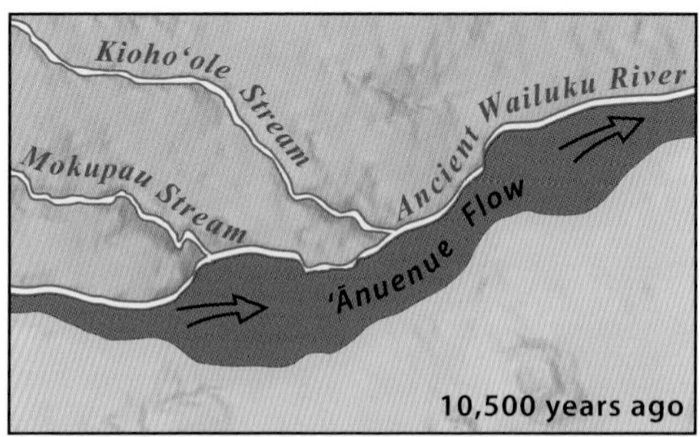

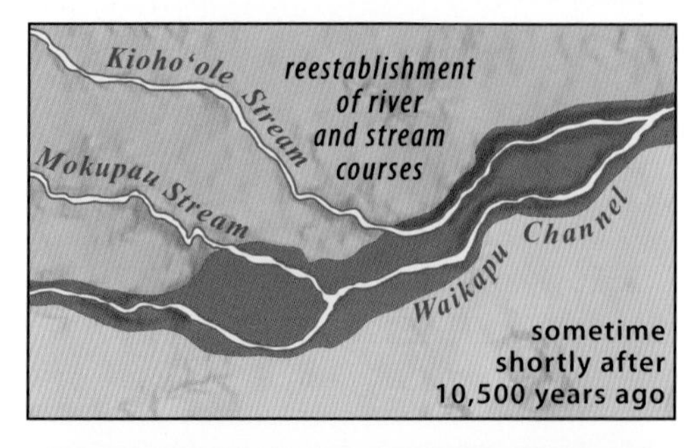

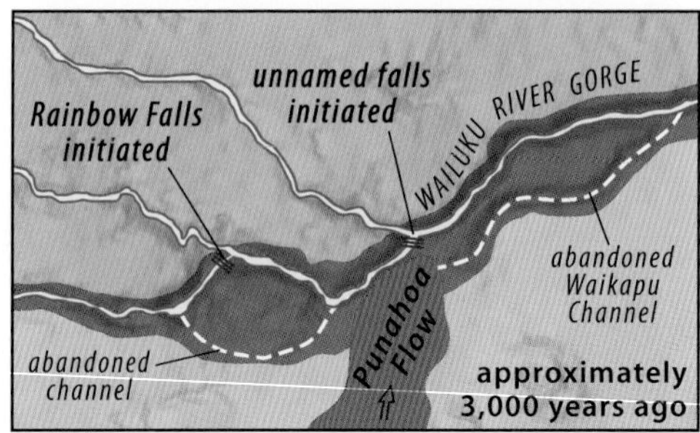

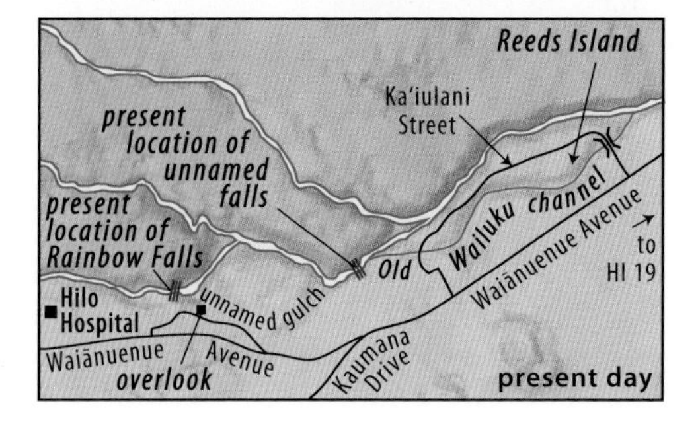

Development of the Wailuku River gorge and related features in Hilo over the past 11,000 years.

a few hundred feet farther upslope into the Reeds Island neighborhood and looking to the right of the road. Please be mindful of private property.

Rainbow Falls (Waiānuenue) tumbles 80 feet down into a large plunge pool. View it from the overlook at Wailuku River State Park off Waiānuenue Avenue, which intersects HI 19 just south of the Wailuku River bridge. As at Reeds Island downstream, a large abandoned channel of the ancient Wailuku River 30 to 50 feet deep and 1,000 feet long slices through the jungle just north of the overlook. Look for it to the right of the paved upper lookout, through the trees. It borders a terrace similar to Reeds Island but uninhabited, measuring 700 by 1,700 feet, with its long axis aligned along the river.

Rainbow Falls pours over a rim of the same ʻĀnuenue Flow that nearly filled the ancient gorge of the Wailuku and—as at Reeds Island—displaced the river to the southern flow margin, carving the now abandoned channel. When the Punahoa Flow entered the channel downstream a half mile from here, 3,200 to 3,100 years ago, it caused the river to back up and—perhaps again during a period of flooding—jump a low divide separating it from another Mauna Kea tributary, the Mokupau Stream. The river then began cutting its modern gorge along the northern edge of the ʻĀnuenue Flow through older Mauna Loa and Mauna Kea lavas to sculpt the modern terrace.

Rainbow Falls probably started when the Wailuku River first jumped to join Mokupau Stream. If true, this means that Rainbow Falls is probably no more than around 3,000 years old, and it cut headward back up Wailuku gorge from the Mokupau Stream initiation point at a rate of 100 to 125 feet every 1,000 years. During this time the Wailuku has also eroded downward another 15 to 20 feet below the level of its former, pre-Punahoa channel.

The power of undercutting, essential to causing headward erosion of the falls, is expressed by the huge grotto beneath the lip. This water-cut cavern measures 250 feet wide by 130 feet deep and has a ceiling 100 feet above the rubble-filled floor.

The 10,000-year-old ʻĀnuenue lava forms the massive lip of Rainbow Falls along the Wailuku River in Hilo. Multiple episodes of erosion and filling by lava flows have shaped the river gorge. Springs pour from rock layers exposed in the nearly vertical southern wall to the left.

During floods, churning water picks up the rubble and other sediment and carves deeper into the old flows forming the grotto walls. Only then, when floods come tumbling over the falls, can the awesome power of this river be appreciated. The name Wailuku translates to "Waters of Destruction," and more freshwater drownings of swimmers have occurred here than along any other watercourse in Hawai'i.

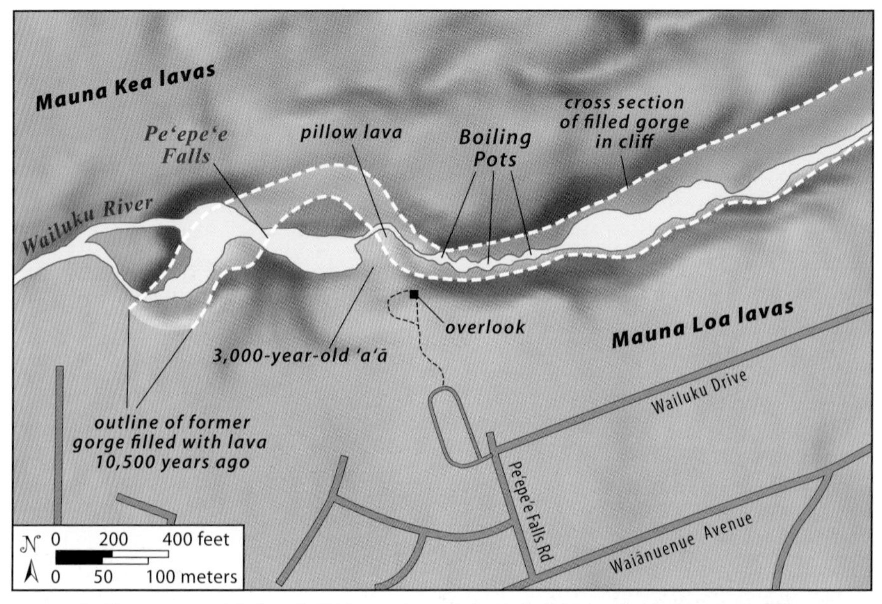

Boiling Pots and Pe'epe'e Falls can be reached via Waiānuenue Avenue in Hilo.

Boiling Pots eroded in columnar basalt along the Wailuku River at the western edge of Hilo.
—Photo by James Anderson

A short drive upstream of Rainbow Falls is Pe'epe'e Falls and the Boiling Pots overlook. Here, the Wailuku River has carved out a spectacular series of plunge pools that appear to "boil" when the water is high. This section of the gorge exposes some of the best examples of both pillow lava and columnar jointing in Hawai'i.

Kaūmana Caves

Follow Kaūmana Drive into the western edge of Hilo to visit a large lava tube formed in the 1880–1881 lava flow—Kaūmana Caves. To reach the caves from Puainako Street (HI 2000), turn north (right) on Wilder Road, then right again on HI 200 (Kaūmana Drive) for about 1 mile. Dense forest now covers most of the flow in this county park, but you can see outcrops of pāhoehoe well exposed around the lava tube skylight, through which a steep staircase descends to the main level of the cave below.

Kaūmana Caves in Hilo. Top: large chamber with secondary roof, partially broken into slabs, topped with a smooth, red-oxidized coating of lava. Several large lava balls wedged in the back of the chamber when lava backed up after the main skylight collapsed. Bottom left: small, irregular stringers formed by trains of bubbles of partial melt that oozed from the walls when the lava level dropped near the end of the eruption. Bottom right: shark tooth texture where the tube narrowed and lava splashed and dripped from the walls.

Two miles of cave passage may be explored here. Unfortunately, vandalism mars much of the easily accessible passages, detracting from this adventure. Still, the geologic features are well worth seeing. You'll need a good flashlight or two, and we recommend a hardhat and long pants. When you reach the bottom of the stairs, follow the left passage upslope for 700 feet. Beyond the low ceiling at the back of the entrance chamber, the passage opens up nicely and it is possible to walk comfortably most of the way.

The walls of the cave show a variety of lava tube features. Stunted stalactites, shelves marking the progressive drop in level of lava filling the main passage, entrail-like ooze-outs of partial melt from smooth shelf-stone, and white coatings of gypsum on the walls abound. In places, you can see a keystone-shaped geometry, evidence that a large amount of thermal erosion, or downcutting by the flowing lava, took place as the lava tube evolved.

The section of Kaūmana Caves downstream from the entrance also shows some interesting features, including nicely exposed secondary crust formation. In fact, a very low-roofed passage goes under the entire open skylight. Intense red oxidation around the large skylight opening and the smooth lava coating some of the boulders are evidence that the skylight began collapsing while the lava tube was still active. More crawling and clambering is needed to navigate the branching passages in the lower section of the cave.

You are never far below the surface here, as shown by the dangling of tree roots that have worked their way through fractures in the ceiling. Blocks on the floor indicate past ceiling collapses, but these took place primarily as the eruption came to an end and the rock was cooling and contracting. Kaūmana Caves is a remarkably stable feature at present. Still, avoid exploration during periods of prolonged heavy rain. Waterfalls gush from some fractures in the walls, and passages can partly flood.

HI 200 (SADDLE ROAD)
Hilo—HI 190 in North Kona

53 miles

See map of full route on page 223.

Multiple broad passes—saddles—exist on the Big Island, each separating adjoining volcanoes. But if you tell a Big Islander that you're planning to cross "the Saddle," your listener will instantly recognize that you mean the wide divide between Mauna Loa and Mauna Kea, formally called Humu'ula Saddle. The crest of Humu'ula Saddle, at 6,200 feet, lies 1.5 miles lower than the summits of the two massive volcanoes to either side—an imposing elevation difference. HI 200 (Saddle Road or Daniel K. Inouye Highway) is a well-graded commuter highway enabling transit between Hilo and Kailua-Kona in about an hour and a half.

No towns or permanent communities exist in the wild volcanic landscapes at the heart of the island. Slopes on the wet, windward (eastern) side of the Saddle are heavily forested, largely the domain of pig hunters, hikers, and ecologists. The leeward (western) slopes, in contrast, are dry and open grasslands and shrublands, largely grazed—or overgrazed—by cattle, wild goats, and sheep, and used for military training. While the windward side of the Saddle receives in places almost 300 inches of rain a year, the leeward gets by with less than 10 percent as much rainfall on average. In any event, some of the most interesting geology in Hawai'i can be viewed up close and from the vistas along this road, weather permitting. This route also provides access to the roads up Mauna Loa and Mauna Kea, leading to the island's tallest summits.

Hilo to Humu'ula Saddle

Milepost numbers rise in sequence from Hilo heading west. Zero your odometer at the beginning of HI 200 in Hilo where it intersects Komohana Street near the University campus. Mileposts are incomplete but are helpful for locating features along many stretches indicated below.

Forest cover develops quickly over fresh flows and ash deposits at low, windward elevations, making it impossible to distinguish individual geologic deposits—even those that have erupted historically. Vegetation changes as the road reaches higher elevations of around 4,000 to 5,000 feet, reflecting cooler climates. Contrasts in the expression and types of vegetation may be correlated roughly with ages and types of flows at these mid-elevations.

For the first 6 miles, HI 200 winds gently upslope across various prehistoric Mauna Loa flows from 10,000 to 1,200 years old. You cannot distinguish these lavas in the landscape because of the dense greenery. Between miles 6 and 8, the forest opens onto grassy fields—good grazing land—nourished by prehistoric ash deposits, probably from Mauna Kea. At milepost 8 the highway crosses onto the 1880–1881 flow, and the grassland ends. At milepost 9.5 the road leads onto slightly older 1855–1856 'a'ā, which it follows for the next 10 miles. You still cannot tell that you are traveling over fresh lava here because of all the trees and ferns.

Around milepost 19, around 5,200 feet above sea level, begin to look for outcrops of lichen-coated 1855–1856 lava in the thinning forest. The 1855–1856 eruption was truly formidable. Lasting for 450 days, it was the longest rift eruption since 1832. The lava fountained from a fissure that opened down the northeast rift zone of the

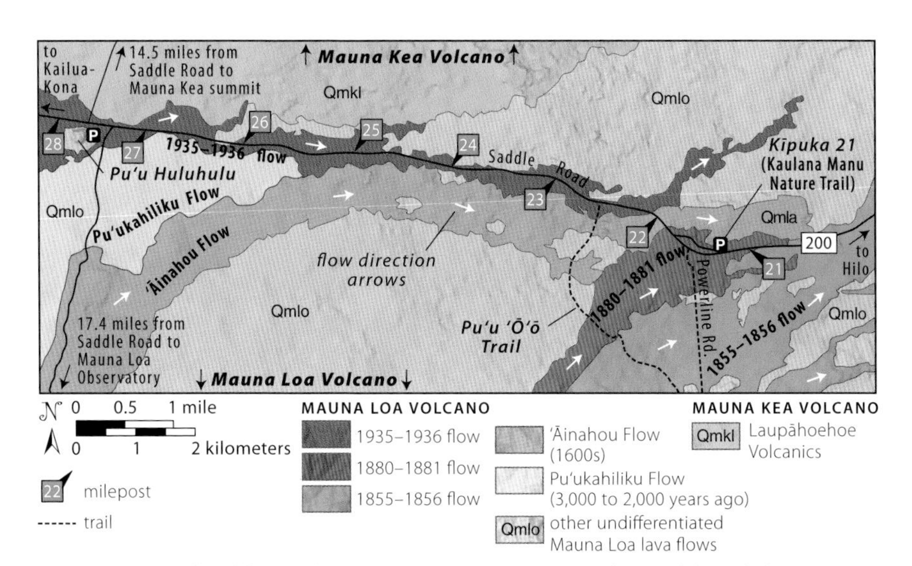

Map of Saddle Road on the upper, windward side of Humu'ula Saddle.

volcano between 12,000 and 9,200 feet of elevation. Discharges waxed and waned erratically. It may be that two different sources of magma from the volcano were involved based upon the changing chemistry of erupted lava. If it occurred today, a similar eruption would have a serious impact on the island economy.

Between mileposts 19 and 20, the road passes the pullout for the Kaūmana Trailhead. This pretty, 2-mile-long footpath leads across gently sloping areas of 1855–1856 pāhoehoe, past kīpukas of 3,200- to 3,100-year-old Punahoa lava and soil now overgrown with native trees and large ferns. Kīpukas are patches of older land surrounded by younger lava flows. The contrast in vegetation is sharp where the 1855–1856 lava and Punahoa Flow come together. The trail leads to another, less well-marked pullout along the highway downslope. The walk down and back is easy, though it may take a couple of hours.

Near milepost 21 the highway crosses from 1855–1856 lava onto less vegetated, darker 1880–1881 ʻaʻā. The parking area immediately west of milepost 21 is for the Kaulana Manu Nature Trail, a half-mile interpretive walk through densely forested Kīpuka 21. The trail begins on 1880–1881 ʻaʻā, but drops into an older landscape of mostly 4,000- to 3,000-year-old pāhoehoe. Note the striking change in vegetation marking the edges of the flows. At this elevation the ecology, including all larger plants and birds, is dominantly native, unlike lowland areas. Kīpukas are important in Hawaiʻi as places where isolation favors rapid evolution of new life forms. They may be thought of as ever-changing islands within islands. They also are natural sanctuaries for rare and endangered species that otherwise might disappear in more homogeneous environments.

From Kīpuka 21, HI 200 continues over 1880–1881 lava almost to milepost 22, where the asphalt gently dips into a young ʻōhiʻa forest springing from 400-year-old ʻĀinahou ʻaʻā, also from Mauna Loa. Crossing the kīpuka near the Saddle House Road turnoff, the highway ramps up onto lava of the 1935–1936 eruption, following it entirely to the crest of the Saddle. The ʻaʻā field to the right (north) of the road

Compression ridges near the terminus of the Mauna Loa 1935–1936 ʻaʻā flow along the Saddle Road.

shows well-developed compressional folds resembling the rumpling seen near the ends of glaciers and many landslides. These lava ridges are most easily seen in the oblique sunlight of early morning or late afternoon. They form roughly at a right angle to the direction of flow, similar to the ropes on a pāhoehoe flow.

To the south of the road between mileposts 22 and 23, a pullout marks the start of the Pu'u 'Ō'ō Trail, which enables hikers to explore a landscape of kīpuka and younger lavas ranging in age from 3,500 years to historical flows of 1855–1856, 1880–1881, and 1935–1936. The scenery nicely shows correlations between ecology, soil development, and ages of flows. The trail was established by sheep and cattle drovers more than a century ago, however, so the meadowy first part, though pleasant, was heavily grazed and ecologically scarred. After about 3.5 miles, Pu'u 'Ō'ō Trail intersects unimproved Powerline Road, leading back to the highway near Kīpuka 21 to complete a loop about 8 miles long. Powerline Road passes the entrance to Emesine Cave along the way. This lava tube in 1880–1881 lava is at least 10 miles long and is related to the same channel system that produced Kaūmana Caves near Hilo, 20 miles to the east. You may adventure 1 mile downslope underground to another skylight, but this is an arduous and challenging excursion requiring good preparation. A permit for entry is required from the Department of Land and Natural Resources to help protect fragile features and cave ecology.

Saddle Road follows the centerline of the 1935–1936 flow west past milepost 22. Builders of the route favored construction across 'a'ā, at least at higher elevations. 'A'ā is largely "self-granulated" to begin with and can be more easily crushed to make roadbeds, in contrast to solid pāhoehoe flows.

The 1935–1936 eruption has the distinction of being the first ever in the United States that anyone tried to control, in this case to protect the port of Hilo. Thomas Jaggar, director of the Hawaiian Volcano Observatory, requested help from the US Army Air Corps to bomb the flow's main lava tube near its source. The idea was that the lava would spread out upslope through a shattered artificial skylight rather than continue to feed the flow front threatening the port. One of the planners for this operation included a future American general, George Patton. Biplanes dropped 600-pound bombs that succeeded in breaking open the main lava tube, sending spectacular sprays of molten lava into the air—some of which burned holes in aircraft wing fabrics! The eruption ended quite naturally a week later, however, and the amount of blasted crust was probably never great enough to have clogged the flow's primary channels. Field inspection of one bomb impact point—a direct hit—shows little outpouring of lava. Nevertheless, at the time the experiment wasn't considered a complete failure. The 1935–1936 lava flow did not come much closer to Hilo than milepost 22, as indicated by the lava compression ridges mentioned above. A second attempt at bombing an active flow took place in 1942 during another Mauna Loa eruption, also with ambiguous results.

Between mileposts 23 and 24, the Saddle Road approaches the contact with Mauna Kea lava to the north (right). The contrast between the dark, fresh-appearing 1935–1936 Mauna Loa flow surface near the road and the lush, grass-covered hawaiite flows from the few-thousand-year-old Mauna Kea cinder cone eruptions upslope to the north is telling. The Mauna Loa lava flowed in a more fluid, less viscous way than the alkali-rich hawaiite flows, which form a characteristically hummocky terrain. They would be covered with forest in this area except for heavy grazing by introduced livestock.

Between mileposts 25 and 26 look for the transition in the 1935–36 lava from ʻaʻā to pāhoehoe, an excellent example of how a single flow can present two very different appearances. Loss of gases, cooling, and gradual steepening were probably responsible for this change as the lava traveled downslope.

Pu'u Huluhulu

Just west of the Mauna Loa Road (see separate road guide) on the left (south) side of the highway is a heavily forested, partly quarried Mauna Kea cinder cone now surrounded by younger Mauna Loa lavas. The cone, Puʻu Huluhulu, "very hairy" or "forested hill," is probably a few tens of thousands of years old based on the ages of similar cones in the vicinity. A parking area near the north base is the start of a walking path to the summit.

Look near the northwestern base of Puʻu Huluhulu outside the fence for remnants of a stone wall built by Japanese ranch hands in the late nineteenth century. It was partially covered by lava during the 1935–1936 Mauna Loa eruption. You can see the original height of the wall where it continues into a kīpuka of older, brownish pāhoehoe just west of the cinder cone. As you walk along the edge of the wall, notice that it lies in a trough in the lava. It almost appears as if something shoved the wall down into the flow, but in reality the lava inflated around the wall as it cooled and hardened. After the lava pooled in the Saddle and a hard crust formed across its surface, still-molten lava continued to flow beneath the crust, lifting and splitting it as though it were rising bread dough. You can judge the amount of the rise as you walk along the wall. In a few places the lava broke out and flowed over the structure, then

A stone wall constructed to control livestock was partly covered by lava in 1935–1936 a short distance west of the Pu'u Huluhulu parking area. Pāhoehoe encountered the wall and inflated against it until it overtopped (right side of the photo), and then filled in and inflated along the other side.

inflated on the other side of the wall, but in most places the inflated crust remains standing above the top of the wall.

The quarry in the western flank of Pu'u Huluhulu exposes the interior structure of the cinder cone. A basaltic dike forms a narrow panel slanting up the right side of the quarry. Patches of dike rock also may be seen crossing the slope near the top of the quarry. The dikes evidently erupted in a few places, locally mantling the slopes of Pu'u Huluhulu with fresh veneers of lava. A very thin 'a'ā flow coats the slope along the left side of the quarry wall. It must have erupted well after the cone formed because it covers a pale yellow-brown paleosol—an ancient soil horizon that probably took centuries to form. A reddened baked horizon immediately underlies the lava, indicating that the lava oxidized the much cooler paleosol as it flowed over the top. Similar flows that erupted after the cone formed drape the southwest and eastern slopes of Pu'u Huluhulu.

The quarried western end of Pu'u Huluhulu reveals several thin dikes that cut through the older cinder cone. The dikes, slanting up at the far right and at the top of the cone, are marked by arrows. On the left side, a small lava flow erupted from the flank and flowed over a weathered, yellow paleosol.

The astonishing thing about these dikes and flows is that they appear to be entirely unrelated to the eruption that built the cinder cone. They have a composition typical of Mauna Loa, not a Mauna Kea cinder cone like this. How did they get here? There is no evidence of a prehistoric Mauna Loa eruptive fissure in the vicinity that may have cut the Mauna Kea cone, though one certainly could have been buried by the younger flows nearby. Perhaps pioneering Hawaiian geologist Gordon Macdonald's thoughts were correct: molten lava from some ancient eruption high on Mauna Loa's slope was moving beneath a surface crust, which put it under considerable pressure when the flow encountered the base of Pu'u Huluhulu. The pressure drove sheets of molten lava through the flow crust into the loose cinder and cooled to create the dikes. In a few places the lava erupted—or, rather, reerupted—from the flanks of the cone!

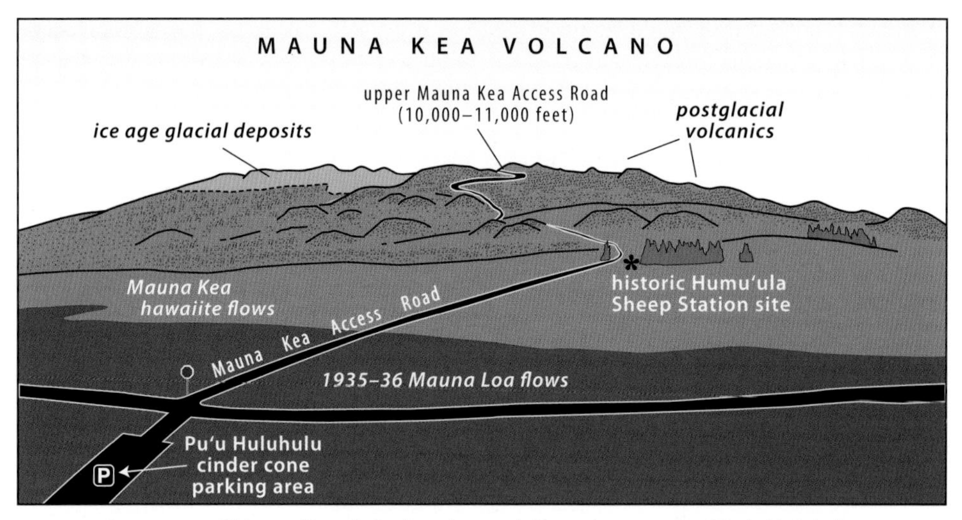

Panorama of Mauna Kea viewed to the north from the summit of Pu'u Huluhulu.

The dirt road through the quarry continues up to the summit of Pu'u Huluhulu, through some of the only surviving koa forest in the area. You can view the partially buried stone wall extending for hundreds of feet across the lava field below. The summit area of Pu'u Huluhulu is also an excellent place to enjoy the southern face of Mauna Kea, packed with geologic features.

Most impressive are the dozens of cinder cones formed in Mauna Kea's postshield Laupāhoehoe stage—Pu'u Huluhulu's kin. The closest group of cones a few miles to the north near the radio tower are all between 40,000 and 20,000 years old. The younger cones that rise farther upslope include Pu'u Kole, Pu'u Loaloa and Pu'u Huikau. Ash, cinder, and alkalic lava burst from these vents only about 4,500 years ago. An even younger group of cones, products of Mauna Kea's most recent eruptions, clusters on the upper mountainside around the 11,000-foot elevation. They may have formed about 3,300 years ago, though this dating is uncertain. In any case, future cinder cone eruptions in this volcanic field seem highly likely, and perhaps the next eruptions will take place even farther up the mountainside, continuing the trend.

If clouds aren't in the way, you can see light-gray material mantling Mauna Kea's upper slopes. The cover forms a somewhat smoother terrain than the eroded, weathered lava and cinder beds downslope. This debris is composed of rock fragments and fine, gritty sediment left from glaciers that covered the top of Mauna Kea during past ice ages. The last glacier melted around 20,000 years ago, though some of the debris you see upslope is much older than that. Note that the younger cinder cone eruptions have buried the light-gray glacial debris on the southeastern (right) flank of the mountain. Not all the cones and flows in the summit area postdate glacial activity, however. Older ice age cinder cones are not as dark as the postglacial group but have a paler, brownish-red color.

Looking south from Pu'u Huluhulu toward Mauna Loa, you can see spatter ramparts and small cones with steep sides, all in the northeast rift zone east of the summit caldera. Flows of diverse age stand out on the high, barren slopes of the shield in a patchwork of color from light brown to dark gray. ʻAʻā flows are darker than pāhoehoe, and all flows become lighter in color and browner with age. About three-quarters of these flows are less than 4,500 years old.

Saddle Road West of Puʻu Huluhulu

HI 200 continues crossing Mauna Loa's 1935–1936 pāhoehoe west of Puʻu Huluhulu for about 5 miles, where the road turns sharply north toward the steep face of Mauna Kea. Here the road runs parallel to a line of five cinder cones of the Laupāhoehoe Volcanics. The road turns west again midway through the chain, and you soon reach Gilbert Kahele (Mauna Kea State) Recreation Area, a good place to take a break and reflect on the scenery.

The massive bulk of Mauna Kea looms steeply above Kahele Park, cut deeply by Pōhakuloa Gulch to the northeast. This flank of Mauna Kea is much steeper than other sides of the volcano, and it exposes the widest terrain of older shield-related lavas this high on the mountain. The steepness approximately matches that of the faulted landscape on the southern flank of Kīlauea, and it could be that Mauna Kea's shield also collapsed southward before Mauna Loa became large enough to support it. Proving this would be difficult, however, because any landslide debris field related to such collapse has since been deeply buried by the younger volcanoes to the south. In any event, younger postshield eruptions have largely mantled this old slope.

Pōhakuloa Gulch owes its great size to the torrents of water released by the melting ice in Mauna Kea's glacial cap, which began to rapidly disappear only about 18,000 years ago. The till related to this glaciation forms the smooth, gray band near the crest of the mountain, 3,500 feet above Kahele Park. One long tongue of ice extended almost 1 mile down the flank of Mauna Kea, supplying the water that carved the gulch.

West of Kahele Park, HI 200 passes Pōhakuloa Military Training Area, as the highway closely follows the base of Mauna Kea. A contact with Mauna Loa lavas lies 0.5 to 1 mile south of the road; look to the left of the highway across a high lava plain of Mauna Kea's Laupāhoehoe alkali basalt flows, ash, cinder, and sediment. The summits of older Mauna Kea cones poke above the landscape of younger lavas. In this same general area—just south of Kahele Park—spatter ramparts related to an eruption of Mauna Loa 3,000 to 1,500 years ago show that the magma feeder systems of the two volcanoes overlap.

Daniel K. Inouye Highway to HI 190

The Saddle Road branches at milepost 41.5, but both routes reach HI 190 (see map on page 207). If you continue straight west past the Waikiʻi intersection, the jumbled cone cluster of Puʻu Keʻe Keʻe comes into close view just southwest of the road. A set of at least seven closely spaced, explosive eruptions built these cones, possibly over a period of many months to several years. Puʻu Keʻe Keʻe is one of the Laupāhoehoe vent areas most distant from Mauna Kea's summit, 11 miles to the northeast.

Rounding Puʻu Keʻe Keʻe, the highway provides outstanding views of both Hualālai Volcano and Mauna Loa. The contrast in their shapes is telling; one volcano is in its prime, the other dying.

Around mileposts 43.5 to 44.5 the landscape north of the road stretches high up the flank of Mauna Kea in an undulating, grassy terrain of ancient ash dunes capping older alkalic flows. The dunes developed as trade winds blew across the till fields atop Mauna Kea at the end of the last ice age.

At milepost 49.5 the highway skirts a prehistoric Mauna Loa ‘a‘ā flow supporting a young woodland of native ‘ōhi‘a trees. This lava is about 300 years old, but the developing forest "prefers" it because it is better drained than the adjacent, older clay and ash-rich terrain. The rough ‘a‘ā may also be less attractive to grazing animals. HI 200 comes to an end, intersecting HI 190 at a T-junction that overlooks the northwest coast of the island.

Saddle Road to Waiki'i

If you turn north at milepost 41.5, you'll take a scenic cutoff linking Saddle Road with HI 190 just south of the town of Waimea. This narrow, paved route winds up and over the high shoulder of Mauna Kea, cresting at 5,300 feet, 3 miles north of the intersection.

On the crest, between mileposts 45 and 46, a deep roadcut reveals the well-layered interior of an ancient dune composed of ash and fine till, blown here from the summit of Mauna Kea as the volcano lost its ice cap thousands of years ago. Dunes actually blanket the slope all around you, but grasses have stabilized them in our modern, warmer climate. About 5 miles north of the intersection, near the community of Waiki‘i, the road leaves the area of dunes and crosses onto older flows of the Hāmākua Volcanics, hawaiites fed by cinder cones scattered across the upper western flank of the mountain. One such cone, Māhoelua, rises to the left of the road.

The Waiki‘i cutoff is known for its sweeping vistas. About 8 miles north of the intersection (near milepost 50) is an excellent view of Kohala, Waimea, and the North Kona coastline. On clear days, 10,000-foot-tall Haleakalā on Maui is visible 70 miles to the northwest.

The road passes another large Mauna Kea cone about 1 mile before you reach HI 190. It is Nohonaohae, of Laupāhoehoe age.

A roadcut through wind-blown dunes near Waiki'i. View is about 12 feet across.

MAUNA KEA ACCESS ROAD
Saddle Road—Summit

15 miles

The road to Mauna Kea's summit begins right across HI 200 (Saddle Road) from the Pu'u Huluhulu parking area. The paved part winds very steeply past cinder cones and across postshield Laupāhoehoe lava flows of the south rift zone to 9,200 feet, where it levels off at Halepōhaku and the Ellison B. Onizuka Center for International Astronomy. Past this point, four-wheel drive is needed to follow the unpaved portion up to the summit area, not so much to climb as to safely navigate the downhill return. Zero your odometer at the intersection with the Saddle Road.

The road begins by crossing a grassy landscape of lumpy alkalic hawaiite lava before steeply climbing. At mile 2.1 it passes the unmarked intersection with Keanakolu or Mānā Road, a dirt four-wheel-drive road that traverses the eastern slope of Mauna Kea all the way to Waimea (see road guide for Mānā Road). The first few miles leading to a cinder quarry may be followed by ordinary passenger vehicle in dry conditions. Here you can see up close some of the younger cinder cones associated with the Pu'u Loaloa postshield eruptions of 5,600 to 4,500 years ago. Various gullies along the way reveal cinder beds exposed in cross section. A permit is required to explore the surrounding landscape by foot, however.

At mile 2.9 on the Mauna Kea Access Road, the road cuts through hawaiite flows adjoining the first prominent roadside cone on the way up to Halepōhaku. At mile 4.2 the road crosses a cattle guard. Upslope from here you can see switchbacks of the unpaved section high above; they cross the youngest volcanic terrain on the mountain.

Two prominent cinder cones appear on the eastern skyline about 5.5 miles from the Saddle Road. Pu'u Kole, which erupted 4,500 years ago, appears to be the younger. In front of the cones you can see an alkalic basalt flow that erupted about 5,300 years ago from the low mound about a half mile up the slope.

At mile 5.7 the road passes through the breached, horseshoe-shaped crater of Pu'u Kalepeamoa. The ridge west of the road is the rim of the crater where the trade winds piled cinder high to one side. The cinder contains many fragments of older rock brought up during the eruption—xenoliths—including speckled black-and-white gabbro and green dunite.

The visitor center at Halepōhaku lies on a smooth blanket of young ash and cinder. An unpaved trail leading southwest of the visitor center leads to the highest point on Pu'u Kalepeamoa crater rim. Look for xenoliths along the way, but please leave any that you find for future visitors to enjoy finding: follow the code of roadside field geologists! The xenoliths—dunites, gabbros, and peridotites—appear to come from magma that crystallized inside the volcano at great depths. The view from the top in all directions is truly spectacular. Note looking upslope and downslope that Pu'u Kalepeamoa appears to be part of a chain of six cones formed along a rift in the mountainside radial to Mauna Kea's summit, active more than 10,000 years ago. Did these all form during the same prolonged eruption or series of related eruptions? It isn't possible to say, though the cones look similar in age. The area around the top of Pu'u Kalepeamoa features oxidized mounds of accumulated welded spatter and scattered lava bombs. If you travel no higher up the

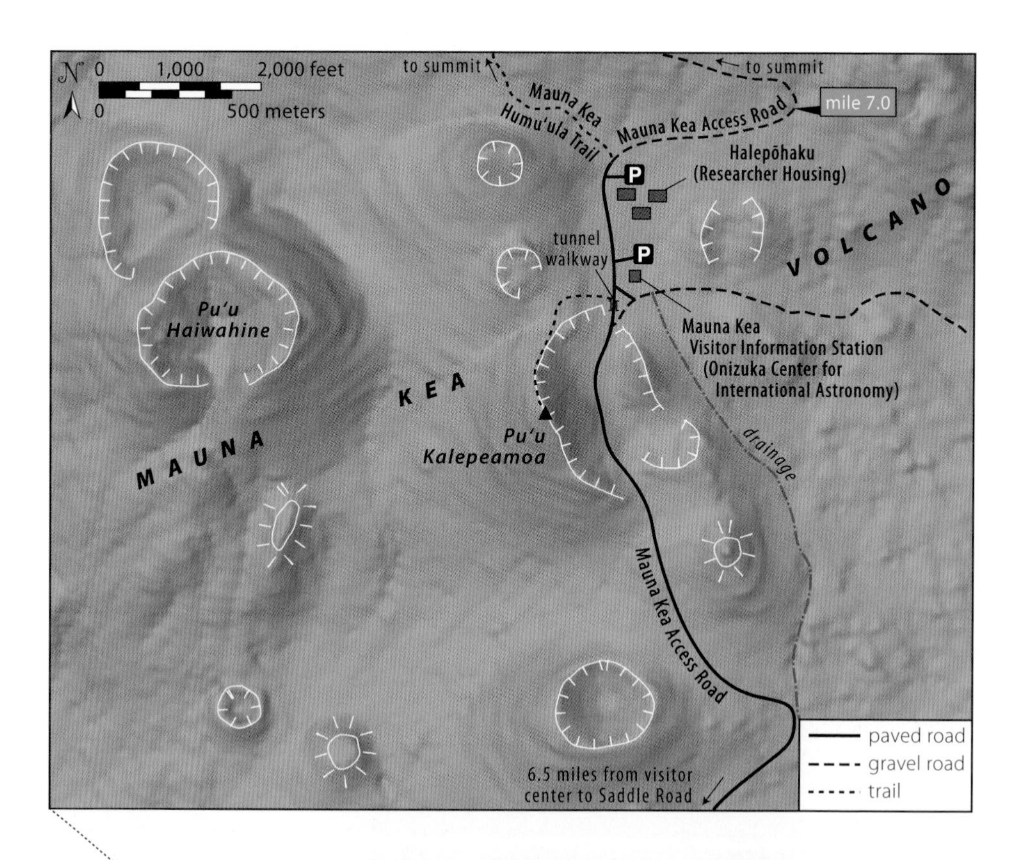

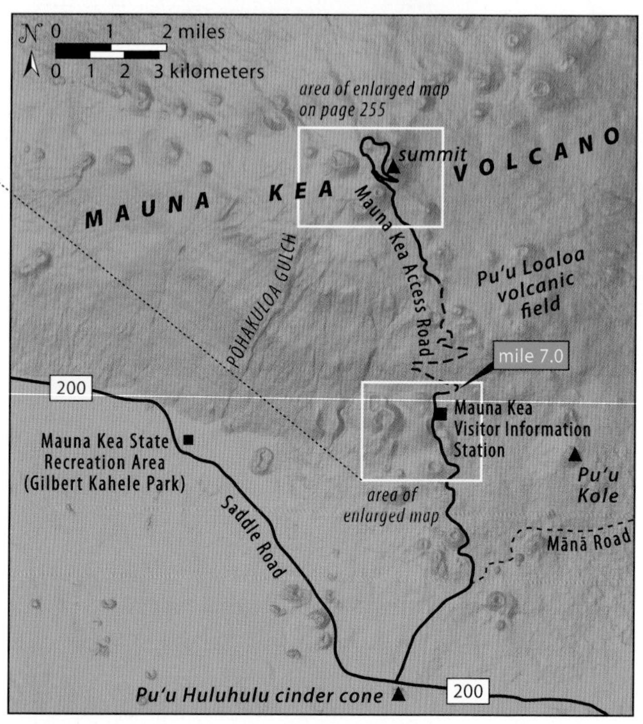

Geologic points of interest in the Halepōhaku area.

flank of Mauna Kea than here, at close to 10,000 feet, you will have reached a good turn-around point.

If you continue upslope on the Mauna Kea Access Road, the next 5 miles past Halepōhaku are unpaved washboard. At mile 6.5 the roadside shows mantles of cinder in wavy beds covering a channeled landscape. These layers dip toward the road. The first big hairpin switchback is reached at mile 7.0. At mile 8.0 around the next switchback, you enter the Mauna Kea Ice Age Reserve.

Looking downslope toward the Saddle from the summit of Pu'u Kalepeamoa cinder cone, near the Mauna Kea Visitor Information Station. Various Pu'u Loa and older Mauna Kea cinder cones dot the lower slopes, while Mauna Loa's northeast rift zone stretches across the southern skyline. The enormity of Mauna Loa can well be appreciated from this perspective.

Looking upslope toward Mauna Kea's top from the summit of Pu'u Kalepeamoa. The chain of ruddy, older cinder cones, of which Pu'u Kalepeamoa is a part, can well be seen.

Cinder and ash layers only a few thousand years old stand out in roadcuts as the Mauna Kea Road switchbacks above Halepōhaku.

Ground-up rocky material related to past ice ages is first reached at mile 9.7 just before the third hairpin turn. Essentially two classes of glacier-generated debris exist: till (also called drift) and glacial outwash. Till consists of dust- to boulder-size angular fragments scraped from the lava by grinding ice, whereas glacial outwash is mostly finer silts, gravels, and sands of this same light-gray material washed downslope by melting ice, either gradually or in sudden outburst floods. You can see an example of the more ordinary kind of glacial outwash on the floor of the small valley near Pu'u Keonehehe'e cone at mile 10.5.

At its climax, the ice cap across the summit of Mauna Kea covered an area of around 10 square miles, though outwash layers cover a much larger area. At most the ice cap may have been about 200 feet thick. From alpine studies done elsewhere, we know that it takes an ice mass of at least 60 feet thick to begin flowing as a glacier, so we know that a lot of grinding and carving of the underlying lava landscape must have happened atop Mauna Kea. In some places, eruptions beneath the glacial cap melted completely through it to build cones isolated in the ice. In other places, the glacial cover didn't melt away completely, but only around the covered vent and flows. This created huge, subglacial reservoirs of hot water mixed with finer silty debris. Tremendous pressure from this water lifted the edge of the ice cap to release great outwash floods.

Glacier-related flood dumps have left chaotically mixed deposits of large rock fragments, some rounded as conglomerates, in deeper canyons high on the mountainside. The huge flood at the coast of the Tropical Bioreserve in Onomea Bay just north of Hilo might also be related. Geologists know of at least six subglacial eruptions beneath Mauna Kea's former ice cap. It's possible that many others took place in earlier times.

Dated lava flows and charcoal from ash deposits suggest that in the last ice age the frequency of Laupāhoehoe eruptions was at a rate of six or seven per 10,000 years.

About twice as many have taken place since the ice age finally ended about 11,000 years ago. Perhaps as a warming climate melted Mauna Kea's ice cover, the resulting reduction of pressure on the mantle below allowed magma to rise to higher levels in the volcano, favoring an increase in eruptions. Similar correlations between ice cover and eruptive frequency have been reported in Iceland and the Andes.

The glacial till and outwash deposits on Mauna Kea record three episodes of glaciation. The deposits include the Pōhakuloa Drift laid down 150,000 to 100,000 years ago, the Waihou Drift accumulated between 100,000 and 55,000 years ago, and the Makanaka Drift deposited from 55,000 to 20,000 years ago. Except, perhaps, for the Pōhakuloa Drift, each of these episodes corresponds to the Wisconsin ice age of North America, from about 125,000 to 11,000 years ago. The lowest elevation that each drift occurs does not correspond to the exact elevation of the lowest extent of the glaciers because the volcano has been slowly sinking during all this time.

Most of the till along the road is the Makanaka Drift. Scattered pale-gray till drapes the dark-gray or brown slopes of cinder cones farther up the slope past mile 10.5. Glaciers eroded some of these cones almost beyond recognition. At mile 11.4 where pavement returns, look for places to the right of the road where moving ice

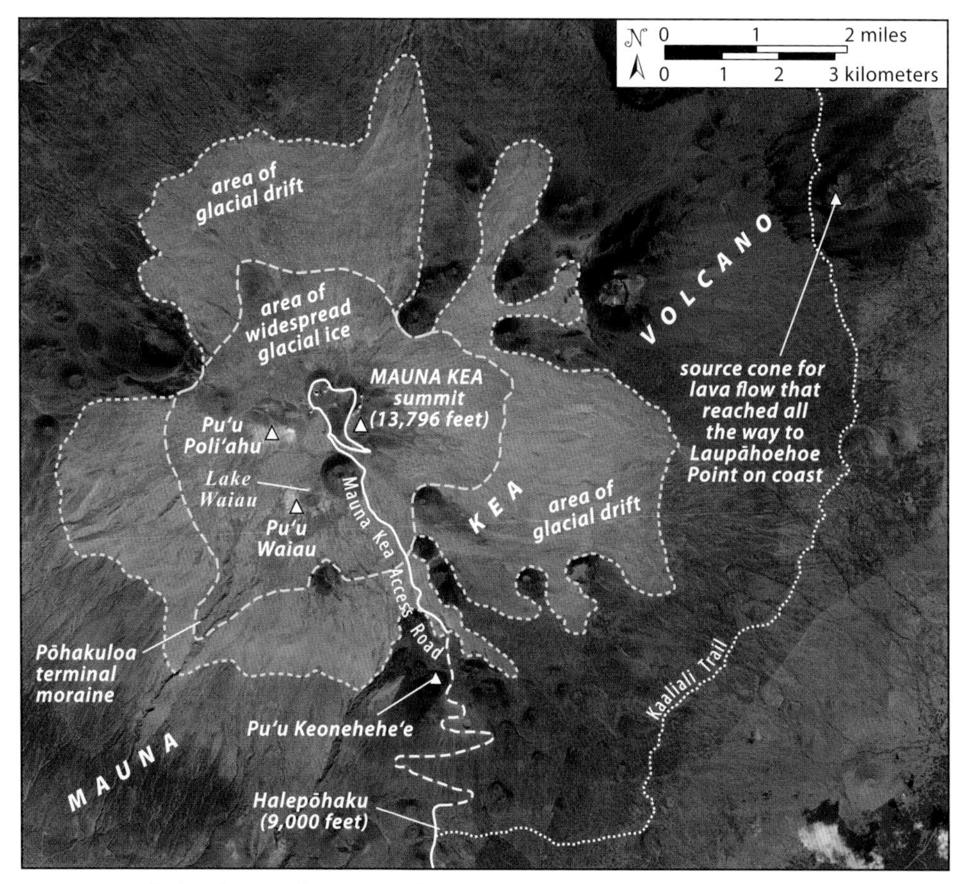

Glaciated parts of Mauna Kea's summit and areas covered by glacial debris.

sculpted the rough lava into smooth, gentle lumps. To French geologists familiar with these features in the Alps, these resemble the broad backs of sheep. They named them roche moutonnée.

Watch at mile 11.6 for scattered piles of basalt rubble on the near horizon to the left of the road. Ancient Hawaiians left these as they quarried the glassy subglacial lava, the raw material for chopping tools. Mauna Kea's summit was a Neolithic mining area and, no doubt, hazardous thanks to the remote, inhospitable terrain. Federal and state laws strictly protect such archeological sites.

Glacial till, ground moraine, and outwash carpet the mountainside with light-gray debris where the road begins to climb less steeply in the summit area. Older ice age cones are strongly oxidized and show remarkably little evidence of erosion.

Roche moutonnée with its surface scratched by rocks embedded in the glacier. The scratches show the direction the ice flowed across the rock.

About 1 mile farther northwest is Pu'u Waiau, one of the older cinder cones that grew here during the past ice age. Steam and hot water percolating through the cone near the end of its eruption altered the cinder, creating the light-colored yellow-brown patches. The alteration products include clay, which makes the cinder impermeable and thus increases runoff from rain and melting snow. Hence the flanks of Pu'u Waiau have more gullies than the more permeable flanks of neighboring, less altered cones. Northwest of Pu'u Waiau is Pu'u Poli'ahu, another noticeably altered cinder cone.

Lake Waiau

At mile 13.2 you reach the parking lot at the trailhead to Lake Waiau. The lava flow along this section of road originated at Pu'u Wēkiu, the large cinder cone ahead with observatory domes near the top. In many places, long tracks of parallel lines and mosaic zones of fracturing mark the flow's surface. These indicate that the lava flowed beneath the ice, melting it and wedging along. The overlying ice created the peculiar fractures as it quenched the flow. Then the ice dragged particles of grit and other till fragments across the lava, polishing its surface and etching long grooves and scratches.

From the parking area, a trail leads a half mile west into the pass between two cinder cones, Pu'u Waiau to the south and Pu'u Haukea to the north. The path winds up the slope from the steep, lobate edge of a lava flow that erupted from Pu'u Haukea 40,000 years ago. The peculiar fracture patterns along the edge of the flow suggest that the lava banked up against ice as it cooled and hardened.

A few hundred yards beyond the pass the trail ends at the shore of Lake Waiau, at the bottom of Pu'u Waiau crater. It is one of the few natural bodies of freshwater

Hackly lava along the trail to Lake Waiau shows distinctive fracturing that forms as a flow moves beneath glacial ice.

The crater of Pu'u Waiau, one of the oldest cinder cones atop Mauna Kea, showing an ancient rock glacier and the lake, a body of water only about 300 feet across when full.

in the Hawaiian Islands and at 13,160 feet is the highest lake (or pond, actually) in Hawai'i. The cinder piled high along the southern rim of the crater tells of strong winds from the north as the cinder cone grew. An inactive rock glacier with lobate margins fills a swale against the crater wall just south of the lake. You can recognize this mixture of rock, once embedded in ice that flowed toward the lake, as a pad of hummocky, light-gray debris.

If the bottom of Lake Waiau were unaltered cinder, it would not hold water. But the floor of the crater contains beds of impermeable clay weathered from volcanic ash. Circulating hot water and steam may have helped by altering the cinder beneath the ash. Lake Waiau occasionally overflows through the notch in the western rim of the crater.

Groundwater fed by the lake may also seep out high on the nearby southern slope of Pu'u Waiau, contributing to development of the set of shallow drainage channels seen on that outer flank. Glance back and look for them on your way back to the parking area. Similar seep-fed drainage channels are observed in many Martian landscapes related to summer melting of ground ice.

The embankment of rough lava along the north shore of Lake Waiau is part of the flow that erupted from Pu'u Haukea, the adjacent cone to the north. It poured across the low northern rim of Pu'u Waiau crater and then stopped. The cavernous voids, mosaic fractures, and lava pillows again indicate that it banked up against ice that has since melted away. Look for many inclusions of dark and coarsely granular gabbro and green dunite in this lava.

When light is low on the cindery slopes, you can also look for coarse fragments of cinder aligned in neat, evenly spaced rows a few inches apart, all stretching downslope. These cover many acres of the subarctic summit region of Mauna Kea. Just how they form is uncertain, although similar forms of patterned ground are not uncommon in other regions with permafrost. Some geologists suggest that a combination of freezing and thawing of water in the shallow ground—known as frost heave—and gravity sorts the cinder into neat rows.

Mauna Kea Summit

The road ends in the observatory complex next to the rim of Pu'u Wēkiu crater. Spindly bombs and blocks embedded in cinder layers are exposed in roadcuts along the way. The summit is on the high eastern rim of the crater, a restricted zone about 600 feet southeast of the parking area.

On a clear day looking north, you can see Kohala Volcano at the north end of the island, the misty blue profile of Haleakalā rising out of the ocean beyond, and on rare occasions even the islands of Lāna'i and Kaho'olawe. To the south, the summits of Mauna Loa and Hualālai stand out. The summit of Mauna Loa appears beveled; you are seeing the northeastern rim of Moku'āweoweo, its giant summit caldera.

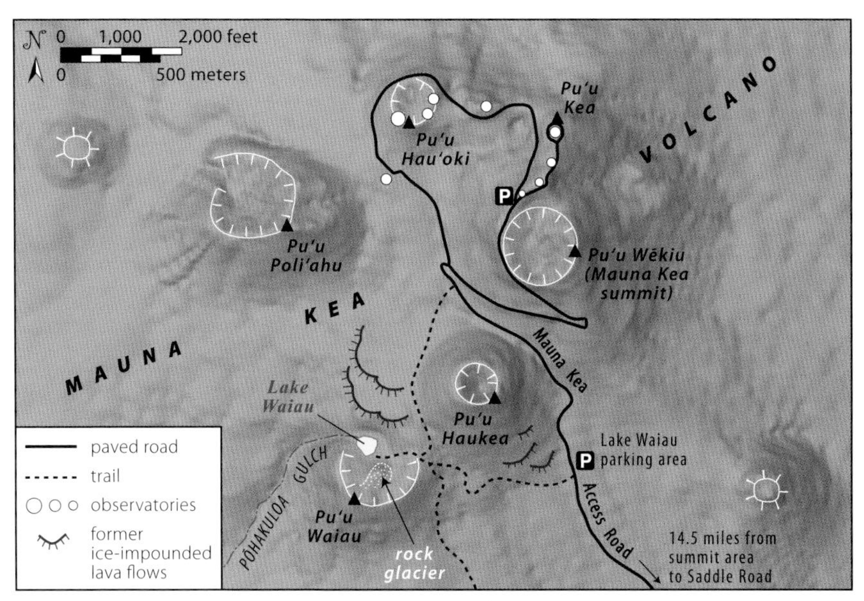

Mauna Kea summit area.

View to northeast from road's end at the summit of Mauna Kea. Cinder cones shown are ice-age vintage. Younger cones only a few thousand years old lie just out of sight downslope, including one that sent a lava flow all the way to the coast at Laupāhoehoe Point.

At sunset the afternoon clouds that generally hide the Hāmākua and Hilo sides of Mauna Kea provide a striking base for an optical effect: the huge, pyramidal shadow of Mauna Kea spreading toward the east.

MAUNA LOA OBSERVATORY ROAD

17.4 miles

This 17.4-mile drive up Mauna Loa's north flank is well paved and graded, but narrow; in many stretches it is wide enough for only one vehicle. Numerous pullouts allow cars to pass one another without much trouble; downhill vehicles are generally given right of way. Use your headlights and drive with care because of sudden blind curves. To follow this road, zero your odometer at its start on HI 200. There are no mileposts on the way up.

Skirting the east side of Pu'u Huluhulu cinder cone, the Observatory Road leaves 1935–1936 pāhoehoe and begins crossing the weathered, lightly vegetated surface of the Pu'ukahiliku Flow, erupted from Mauna Loa 3,000 to 2,000 years ago. Up ahead, the bumpy profile of Mauna Loa's cone-studded northeast rift zone forms a gently sloping horizon, 12 miles away at its closest point. The striking patchwork of different-colored flows draping the flank of the huge shield correlates with flow types and their relative ages. 'A'ā flows are darker and less glassy than pāhoehoe, which is the dominant lava type. Lava flows tend to become duller and browner with time as they slowly oxidize in Earth's atmosphere. Geologists Frank Trusdell and Jack Lockwood have determined that these older, browner flows have ages in the range of 4,500 to 1,500 years. Roughly half of Mauna Loa's surface, seen where you begin your ascent, has formed since 1843.

At mile 1.3 the road leaves the Pu'ukahiliku Flow and climbs onto finely granulated, 400-year-old 'Āinahou 'a'ā. Because it was easier to build the road across an 'a'ā base, the road follows this tongue of lava a long way upslope. At mile 2.1 and approximately 6,820 feet elevation, the road crosses a large channel in the 'Āinahou Flow with considerably larger fragments of rubble. Some crudely formed, large accretionary lava balls lie scattered on its banks.

Just past the 7,000-foot elevation, marked by paint in the asphalt at around mile 3.4, the 'Āinahou Flow transitions from 'a'ā to pāhoehoe patchy with low, light-green vegetation. This location is just above timberline for forest trees in this area, so you can view the continuation of the pāhoehoe well upslope. It is flanked to the right by a spatter rampart marking the source of the Pu'ukahiliku Flow. The white dome on the mountainside just right of the Pu'ukahiliku rampart is a vinyl-covered structure measuring 35 feet across and 20 feet high. This scientific field site is the Hawai'i Space Exploration and Analog Simulation for testing human behavior during extended travels in space. Crews have spent as many as eight months inhabiting this facility, designed for an extraplanetary or Mars-like setting. On the left the 'Āinahou Flow is bordered by a narrow dark band of 1880–1881 'a'ā.

At mile 4.0 (elevation 7,120 feet), the road crosses onto a strip of 'a'ā erupted in 1899. A gravel road branches straight ahead where the paved Observatory Road bends sharply to the left. Called the Hilo–Kona Road, this represents an abandoned attempt to build a cross-island highway in the 1950s. It terminates at Pōhakuloa US Army Training Area fencing.

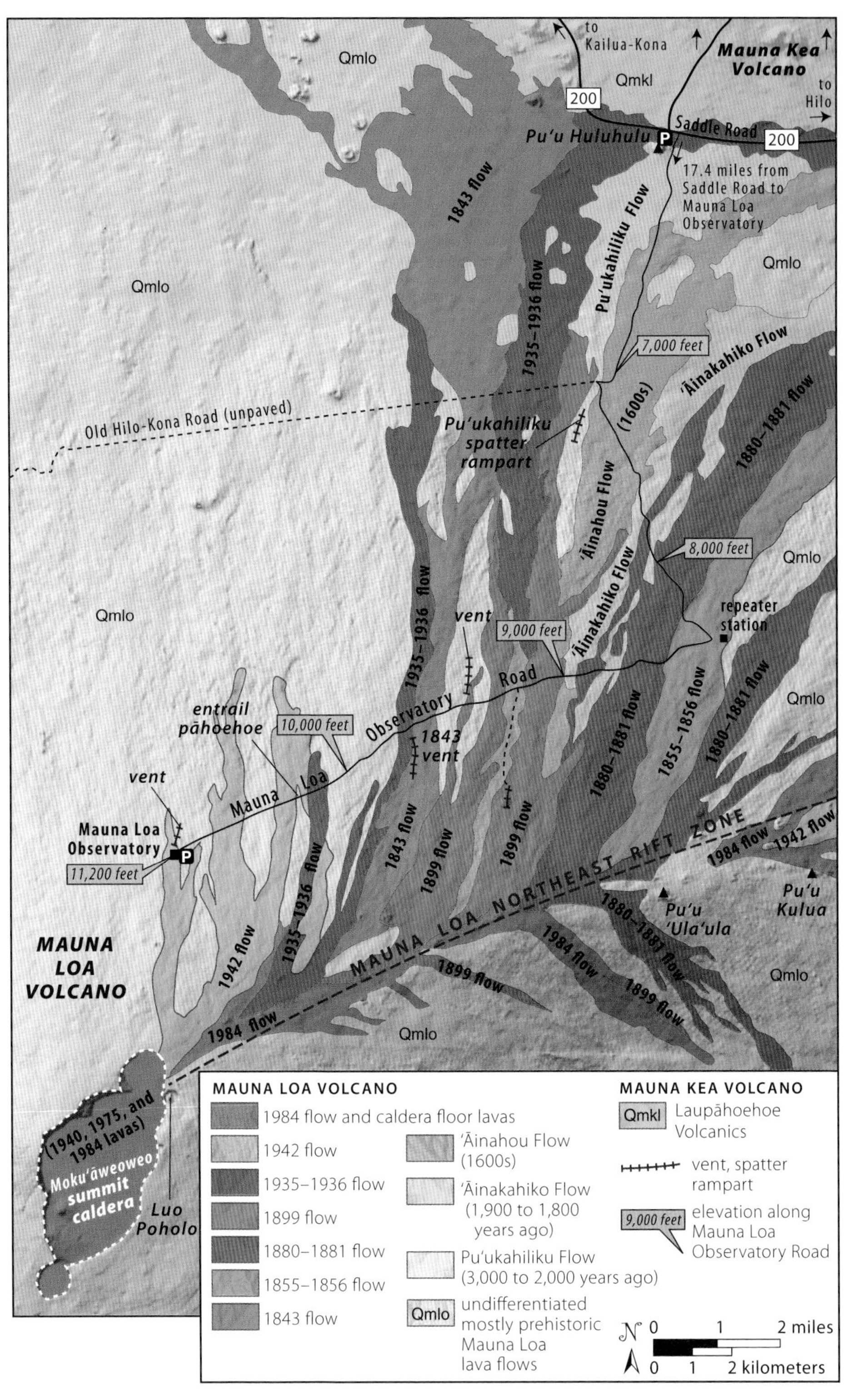

Lava flows crossed by the Mauna Loa Observatory Road. Other historic lava flows not crossed by the road are not included.

At mile 4.1 a large pullout next to the Mauna Loa Observatory Road allows you to park and explore the adjacent, much weathered Pu'ukahiliku spatter rampart. This vent does not lie within the trend of the northeast rift zone but is part of the fissure series that radiates across the northern flank of Mauna Loa. As Mauna Loa swells with magma, lava may burst from any number of flank fissures on this slope, though only a small fraction of the volcano's eruptions occurs this way.

Leaving the spatter rampart, the road returns to crossing 'Āinahou pāhoehoe. At mile 5.3 it climbs onto 1880–1881 lava, then at mile 5.9 it begins passing through a kīpuka of mixed 'Āinakahiko pāhoehoe and 'a'ā from 1,900 to 1,800 years old. The road returns to 1880–1881 lava at mile 6.3. Just before reaching that contact, look for a small lava tube cave entrance in 'Āinakahiko lava just north of the road on the Mauna Kea side.

Just below the 8,000-foot elevation at mile 6.8 and still within the 1880–1881 flow, the Observatory Road intersects the turnoff to the Mars-simulation habitat, which is neither visible from here nor open to the public. Several large kīpuka of 'Āinakahiko lava crop out across the 1880–1881 flow field.

At mile 7.0 look for excellent examples of accretionary lava balls scattered across 1880–1881 'a'ā, associated with a wide, mostly clogged lava channel. The flow transitions to pāhoehoe at mile 7.2. For the next half mile or so, the road winds upslope to reach the 1855–1856 flow, featuring a prominent lava channel at mile 7.7. 'Āinakahiko and other prehistoric kīpukas continue to appear in the landscape.

Distinguishing the 1855–1856 flow from that of 1880–1881 is possible at lower elevations by looking at differences in vegetative cover, as along the Pu'u 'Ō'ō Trail accessed along HI 200. But at this elevation these two young flows appear to be one and the same. This is true for all historically erupted flows this high on the mountainside: vegetative recovery and weathering are just too slow. Sometimes subtle differences in chemistry are the only way to tell them apart. The patchwork of flows and kīpukas also becomes much more complex the closer you get to the summit. From this point on, we can point out only major, clearly recognizable features in the landscape.

At mile 8.3 note how the 1855–1856 pāhoehoe takes on a slabby character, a good example of this lava type. At mile 8.5 next to the repeater station, the Observatory Road takes a sharp bend to the right (west). You are almost halfway to the end of the road. At this elevation of 8,500 feet, the surrounding 1855–1856 lava appears quite fresh. Upslope, a prominent lava tube skylight just south of the bend is mostly choked with rubble. The two prominent cones marking the northeast rift zone upslope are Pu'u 'Ula'ula ("Red Hill") on the right and Pu'u Kulua to the left and at a slightly lower elevation. Both cones are less than 1,000 years old.

Between miles 9.1 and 9.4 the road leaves the 1855–1856 flow and crosses another stretch of 1880–1881 lava. Look for spectacular patches of entrail-form pāhoehoe mixed with 'a'ā around mile 9.7. The road crosses another large 'Āinakahiko kīpuka from mile 10.0 to 10.2 and again from mile 10.7 to 10.8.

At mile 10.9 just past the 9,000-feet elevation marker, the road crosses onto 1899 lava. Watch for the gated gravel road and windsock to the left just past this point to see various cinder cones and spatter ramparts upslope. You are approaching Mauna Loa's northeast rift zone.

Between miles 11.6 and 11.8 the road crosses a mix of 'Āinahou 'a'ā and pāhoehoe about 400 years old. Look for a low, smooth-skinned mound in a cluster of rounded

Entrail pāhoehoe is common on the steeper flanks of Mauna Loa along the upper Observatory Road.

lava balls in the ʻĀinahou lava just to the right (north) side of the road. The thick, fresh looking ʻaʻā just past this point erupted in 1899. The mound formed where lava gushed from a lava tube skylight, flushing out lava-coated blocks of rubble. Possibly the conduit became clogged from a ceiling collapse a short distance downslope, or a great surge of lava issuing from the vent pulsed through the tube. Either way, this site is spectacular and worth the stop. Be careful approaching the skylight opening: it is quite deep and steep-sided. A series of similar "eruptive" skylights and huge collapse pits can be discovered downslope for hundreds of yards. The lava is shelly in places, so take care if you explore further.

Continuing upslope, the road shortly crosses the 1899 lava back onto weathered surfaces of older, buff-colored flows. At mile 12.3 a gravel road branches off to the right, next to a pullout, and leads one-third mile downslope to a quarry cut into a prehistoric spatter rampart. On the way down, the road crosses ʻaʻā of the 1843 eruption with abundant small clusters of olivine and feldspar crystals, best viewed with a pocket magnifier. The spatter rampart is probably 1,500 to 750 years old, based on its present color and state of weathering and the radiocarbon dating of similar-appearing surfaces elsewhere.

Between miles 13.0 and 13.2 around elevation 9,700, the road crosses a strand of 1935–1936 ʻaʻā. Just beyond this point, it drops back down onto late prehistoric pāhoehoe, and another gravel track leads upslope past a gate on the left (south) side. This track ends a few hundred yards uphill at a prominent, well-oxidized cone that grew during the 1843 eruption. The 1935–1936 ʻaʻā partly fills a well-developed channel in the pāhoehoe to the right of the road, showing well the contact relation between these flows. Look for a small lava tube entrance nearby.

Lava balls flushed out of a deep skylight pit during eruption of a 400-year-old pāhoehoe flow along Mauna Loa Observatory Road.

The Mauna Loa Observatory Road passes through a patchwork of flows of differing types and ages between 9,000 and 10,000 feet. This view is seen looking upslope between the 1899 and 1843 flows, mentioned in the text (at approximately miles 12–12.2). The ages of some prehistoric lavas can only be inferred based upon their color and degree of oxidation; charcoal for precise carbon-14 dating is not available at this high elevation.

Approaching mile 13.5 look left (south) to see the pale-reddish 1843 cone. Like other vent structures near the road, this one resulted from a radiating flank eruption away from either of Mauna Loa's rift zones. At mile 15.6 a spectacular display of entrail pāhoehoe marks a kīpuka surrounded by well-granulated 1942 ʻaʻā—the youngest lava crossed by the Observatory Road.

The road finally ends at a small public parking area at mile 17.4 at elevation 11,200 feet. This area adjoins the Mauna Loa Observatory and the trailhead for a roughly 13-mile round-trip hike to the summit of Mauna Loa. A long, artificial barrier of lava riprap protects the observatory from lava pouring downhill, but it is not entirely immune from eruptions. Like all other locations on the upper north flank of Mauna Loa, new fissures can open underfoot. One prehistoric vent, including spatter and slabby pāhoehoe, trends downslope right next to the parking area. Much of the spatter rampart has been bulldozed, but you can still make out some details. The views of Mauna Kea may be especially spectacular as you drive back down to HI 200.

PUNA DISTRICT AND KĪLAUEA

HI 130
Keaʻau—Pāhoa—Kalapana

23 miles

Begin your exploration of the Puna District at the intersection of HI 11 and HI 130 near the small town of Keaʻau a few miles south of Hilo. The Puna District, a landscape of dense forests, young lava flows, and stunning tropical coastlines, is also where volcanic eruptions in Hawaiʻi have been most damaging owing to large numbers of residents living on a very active volcano—Kīlauea.

South of the HI 11 junction, HI 130 crosses ash-covered, weathered lava that erupted from the northeast rift zone of Mauna Loa within the past several thousand years. Enough soil has developed on it to support agriculture such as macadamia nut and papaya orchards.

About 2 miles south of the junction, the highway passes onto a large swath of inflated pāhoehoe lava flows that were erupted from Kīlauea's summit in the ʻAilāʻau eruption between roughly AD 1410 and 1470. Very little soil has developed on this young terrain, so it supports few crops and only sparse forest. Isolated patches of taller trees and denser plant growth mark kīpukas, areas of older flows not covered by the younger flows. Look for these near the town of Pāhoa.

From October to December 2014, lava flows from Puʻuʻōʻō, a vent on Kīlauea, threatened Pāhoa. To see the lava, take the first right at the roundabout as you enter Pāhoa and continue past the shopping centers and then turn right on Apaʻa Street. Continue for 0.8 mile until you reach the Pāhoa Transfer Station. Slow-moving lava encountered the back of a fenced wall around the transfer station and inflated up against it. The inflationary pressure pushed through part of the fence, while breakouts of lava also burned through and filled the truck pull-in, which has since been cleared. Ropy lava cascades drape the back walls. The rest of the infrastructure survived. Just across the street, though, a house was destroyed; its rusting roof may still be visible.

A little farther along Apa'a Street you can see lava flows that crossed the road. This area, mostly private property, has many tree molds and embedded metal fences or structures that were uplifted when the lava flow inflated. This pāhoehoe has abundant crystals of olivine and plagioclase.

In the town of Pāhoa, take time to visit the Pāhoa Lava Zone Museum on the main street near the east end of town. This small but richly displayed museum provides

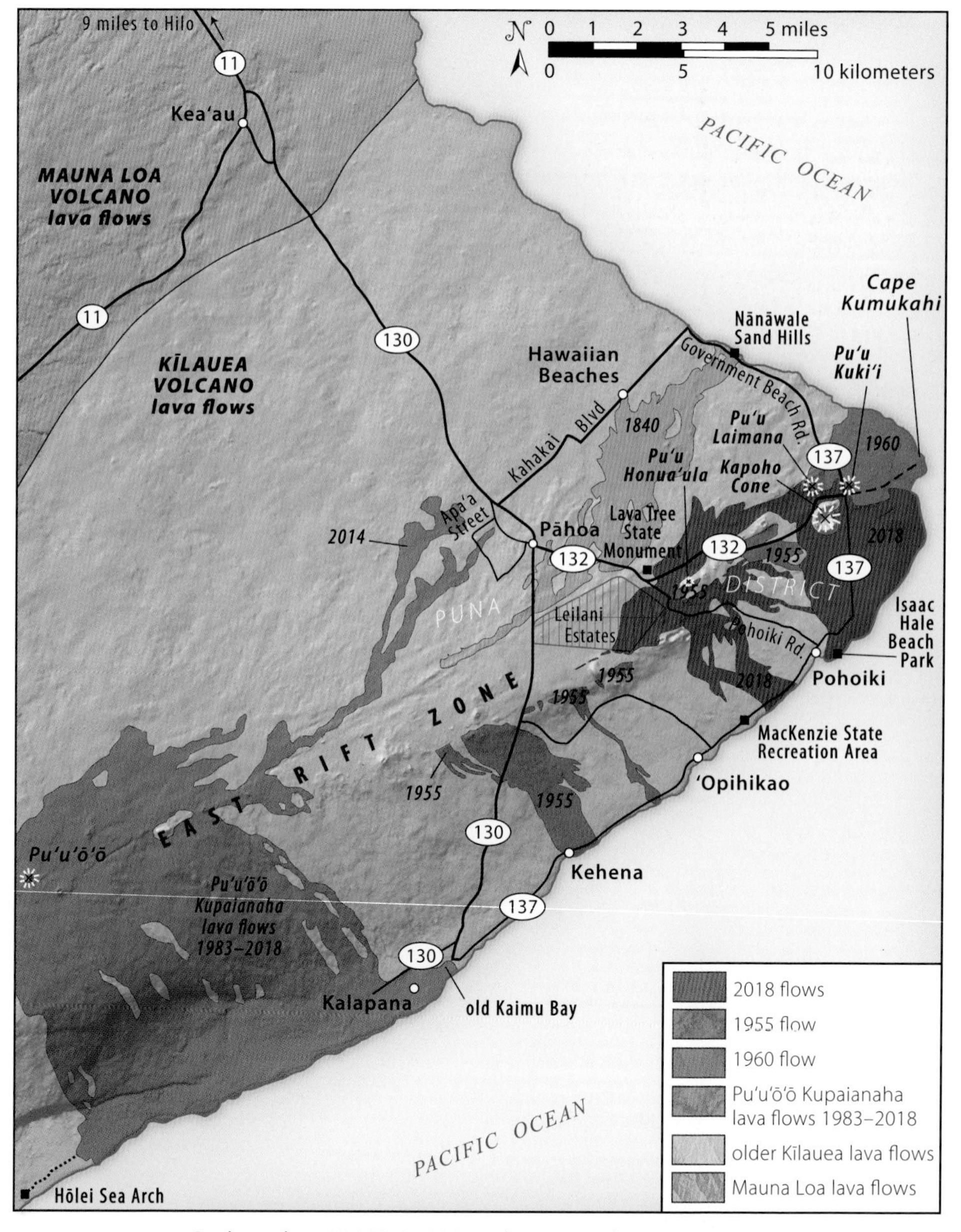

Geology along HI 130, HI 132, and HI 137 in the Puna District.

THE 2018 KĪLAUEA ERUPTION

The 2018 eruption from fissure 8 vent in Leilani Estates built a cone (Ahu'ailā'au) and produced a massive river of lava. —Photo by C. Parcheta, US Geological Survey

On April 30, 2018, the Pu'u'ō'ō vent—active for 35 years—collapsed, and magma migrated down the rift zone. Residents began evacuating, and three days later fissures began ripping open right through Leilani Estates subdivision, spewing lava. The initial lava flows were small and composed of a relatively cool, viscous magma that had been stored within the rift zone from previous eruptions. A magnitude 6.9 earthquake generated by southward sliding of the south flank of Kīlauea on May 4 shook the islands and probably helped open the rift zone pipeline, allowing more magma to drain from the summit. The eruption paused for a few days as magma moved a little farther down the rift, erupting again on May 12. The lava was hotter and more fluid and quickly increased in volume. Flows from new fissures reached the ocean to the south on May 23. This hotter, more fluid magma began reoccupying earlier eruptive fissures.

On May 27 fissure 8 reactivated and quickly became the main vent. Fountains of foamy tephra blew 250 feet high, and a massive river of lava gushed from the vent. An 'a'ā flow plowed northeast and within a few days went around Kapoho Cone and obliterated the beautiful subdivisions of Beach Lots and Vacationland. The active channel was up to 1,300 feet wide, and lava near the vent flowed at 18 miles per hour. About 850 acres of new land built off the coast as the lava spread along the shore, halting just as it reached Pohoiki. On August 4, the lava stopped issuing from the vent. This was the largest eruption in the lower east rift zone in at least 200 years, with 24 fissures and 700 homes destroyed.

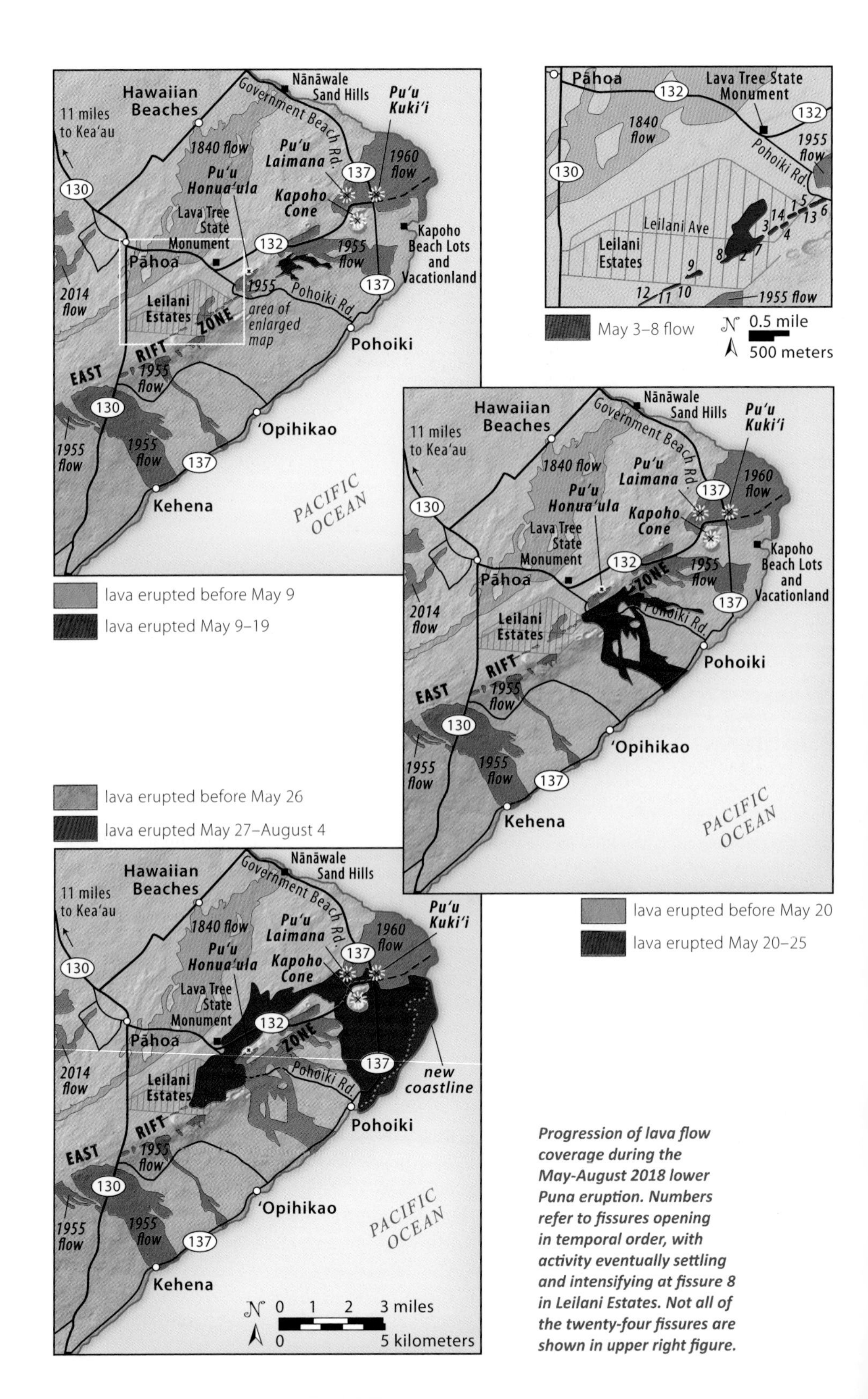

Progression of lava flow coverage during the May-August 2018 lower Puna eruption. Numbers refer to fissures opening in temporal order, with activity eventually settling and intensifying at fissure 8 in Leilani Estates. Not all of the twenty-four fissures are shown in upper right figure.

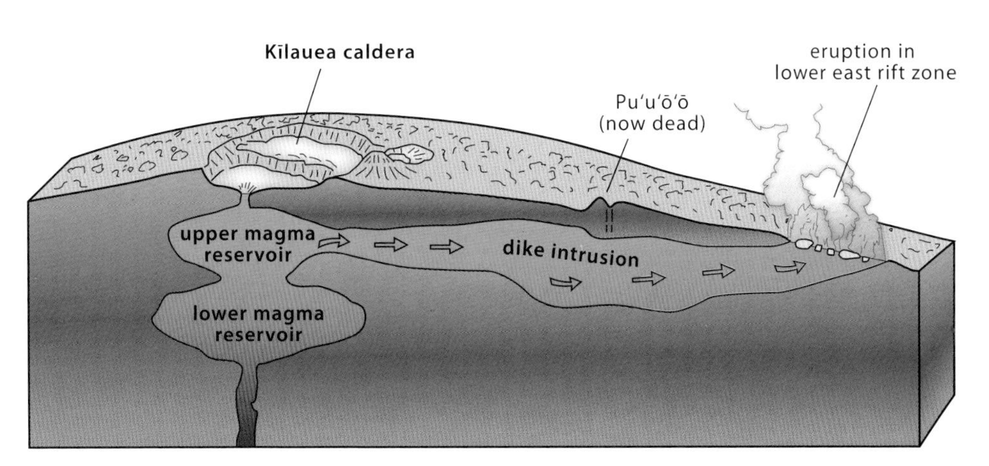

Magma intruding via an enormous dike from Kīlauea's summit reservoir fed the eruption that broke out in lower Puna in 2018.

The eruption in the lower east rift zone in turn caused collapse within Kīlauea's summit caldera 25 miles away. Halema'uma'u Crater at Kīlauea's summit had hosted an active lava lake from 2008 to 2018; as magma was withdrawn, the lava level quickly dropped, sometimes accompanied by ash-laden explosions as rocks fell into the deepening crater. Thousands of earthquakes accompanied the gradual, stepwise down-dropping of the caldera floor over the next three months. When it ended in August, the deepest part of the crater was 1,800 feet below the rim and the caldera landscape had been dramatically reshaped.

an excellent overview of recent Kīlauea eruptions, volcano-related cultural tradition, and local volcanic rocks, many surprisingly beautiful and interesting.

South of Pāhoa, HI 130 ascends the forested flank of Kīlauea's east rift zone. Two miles south of the intersection with HI 132, watch for the entrance to Leilani Estates on the left side of the road. You may turn left and follow Leilani Avenue until it dead-ends. This vantage point permits a view of the massive lava flow that poured through the local neighborhood in the summer of 2018, but note that growth of vegetation obscures much of the view even just a few years later. You may explore side streets to the south to get a look at the principal vent of the eruption, but be respectful of private property and "no trespass" signs. A 150-foot-tall cinder and spatter cone grew over the main vent, which was designated "fissure 8" during the eruption because it was the eighth vent to activate during the three-month-long outbreak. It has since been named Ahu'ailā'au or Altar of the God 'Ailā'au, the Eater of Forests, by Hawaiian elders. Smaller fissures cut through the subdivision in this area, leaving steaming cracks, spatter, and sulfur deposits. As you return to the HI 130 intersection, look straight across the highway from the subdivision entrance to see an exposure of a spatter rampart from a large eruption circa 1790.

The eastern end of the dike that intruded in 2018 opened cracks across HI 130 between mileposts 14 and 15. These fractures may still be apparent from dead trees, rising steam, and plates in the road.

Lava channel crossing Leilani Avenue during the 2018 eruption.
—Photo courtesy US Geological Survey

Past milepost 15, the highway begins a gradual descent from the crest of Kīlauea's east rift zone to the south coast. Around milepost 16 the highway crosses a huge ʻaʻā flow full of glassy crystals of green olivine that erupted in 1955. It is now heavily overgrown with lichen and native ʻōhiʻa trees—rapid colonizers of fresh lava flows. It erupted from the pair of large cinder and spatter cones upslope of the highway, mostly obscured by new forest. The 1955 eruption was the first volcanic outbreak to strike lower Puna in 115 years and so came as a shock to local residents. However, this and a subsequent eruption in 1960 did not prevent the development of new subdivisions in the area soon afterwards, including Leilani Estates.

Not far beyond milepost 21, HI 130 continues—unpaved—onto pāhoehoe flows from the Puʻuʻōʻō eruption that covered the town of Kalapana in 1990. These are sheet flows: flat-topped, inflated pāhoehoe flows first recognized and documented here in 1990. At milepost 21.5 the road goes through a kīpuka of older ground in the 1990 flows. Beyond this kīpuka is the massive Puʻuʻōʻō flow field, pāhoehoe lava erupted between 1983 and 2018 that covered 7.5 miles of coastal highway.

Many new homes have been built on the 1990 lava, although a few were destroyed when lava returned to the area in 2010–2011. There are excellent views upslope to the long ridge of the east rift zone where Puʻuʻōʻō and other vents sit out of sight. Decades of young lava flows drape the steep south-flank fault scarps. Inflationary structures are visible in the pāhoehoe flow field that composes most of the coastal plain below, particularly where the road cuts through tumuli. Seaward of the road, a line of trees visible in the distance marks the top of what was once a nearly 50-foot graben cliff, but lava flows filled in the upslope side so that it is nearly level today.

As of 2022 the accessible road ends at about milepost 23, but in the future HI 130 may continue into Hawaiʻi Volcanoes National Park as it did before the Puʻuʻōʻō eruption covered the road. In the meantime, head back along HI 130 until you reach the turnoff to Kalapana/HI 137 and head right. This short connector will take you to coastal HI 137. The intersection looks out over what was once Kaimu Bay before lava flows filled it in 1990. This is the largest mass of new land built by lava flows during the Puʻuʻōʻō eruption—about 500 acres.

HI 137
Kalapana—Isaac Hale Beach Park

11 miles

See map on page 262.

From near the end of HI 130, turn toward Kalapana. A short connector brings you to HI 137. The road to the right passes a few houses from Kalapana that survived the 1990 eruption of Puʻuʻōʻō and dead-ends within a quarter mile. Park here to take a 0.3 mile walk out to a fresh black sand beach, across 1990 lava flows that filled old Kaimu Bay. New land is state property, although residents have made a trail that covers many features but makes walking easier. Watch for classic inflated pāhoehoe structures like large, steep-sided tumuli, squeeze-outs, and cracks with multicolored stripes.

Near the shore, the first black sand you encounter is coarse and glassy, created when lava was quenched as it flowed into the sea. The sand becomes finer closer to the water, reworked by the waves. The condition of the beach varies widely depending on recent storm activity; sometimes it is sandy, sometimes cobbly or bouldery. All the trees at the beach were planted in an effort to restore the magnificent palm grove that grew in Kalapana before the 1990 eruption.

Zero your odometer at the intersection before heading northeast, following HI 137 along the coast. In the first couple of miles the road crosses a flow of basalt lava with an ʻaʻā surface that erupted around 250 years ago. Forest recovery is impressive though not yet complete despite heavy rainfall in this area.

Just east of Kehena and 3.4 to 4.5 miles north of Kalapana, HI 137 crosses rough lava that erupted in 1955. New houses are scattered across the flows. A short distance farther east the road crosses a much narrower tongue of ʻaʻā lava that also erupted in 1955. Approaching the community of ʻOpihikao, the road winds through the dense coastal forest on lava from approximately 1790 and somewhat older eruptions.

MacKenzie State Recreation Area is an excellent place to watch heavy surf pound resistant 1790 ʻaʻā flows exposed in a shoreline cliff. Ironwood trees, introduced to Hawaiʻi from Australia in 1872, carpet the flow surface with thick beds of soft needles. It is a great place to picnic and bask in a sedative trade wind. A large overgrown skylight opens a few hundred yards to the northeast, and the huge 2018 ʻaʻā flow lies beyond that.

At 0.2 mile past MacKenzie State Recreation Area, HI 137 crosses the first of three massive ʻaʻā flows erupted in late May of 2018. The road passes through two kīpukas of forested 1790 lava in between the newer flows. The top of the third ʻaʻā flow provides a good view upslope to the eruptive vents—cones and spatter ramparts—on the east rift zone. The fissure 22 cone, a rare steep-sided, symmetrical cone created by pulsating bursts of lava, dominates the horizon.

At Pohoiki Road, 10.9 miles from HI 130, turn right toward Isaac Hale Beach Park, once a small, stony beach and a landing for small boats. A large, new black sand beach was deposited here in 2018 as lava poured into the ocean a short distance to the northeast and shattered into glassy bits that were swept downcurrent. The beach contains not only sand but also plentiful water-worn cobbles, many of which contain numerous shiny crystals of olivine.

Swimming is not recommended at this popular weekend spot due to strong currents and rough seas. A series of warm pools have developed on the inland side of the beach—the pre-2018 shoreline—including at the now-blocked concrete boat ramp. The water quality is unhealthy, however, thanks to poor water circulation and dense algal growth.

Walk to the edge of the huge 2018 ʻaʻā flow that covers part of the park grounds at the northeast end of the beach. The flow covered more than 5 miles of coastline and added over 850 acres of new land to the island, some of which has since sloughed off into the ocean. Pohoiki was the only boat launch along this entire southern coast and very important to the local fishing economy. Many believe that boat access will eventually need to be restored.

As of 2022, both HI 137 and Pohoiki Road are partly covered by 2018 lava a short distance north and west of Isaac Hale Park. These routes may soon be restored, however. In the meantime, you will need to backtrack to Kalapana to return to HI 130, then return upslope to the intersection with HI 132 in Pāhoa.

When Pohoiki Road is reopened, you will be able to drive directly upslope from the HI 137 intersection at Isaac Hale through about 2 miles of huge, overgrown mango trees, then cross the 2018 flow to reach HI 132 near Lava Tree State Monument, described in the next section. The new road will cross the massive 2018 flow channel and skirt the geothermal power plant that survived the eruption and continues to supply much of the island's renewable energy. A reopened HI 137 will also cross the fresh 2018 ʻaʻā coastal plain and intersect HI 132 near Kapoho Cone, also described below.

A black sand and cobble beach at Pohoiki created by lava from the 2018 eruption entering the ocean nearby and shattering into pieces.

As the black sand beach at Pohoiki was created in 2018, it blocked the boat launch and buried the old breakwater (center right). Freshwater seeping in behind the new beach has created brackish warm ponds.

HI 132
PĀHOA—KAPOHO

7 miles

See map on page 262.

A 2.5-mile drive east from Pāhoa down HI 132 brings you to Lava Tree State Monument. The park sits on a pāhoehoe flow that erupted around 1790 from a gaping fissure partially visible where the restrooms now stand. A 0.7-mile trail loops around more than forty lava trees up to 15 feet high. The lava trees formed when the flow first submerged the trunks of ʻōhiʻa trees, hardened against them, then subsided as lava flowed downslope or drained back down the fissure with the waning of the eruption. The trees burned away, leaving their imprints as hollow cavities, or molds, inside the pillars. Plates of lava slope away from the lava trees, dropped as the lava drained from underneath a solid surface crust.

Just east of the entrance to Lava Trees, HI 132 veers left, while straight ahead is Pohoiki Road, which used to—and may again—continue on to Pohoiki/Isaac Hale Park. The massive lava channel from the 2018 eruption crossed the road ahead. Set your odometer to zero at this intersection.

HI 132 climbs onto 2018 lava 0.3 mile east of the Pohoiki Road intersection. Parts of this lava field are very thick and likely to stay hot for years, so steaming may occasionally be visible, particularly on cool, wet days. Puʻu Honuaʻula, several hundred years old, is the tree-covered cone visible on the right (south) side of the road. At its western (right) foot sprawls the only geothermal power plant in Hawaiʻi, which is not visible from the road.

Lava trees at Lava Tree State Monument. Tallest here is more than 10 feet and hollow where the tree inside burned away.

The Hawai'i Geothermal Project drilled its first test well to a depth of 6,500 feet in 1975 and 1976. The temperature of the rock at the bottom was 1,220 degrees Fahrenheit. In 1993 Puna Geothermal Venture plant opened several new wells at the base of Pu'u Honua'ula. In 2005 drilling encountered a partly molten body of silica-rich magma called dacite at about 8,000 feet down. Such highly evolved magma had never been found on Kīlauea before, so its discovery was exciting for geologists. This facility was generating 38,000 kilowatts, a substantial proportion of the Big Island's electricity, before it was shut down during the 2018 eruption. Wells were capped and part of the property was covered with lava, though most infrastructure survived, and the plant resumed production in 2021.

About 1.1 miles east of the Pohoiki Road intersection, HI 132 crosses a 100-foot-wide branch of the larger 2018 lava channel that was flowing toward the north and quickly widens to more than 300 feet. The pullout here is an excellent place to examine the channel walls, some more than 20 feet high. Along the left (north) side of the road, the structure of the walls shows multiple thin layers, each an overflow of shelly lava along the levees of the channel. Many of the inner walls show glassy ledges where lava caught and stuck while flowing; other sections have collapsed, particularly in the main, much larger channel to the south. The channels are filled with the last-erupted, spiny, rubbly 'a'ā. On top of the walls, however, shelly pāhoehoe overflows form a deceptively smooth-looking surface. This billowing lava is glassy, fragile, sharp, and full of hollow pockets that collapse underfoot.

At 1.3 miles is another pullout on the right next to the rusted ruins of a shipping container that was filled and overrun by lava. On the left side of the road a short distance ahead, a skeletal metal roof draped in lava protrudes from the flows, next to a small patch of surviving trees. They sit right beneath a sharp bend in the channel

Two branches of the massive 2018 lava channel split around the island in the center, now cut by the new road.

that was the site of multiple overflows. The road continues into the rubbly main channel, walls visible on both sides, where the lava river headed downslope to join up with the smaller branch already passed.

HI 132 soon bends east and cuts down through thick ʻaʻā and into a large, well-vegetated kīpuka. At about 2.2 miles east of Pohoiki Road, look to the left for a row of spatter cones from now-forested fissures that fed a large eruption in 1955. Vegetated ʻaʻā flows from 1955 cover much of this area. Several dozen homes in the kīpuka survived the 2018 lava flows and related fires but were mostly inaccessible for nearly a year. This region was particularly well known for its papaya fields, but much of that farmland was destroyed. Some fields of papaya and a medicinal fruit called noni survived, and others are being replanted.

At 3.6 miles, HI 132 passes back into 2018 lava where it flowed around Kapoho Cone. A pullout at 3.8 miles gives a good view of the thickly forested cone on the right. The lava flows in this area spread widely from the main channel off to the north (left) as it stagnated and broke out in June and July of 2018.

A little farther, the road cuts through an ʻaʻā channel formed when the lava changed direction and began flowing around the west side of Kapoho Cone in July 2018. These are the flows that reached Pohoiki right as the eruption ended. Just ahead, look for a house on the left, with only minor damage, that survived in a tiny kīpuka. Local residents dub it the Miracle House. Past this point, the drive reveals spectacular sections of lava channel, massive accretionary lava balls, ʻaʻā flow front intermixed with shelly overflows, and the burned trunks of huge dead albizia trees.

On the north side of the road at 4.6 miles is a private access road to Puʻu Laimana, a heavily quarried cone made of ash, cinders, and spatter. Puʻu Laimana grew in one month during the 1960 eruption. Lava occasionally fountained to 1,700 feet and was even visible from Hilo, 20 miles to the northwest. This outbreak buried the sugarcane plantation village of Kapoho, which had been damaged when an eruption also threatened to break out here in 1924. The stunted remains of Puʻu Laimana are mostly hidden in ironwood trees to the north. Watch out for large, fast-moving trucks carrying red cinders from a quarry on the cone.

A wide pullout at 4.8 miles puts you in the middle of the channel that fed lava around the eastern side of Kapoho Cone, with channel walls visible on both sides of the road. You can get close to the walls on the left side of the road to see the shelly overflows at the top. On the right side of the road, large broken slabs of spiny lava dominate the flow. The stretched, spiny surface texture indicates that the lava was quite viscous. It represents the last outbreak of lava after the main channel diverted to the west side of the cone. The total thickness of the 2018 lava flow along this section is 60 to 90 feet. Steaming here was not uncommon under wet conditions as recently as 2022. Look back upslope to see the well-defined channel skirting the line of tree-covered prehistoric cones that mark Kīlauea's lower east rift zone.

Kapoho Cone, originally about 350 feet high, dominates the skyline southwest of the pullout. The cone grew during a violent series of steam blasts from magma intruding shallow groundwater or interacting with shallow seawater through an open vent in the east rift zone sometime between 600 and 350 years ago. The crater is shaped like a horseshoe open to the east. A line of four small explosion pits used to cut across the floor of the crater, one of which hosted Green Lake, the largest freshwater lake in Hawaiʻi. Lava evaporated the lake and buried the pits in a matter of hours on June 2, 2018, surrounding the cone on its way toward the shore.

The smaller, well-wooded ash and cinder cone that rises northeast of Kapoho Cone is 1,000- to 400-year-old Puʻu Kukiʻi, or Puʻu Kukae. An important fifteenth-century religious site, Kukiʻi Heiau, was established in this area.

HI 132 ends at a right-angle bend between Kapoho Cone and Puʻu Kukiʻi. This corner is a good place to stop. An old gravel road to the east leading to Cape Kumukahi is accessible with a short walk directly eastward over the new lava, just to the right of Puʻu Kukiʻi (this may be reopened as a new through-going road). As of 2022, it is a 1.5-mile hike to the cape, which marks the easternmost point of land in the Hawaiian Islands. A large flow of ʻaʻā added new land to Cape Kumukahi in 1960, just missing the Coast Guard lighthouse that stands at the end of the road.

As you approach the lighthouse, head right to see a new black sand beach deposited in a cove during the 2018 eruption. Exposures of lava cliffs along the back of the beach show coral growth, so were likely often underwater before the beach was created.

Nānāwale Sand Hills via Government Beach Road

From the corner where HI 132 ends, zero your odometer and turn left to head north on Government Beach Road. The road heads across the 1960 lava flow, partly covered with tephra from Puʻu Laimana and shaded by ironwood trees. The road then winds as a narrow, paved route through the lush Nānāwale forest growing atop 1,500- to 250-year-old lava flows.

About 4 miles north of the corner, the road passes onto an overgrown pāhoehoe flow that erupted in 1840. Activity began in the upper east rift zone, 20 miles to the west, too far away to disturb people in Puna. The action moved closer when new vents

Nānāwale Sand Hills, an eroded littoral cone formed by steam explosions as lava from the 1840 eruption of Kīlauea flowed into the water.

near Pāhoa erupted a flow that moved as fast as 5 miles an hour, mostly through dense forest. It came on a Sunday when people were gathered for church. Panic ensued as lava destroyed whole villages. Many residents barely escaped. According to the Reverend Titus Coan, the intense night glow from molten rock and burning forest could be seen from 100 miles at sea. People 40 miles away could read by this light.

A small and sometimes muddy pullout and trail about 4.2 miles north of the Kapoho–Pu‘u Kuki‘i intersection provides access to Nānāwale Sand Hills. These two large, layered littoral cones grew where lava poured into the ocean for about three weeks in 1840 and steam blasted the lava into tiny bits. Olivine crystals are abundant in the sand and spatter of the cones. Wave erosion and seaward landsliding have consumed about half the mass of the cones, which initially stood 300 feet high. From here the road skirts the scenic rocky coast for another mile to meet Kahakai Boulevard in the Hawaiian Beaches community. Turn left at the intersection for a speedy return to Pāhoa across forested prehistoric lava flows.

HI 11 (HAWAI‘I BELT ROAD)
Hilo—Volcano Village—Wai‘ōhinu
64 miles

From Hilo, HI 11 heads south across forested Mauna Loa flows to Kea‘au, then turns southwest, crossing the summit of Kīlauea—Mauna Loa's junior neighbor—about 30 miles away. The highway then descends Kīlauea's western flank, skirting the dry southwest rift zone and returning to Mauna Loa near the ancient Nīnole Hills and the island's south point. Short side trips lead to lush kīpukas, young volcanic vents, archaeological areas, and some of Hawai‘i's wildest shorelines.

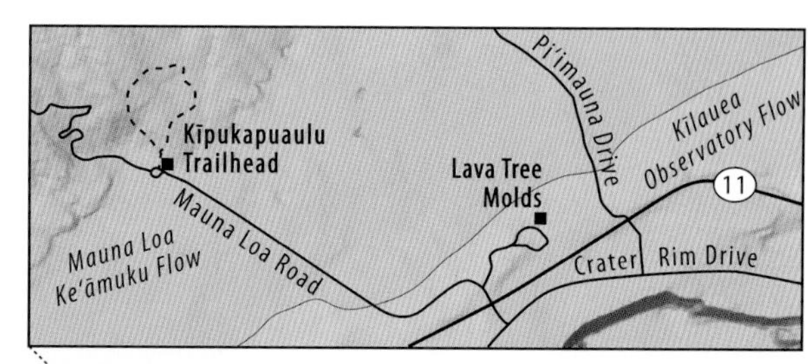

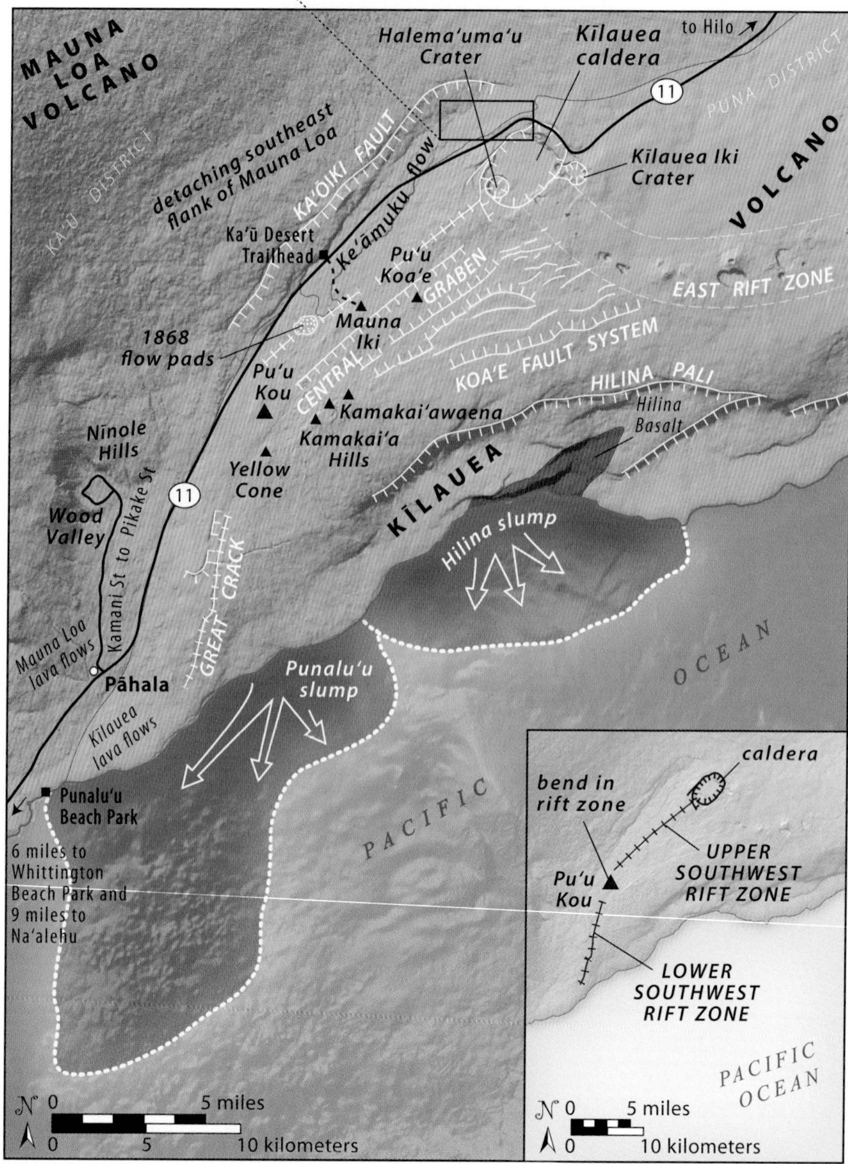

Major geologic features in the Kīlauea southwest rift zone.

The first 5 miles of highway, between Hilo and Kea'au, cross 1,000-year-old Pana'ewa lava, a flow from Mauna Loa's northeast rift zone largely responsible for the development of Hilo Bay. Between mileposts 20 and 21, just upslope from Hirano Store, the highway turns sharply where it passes the boundary between ash-covered, heavily vegetated, olivine-bearing Mauna Loa lavas as much as 6,000 years old and the lightly vegetated, crystal-poor flows that erupted only 600 to 550 years ago from Kīlauea's summit near Kīlauea Iki Crater. From this point to the national park entrance, the highway crosses increasingly thick deposits of ash formed during the explosive eruptions that created the modern Kīlauea caldera, especially during the culminating blast of 1790. Ash weathers more readily than lava, forming fertile soils within a few centuries. The increasing density of forest cover upslope reflects this trend.

A short distance after crossing into Hawai'i Volcanoes National Park along HI 11, you can see Mauna Loa's northeastern flank through an opening in the tree-fern forest. The flank is studded with tiny cones marking vents along the skyline. It is 15 miles away and about 6,000 feet higher than the highway here.

HI 11 rounds the northern perimeter of Kīlauea caldera as the highway heads west from the park entrance intersection. (See the road guide for Hawai'i Volcanoes National Park on page 282.) The highway enters the larger caldera, crossing one of its lesser fault scarps in the area of milepost 29.

Near milepost 30, Golf Course Road (Pi'imauna Drive) heads to the right of the highway up a 10-foot-high fault scarp, which marks the northern boundary of Kīlauea caldera. This scarp probably began forming only a few hundred years ago. HI 11 exits the caldera a short distance west of this location.

At milepost 32 on HI 11, look ahead to view the flank of Mauna Loa as the highway leads into the drier landscape of the leeward flank of Kīlauea. The road continues on the Observatory Flow pāhoehoe for another mile or so.

To the west (right) of the highway just south of milepost 33, look for the thick, fantastically rugged Ke'āmuku Kīpukakulalio 'a'ā flow. This lava erupted from Mauna Loa's northeast rift zone at around 9,000 feet elevation around 575 years ago. The base of the Ke'āmuku Kīpukakulalio Flow marks the contact between Kīlauea and Mauna Loa.

At milepost 34 look to the right of the highway to see the steeply sloping scarp of the Ka'ōiki fault system at the foot of Mauna Loa. The scarp is as much as 300 feet high and stretches close to the contact between Kīlauea and Mauna Loa for almost 10 miles. The last large earthquake to originate here, a magnitude 6.6, struck in 1983. The origin of the Ka'ōiki fault system is unknown. Perhaps it began as landslides in the flank of Mauna Loa before Kīlauea grew large enough to provide support. But that would not explain why recent movement in the Ka'ōiki fault system has been largely horizontal rather than vertical. Perhaps differential swelling of the magma chambers beneath Kīlauea and Mauna Loa creates the forces that move the faults today. Here the road also passes onto another young Mauna Loa 'a'ā flow, the 300-year-old Ke'āmuku Kīpukakēkake Flow.

Just north of milepost 35, look to the west (left) to see cones of Kīlauea's southwest rift zone along the horizon, mostly spatter cones associated with the September 1971 eruption. The ragged summit of much older Pu'u Koa'e pokes up prominently, too.

TREE MOLDS ALONG MAUNA LOA ROAD

Turn onto Mauna Loa Road halfway between mileposts 30 and 31 to do a short loop through some spectacular giant tree molds. The rolling pāhoehoe lava in this area is part of the widespread Observatory Flow series that erupted at Kīlauea's summit 600 to 550 years ago. Collapse of the modern caldera broke up the source vent of these flows, a big lava shield in the vicinity of present-day Halema'uma'u Crater. The impressive depth of the tree molds in the ground, the fossils of giant native koa trees, gives a good sense of the thickness of this inflated pāhoehoe flow.

A picnic area 1.5 miles farther down the Mauna Loa Road is home to Kīpukapuaulu Trail, locally known as Bird Park, a lovely 1-mile hike through an old forested kīpuka of ash-covered Mauna Loa lava. Beyond is the Mauna Loa Strip Road: 10 miles of narrow paved road through koa forest and Mauna Loa lava flows ranging in age from 8,700 to 270 years old. The road gets narrower and rougher after about 2 miles. There are panoramic views of Kīlauea's summit area along the way and from the shelter at the end of the road (elevation 6,662 feet). A short trail permits view of endangered silversword plants growing in patches of soil atop 2,000-year-old lava. The rugged Mauna Loa Trail begins here as well.

Huge tree mold (and new 'ōhi'a tree) in inflated pāhoehoe from Kīlauea summit overflows.

Ka'ū Desert Trail

Between mileposts 37 and 38, a broad pullout to the west (left) marks the beginning of the Ka'ū Desert Trail. The first few miles to the summit of Mauna Iki lava shield are well worth exploring. The round-trip may take a couple of hours.

The stark landscape of this region contrasts greatly with the lush green jungle of Kīlauea's summit and windward side. In part, as elsewhere in Hawai'i, the desert can be explained by the presence of a rain shadow, caused as tradewind-driven storms dump their moisture while blowing across the broad summit. But the Ka'ū Desert still receives 20 to 60 inches of rain a year—as much as in Seattle—so it can hardly be called a true desert. Other factors also contribute to its desolation, including the youthfulness of the volcanic terrain, the highly permeable nature of the lava flows and coarser explosive debris, and the acidic precipitation created as strong volcanic gases drifting west from the summit mix with rain. Only the hardiest plants tolerate these conditions. Technically, this is an acid rain desert and not the arid landscape ordinarily expected from too little rain.

The first half mile of trail crosses younger Ke'āmuku lava featuring excellent examples of accretionary lava balls, the crudely spherical clots of lava as much as 4 or 5 feet across. These structures formed as pieces of lava channel walls miles upslope broke and floated downhill in the active stream of lava. As the clots rolled and tumbled, they acquired coats of fresh lava that gradually increased their diameters and gave them rounded shapes—just like giant snowballs but hot. In places beside the trail, broken surfaces reveal the interiors of lava balls, showing the structure of accreted coatings around original core fragments.

Accretionary lava ball along trail to Mauna Iki, one of many in the Ke'āmuku 'a'ā flow. The broken side reveals concentric layers of lava that coated the boulder. The Ka'ōiki fault scarp is visible in the distance.

Descending the Ke'āmuku Flow on the south side, the trail begins crossing older Observatory pāhoehoe. You can stand right on the contact between Kīlauea and Mauna Loa here. Ash and grit from the powerful Kīlauea summit eruption of 1790 cover most of the flow. The wind occasionally uncovers fossilized footprints of native Hawaiians caught in the eruption. Nearly two thousand have been discovered so far, many of them left by women and children, everyone barefoot. It is difficult to imagine walking across this landscape in bare feet, but that is how most Hawaiians traveled while ashore. If you come across any footprints as you explore, please leave them undisturbed. They are quite fragile, and some have already been badly damaged

Fossil human footprints in 1790 ash near the trail to Mauna Iki.

by past vandals. Look in the 1790 deposit for the rounded pellets of ash that geologists call pisolites, or accretionary lapilli. They may have formed as electrostatically charged ash clumped together inside steamy eruption clouds.

After about 1 mile the trail crosses a slabby flow less than 400 years old with a well-defined channel. After almost 2 miles the trail takes you to the summit of Mauna Iki, a lava shield a couple hundred feet tall and three-quarters-mile wide that grew over eight months during 1919 and 1920. An enormous lava tumulus breaks the flank of the shield along the way up—one of the largest to be observed in Hawai'i. From the summit, the panorama of cones, fissures, and lava flows marks the trend of Kīlauea's southwest rift zone. The three prominent cinder cones 4 miles away to the south are the Kamakai'a Hills, an area of unusually explosive vents in Kīlauea's flank where numerous small but violent eruptions have taken place over the past few thousand years. The two prominent western cones, Kamakai'awaena and Kamakai'a Hill, mark the most recent of these explosions in the early nineteenth century. A small pit crater at the top of Mauna Iki opened when a shallow magma pocket drained away beneath it, perhaps feeding one of the stubby, rough 'a'ā flows that squirted from the flank of the shield as its activity drew to a close.

Nīnole Hills to Punalu'u and Whittington Beaches

As HI 11 heads southwest from the Ka'u Desert Trail, the road drops off the Ke'āmuku Flow near milepost 39, crossing older Observatory pāhoehoe—sourced from Kīlauea's summit—for the next several miles. Numerous shallow roadcuts show fine examples of this inflated pāhoehoe in cross section, including short, crudely formed columnar joints inside large tumuli. To the north the Ka'ōiki scarp gets ever smaller and is all but gone by milepost 43.

The slope of Mauna Loa to the northwest clouds over many afternoons in this area, thanks to strong, steady updrafts off the ocean. Hawaiians had a name for this part of Mauna Loa: 'Āinapō, the Dark Land. Persistent cloudiness contributes to excellent conditions for growing coffee and macadamia nuts on the weathered ash deposits, especially around the town of Pāhala a few miles ahead.

Near milepost 49, HI 11 passes from Observatory pāhoehoe flows onto older weathered ash deposits and lava flows from Mauna Loa. Between mileposts 49 and 57, views inland reveal a rugged cluster of heavily forested knobs and bluffs—the Nīnole Hills. The rough landscape of the hills shows that they are older than the less-eroded landscape around them. Most of Mauna Loa's surface formed in the past 4,000 years, but this part of the southern flank is 200,000 to 75,000 years old.

Many geologists think the Nīnole Hills are eroded remnants of an older version of Mauna Loa. At one time, the southwest rift zone of the volcano may have passed through this area, but a set of major collapses in the western volcano's flank, most recently about 127,000 years ago, caused the rift zone to shift westward 5 to 10 miles, suddenly isolating this side of the mountain from fresh supplies of lava. Mauna Loa has since healed much of the scar left from these collapses, though not enough to rebury the terrain around the Nīnole Hills.

The steep, seaward-facing slopes of the Nīnole Hills also show that a giant landslide must have torn away their southern flanks. The northeast coast of Kohala

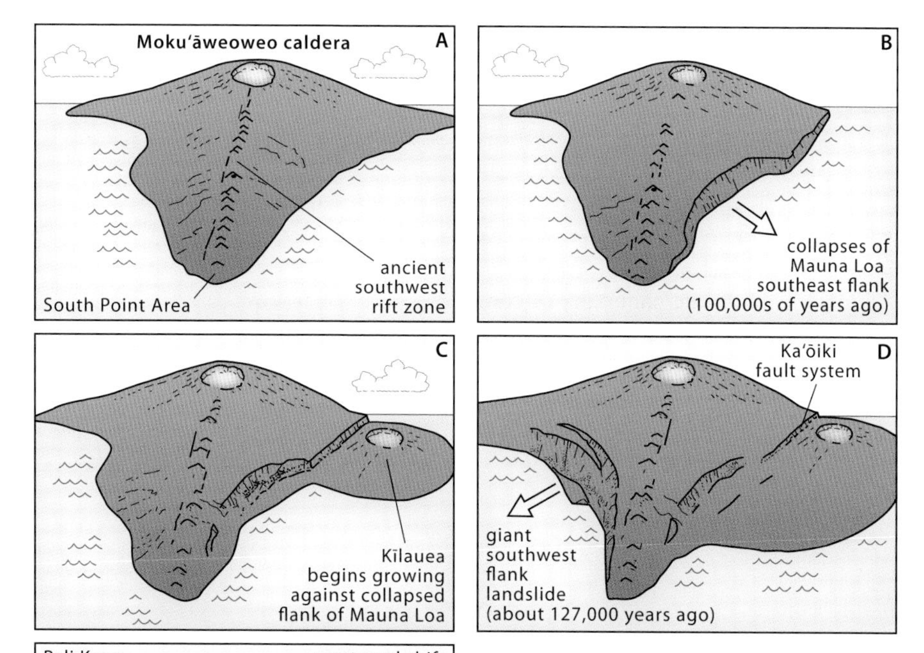

Major changes in the geologic structure of Mauna Loa's southern flanks have taken place in the past few hundred thousand years.

Volcano at the northern tip of the Big Island is unquestionably an eroding slide scarp, with a similar cliff and canyon topography. The difference is that lava erupted from Kīlauea has covered the slide terrain downslope from the Nīnole Hills, whereas evidence of sliding from the north flank of Kohala remains largely exposed and much easier for geologists to study.

Kamani Street, the turnoff to the town of Pāhala near milepost 51, allows you to approach the base of the Nīnole Hills. Drive upslope a short distance through town, turn right at the stop sign onto Pikake Street, and travel another 4 miles east through orchards and plantations largely irrigated by water drawn from wells drilled along ash layers in the adjacent hills. Some well tunnels, constructed nearly a century ago for sugarcane cultivation, extend for thousands of feet into the flank of Mauna Loa. The road ends near a Buddhist temple in scenic Wood Valley, where a huge debris avalanche burst from neighboring, heavily weathered Nīnole slopes during the great earthquake of 1868. The slide mass swept away 10 houses and killed 31 people and hundreds of cattle. It dropped more than 1,000 feet before coming to rest on the valley floor, leaving a muddy deposit tens of feet thick. Today, thick vegetation covers both avalanche and scar.

About 5 miles south of Pāhala, HI 11 passes the Nīnole turnoff to Punalu'u Beach Park. This attractive beach is fringed with palm trees in the finest Hawaiian tradition. The beach's black sand is composed of particles of glassy lava most likely shed from Kīlauea lava flows that entered the sea a few miles away in 1823.

West of Punalu'u between mileposts 57 and 59, the highway crosses a set of Mauna Loa lava flows ranging in age from 3,000 to 900 years, all from the southwest rift zone. The road reaches the foot of a steep slope at Whittington Beach Park. Less

Punalu'u Beach, a pocket of glassy black sand and a popular resting spot for honu (green sea turtles).

Honu'apo estuary next to Whittington Beach Park.

a beach than a scenic, rocky shoreline lined with ancient fishponds, Whittington is the site of the ancient Hawaiian fishing village of Honu'apo. A wetland favored by shorebirds lies next to the picnic area near the entrance.

The tsunami that accompanied the earthquake of 1868 largely destroyed Honu'apo, but in 1883 plantation workers began loading raw sugar onto barges at Whittington, facilitated by a pier. The landing closed around 1940, when improved highways and trucks provided a better method of transportation. In 1946 another big tsunami seriously damaged the pier, though its remains still stand. Exploration of the rocky shoreline nearby reveals spectacular examples of salt corrosion pockets, or tafoni, in the seaside lava. The adjacent sea cliff exposes layers of stacked pāhoehoe flows and the opening to a small lava tube in cross section.

From Whittington Beach, HI 11 winds upslope toward Mauna Loa's present-day southwest rift zone through the towns of Na'alehu and Wai'ōhinu.

WAI'ŌHINU AND THE 1868 EARTHQUAKE

Residents of Wai'ōhinu, a small town just a mile or so upslope from Na'alehu, reported the most severe shaking of the great earthquake of 1868. The epicenter of the quake is thought to have been very close to here. No seismographic instruments existed then, so we cannot determine the exact magnitude of the earthquake. A recent estimate places it around 8 on the Richter scale—enormous! The quake came a day after a small eruption at Mauna Loa's summit, so magma moving inside the volcano may have played a role in triggering the disaster.

William T. Brigham, the first director of Honolulu's Bishop Museum, gave a chilling eyewitness account of events, which began with a series of foreshocks the day after the eruption and culminated with the main shock at 3:40 p.m. on April 2:

> ***March 27, 1868, about half-past five in the morning, persons on the whale ships at anchor in the harbor of Kawaihae saw a dense cloud of smoke rise from the top of Mauna Loa [40 miles away], in one massive pillar, to a height of several miles, lighted up brilliantly by the glare from the crater Mokuaweoweo [sic]. In a few hours the smoke dispersed, and at night no light was visible . . . .***

The shocks commenced early in the morning; the first was followed at an interval of an hour by a second, and then by others at shorter intervals and with increasing violence, until . . . a very severe shock was felt all through the southwest part of the island. From this time until the 10th of April the earth was in almost constant tremor. . . . It is said that during the early part of April two thousand distinct shocks occurred in Kau [sic], or an average of one hundred and forty or more each day . . . .

Every stone wall, almost every house, in Kau was overturned, [and damages were] done in an instant. A gentleman riding found his horse lying flat under him before he could think of the cause, and persons were thrown to the ground in an equally unexpected manner. (Brigham, 1901, 101).

The main quake evidently accompanied slippage along many faults; the southern flanks of Mauna Loa and Kīlauea suddenly jumped seaward across a broad region, possibly as much as a third of the island. Much of Kīlauea's southern coast sank several feet or more into the ocean during the main shock, causing a tsunami that included at least eight surges as high as 20 feet. These wiped out coastal villages and drowned forty-six people between South Point and Cape Kumukahi. The shaking also opened up the magma plumbing system within Mauna Loa. It erupted again on April 7, this time from vents closer to the ocean in the southwest rift zone just west of Wai'ōhinu. Fast-moving lava issued as fountains of lava 500 to 1,000 feet high from a fresh fissure that you can visit in the Kahuku section of Hawai'i Volcanoes National Park, a few miles west of Wai'ōhinu. At least a hundred head of cattle were killed and local residents were forced to run for their lives. The earthquake also coincided with two small eruptions at Kīlauea, a general collapse in the floor of Kīlauea caldera, and the landslide in Wood Valley. Nothing quite so violent or dynamic has occurred on the Big Island since.

HAWAI'I VOLCANOES NATIONAL PARK, KĪLAUEA SECTION

You can begin exploring Hawai'i Volcanoes National Park via a turnoff from HI 11 between mileposts 28 and 29 (about 1 mile or so south of the national park boundary sign). A short drive past the entrance station leads to the visitor facilities, including Kīlauea Visitor Center, Volcano Art Center, and Volcano House Hotel, perched right at the northeastern rim of Kīlauea caldera. Volcanic activity in 2008 and 2018 closed or damaged much of the road and trail system around the caldera, though miles are still open to the public.

A stunning view of Kīlauea caldera may be seen from the patio and bay windows of Volcano House Hotel. Other viewpoints along the north rim worth visiting are Kīlauea and Uēkahuna (Uwekahuna) Overlooks, just a short drive on Crater Rim Drive from the visitor center. For hundreds of thousands of years, Kīlauea's summit area has been an ever-changing panorama of deep coalescing craters, lava lakes, and overflowing volcanic shields. The most recent history of the caldera is a geologic snapshot of this perpetually changing landscape. In fact, if there is one theme that summarizes this national park overall, it is *change*.

Thick ash deposits revealed in lower cliff faces show that a previous version of Kīlauea caldera existed 1,000 to 2,000 years ago, well before the first Hawaiians arrived. This ancient caldera of unknown dimensions completely filled with lava so that the summit region changed into a landscape of several coalescing, broad lava shields. At that time far-traveling lavas draped the flanks, including the Observatory Flows to the southwest and the ʻAilāʻau Flows to the east, reaching as far as 25 miles to the coast south of Hilo. Toward the close of the fifteenth century these giant summit shields began to collapse piecemeal, leading to formation of the modern caldera. The stepwise process of collapse is shown in multiple arcuate rims and sagging blocks enclosing the caldera. Foundering climaxed in a massive pyroclastic eruption in 1790 that killed an unknown number of Hawaiians. Possibly hundreds of the war party of the chief (aliʻi) Keōua died, and this event had a profound impact on subsequent Hawaiian history, enabling Kamehameha's campaign to unify the Hawaiian Islands.

Following this key event, the first non-Hawaiian to visit and write about Kīlauea was the Englishman Reverend William Ellis in 1823. When he stood on the new caldera rim, he gazed into a basin generally deeper than the one you see today, with restless, gaseous lava lakes and spatter cones active across much of its floor. This volcanic activity was continuous and concentrated toward the southwestern side—opposite today's Volcano House. An inner rim, called the Black Ledge, enclosed the lava lakes, indicating that even before 1823 the caldera had partly refilled with lava and collapsed back down again.

The pattern of partial caldera lava refilling abruptly terminated by large-scale collapse repeated in 1823–1832, 1832–1840, and 1840–1868, with lesser episodes of floor collapse occurring in 1895, 1924, and 1960. During the summer of 2018,

A short-lived warm and somewhat acidic lake in the bottom of Halemaʻumaʻu Crater in 2020. The deepest part of the pit filled with lava in 2020–2022. —Photo by M. Patrick, US Geological Survey

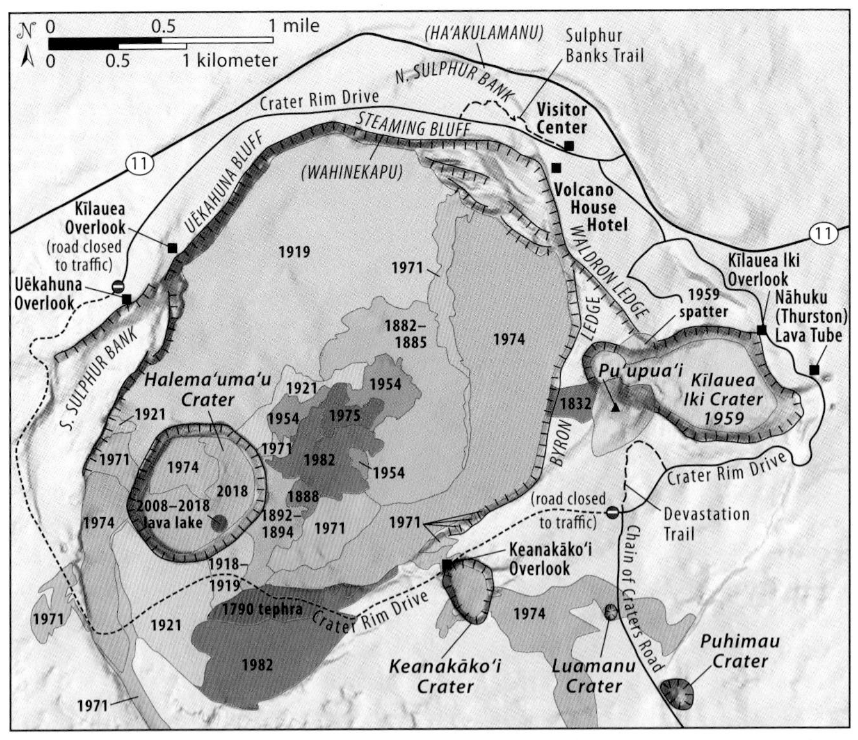

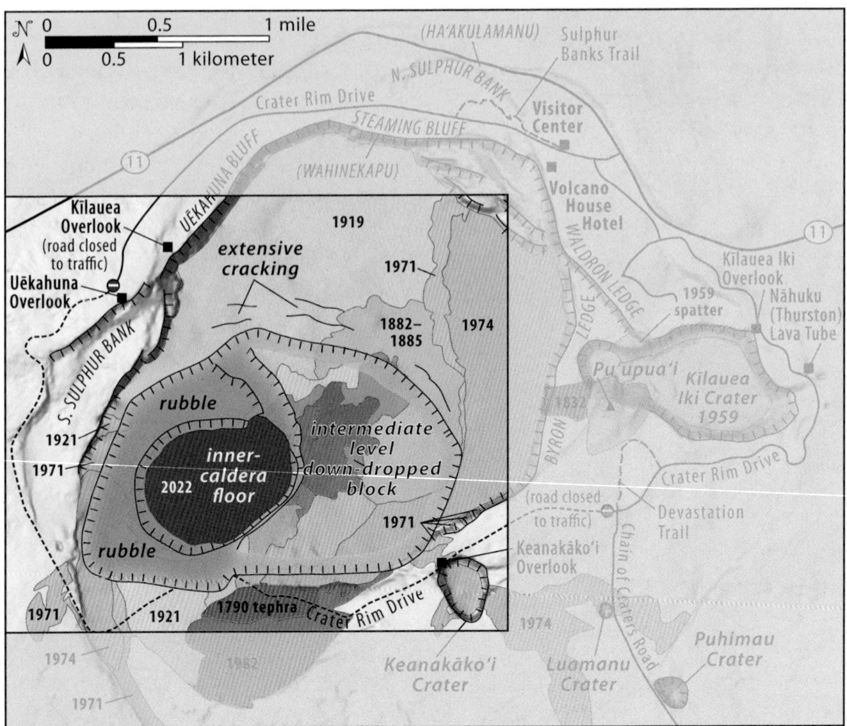

Geologic map of historic flows in the Kīlauea caldera area, before the 2018 eruption and showing conditions as of May 2022. The landscape changes drastically on a human time scale. Dates of historic flows are indicated.

about a third of the caldera floor once again sagged, broke, and sank over a period of two months, reaching a maximum depth of 1,800 feet relative to the preexisting land surface. The collapse was so deep that a small water lake formed at the bottom, the first time in recorded history that this has been observed at Kīlauea. Lava quickly boiled it away in December 2020.

Kīlauea caldera overlies the volcano's active magma reservoir, which is mostly at a depth of 1 to 3 miles beneath the summit. The rate of fresh magma recharge from the mantle source, 25 to 35 miles down, is around 1 cubic mile every decade. As magma pressure builds, liquid lava can erupt. Imagine a long-lasting lava lake as the uppermost portion of a magma reservoir with its roof removed.

Collapses take place when magma is suddenly drawn away, moving into one or both rift zones. Rapid summit drainage often follows earthquakes, as in 1823, 1868, and 2018. The floor of the caldera drops straight down in response, usually along arcuate cracks known as ring fractures that develop over the magma center. The size of the new pit allows a rough calculation of the volume of magma removed.

In a simple model, the amount of lava erupted lower in a rift zone should exactly match the summit collapse volume. But this is not necessarily the case. As much as 40 to 50 percent of the magma moving inside Kīlauea never reaches the surface. It simply remains stored as slowly cooling intrusions, destined to become dikes, sills, and laccoliths—the "skeleton" of the volcano. Occasionally, as in 2018, older pockets of rift zone magma can be intercepted and driven out by fresh batches of molten rock intruding beneath an area.

Today, Kīlauea caldera measures 3 miles long from Volcano House straight across to the opposite wall and is 2 miles wide. The rims nearest the hotel loom as high as 400 feet. The flows on the caldera floor directly below the hotel range in date from 1883 to 1974 and are mostly pāhoehoe. The inner caldera sink on the far side is the center of the most intense degassing and frequent eruption. This depression is the traditional home of the Hawaiian fire goddess, Pele, and named either Halema'uma'u, meaning "(Sacrificial) Fern House," or Halemaumau, meaning "Everlasting House." Early Hawaiians apparently referred to the entire inner caldera as Kaluaopele, "the Pit of Pele."

Light-yellow staining on the far caldera rim above Halema'uma'u is the South Sulphur Bank, first noted in surveys as early as 1842. This persistent area of degassing and sulfur mineralization directly overlies Kīlauea's currently active magma body. The position of this magma body may have shifted a short distance south of the main caldera floor in response to the great 1790 eruption.

Sulphur Banks Trail

Beginning at the western end of the visitor center parking lot near Volcano Art Center, Sulphur Banks Trail leads to Ha'akulamanu, the North Sulphur Bank. The cliff along the right side of the trail is a fault, partially down-dropped during an older phase of caldera collapse. The rotten-egg stench of hydrogen sulfide (H_2S) is sometimes strong here, though it is harmless to casual hikers. Visitors in the nineteenth century used this site as a natural sauna. Look for bright-yellow, needle-like sulfur crystals lining some fumarolic vents.

If this were Yellowstone, you might also expect to see geysers and hot springs, but the water table here is simply too deep to generate those features. Like South Sulphur Bank, North Sulphur Bank might once have at least marginally

The interaction of sulfur gases and water creates a highly acidic environment at Sulphur Banks that breaks rock down to a colorful mélange: white from gypsum and opal, red and orange from oxidized iron, and bright-yellow sulfur crystals.

—Inset photo by T. Elias, US Geological Survey

overlain Kīlauea's summit reservoir. That degassing magma releases both sulfur gases and carbon dioxide, though at different depths. Intensely fractured crust at Ha'akulamanu—revealed by the sheer, hundred-foot-high fault scarp right behind the fumaroles—permits those gases and steam from deeply percolating groundwater to escape. Ground heat is locally so high, as indicated by the scalding temperatures of steam vents, that trees and most other vegetation cannot survive in the meadows. North Sulphur Bank and the nearby steam field have been active for at least the past 180 years and probably much longer.

Steam Vents, Kīlauea Overlook, and Uēkahuna Bluff

Several steaming cracks are easily accessible at the Steam Vents parking area. High heat flow and deep fracturing in this partially down-dropped block of caldera rim boil infiltrating groundwater and create the steam vents and treeless meadows. These cracks do not extend as deeply as those at Sulphur Banks and do not carry any volcanic gases.

A trail leads to the caldera rim at Wahinekapu, the rim of the "Wailing Woman" (or chanting priestess). In addition to an outstanding view of the caldera, active steam vents here parallel the rim, hence the Anglicized name, Steaming Bluff.

Though not as apparent except when it is cool and cloudy, vents also steam on the caldera floor below.

No eruptions have broken out on Kīlauea's north rim for centuries. However, the presence of a forested spatter rampart at the eastern edge of nearby Kīlauea Military Camp shows that eruptive fissures have opened in recent geologic times in what are now developed areas. Southward shift of the magma reservoir during formation of the modern caldera may have cut off the connections enabling volcanic activity on the north side of the caldera—for the time being.

The Kīlauea Overlook and Uēkahuna Bluff provide excellent views of the caldera and the inner Halema'uma'u collapse crater of 2018. A walk along the Crater Rim Trail between the two overlooks takes you through deposits from the 1790 explosive eruptions, ranging from ash to large boulders, and over several small fault scarps. Uēkahuna is the highest point on Kīlauea and has the best views toward the southwest rift zone and Mauna Loa. The remains of the old Hawaiian Volcano Observatory and Jaggar Museum—badly damaged in 2018—still sit on the rim of the caldera at the time of writing (2022).

CRATER RIM DRIVE
VISITOR CENTER—CHAIN OF CRATERS ROAD

This 3.2-mile drive, which heads south from the park entrance station, leads past Kīlauea Iki Overlook, Nāhuku (Thurston) Lava Tube, and the Devastated Area produced by the 1959 eruption. The view from Kīlauea Iki Overlook on a clear day is one of Hawai'i's celebrated panoramas. The viewpoint allows you to peer into the bottom of Kīlauea Iki ("Little Kīlauea"), a crusted-over lava lake that formed in November 1959 about 400 feet below the overlook. A straight foot trail, paler than the lava it crosses, traverses the crater floor and is one of the most popular hikes in the national park. Look for hikers to get a sense of scale for this huge landscape.

Like other parts of the caldera, Kīlauea Iki was an actively growing lava shield in the fifteenth century. The shield remained continuously active for at least sixty years. Its lava flows of the 'Ailā'au series included formation of one of the world's longest lava tubes, Kazumura Cave, with over 60 miles of mapped passages to date. Nāhuku Lava Tube lies near the uppermost entrance to Kazumura Cave but does not appear to connect to Kazumura.

Kīlauea Iki Crater began forming with the general onset of caldera collapse beginning 500 years ago, in two or three stages. The main, deeper, and larger pit containing the 1959 lava lake drops directly beneath the overlook. A western, smaller pit, adjacent to the buff-colored cinder and pumice cone on the far west wall, is somewhat shallower and largely filled with dark 'a'ā. Together the compound, overlapping pits stretch east–west more than 1 mile, with a maximum width of more than a half mile. A narrow, elevated strip of forest-covered land, Byron Ledge or Uēaloha, only a few hundred feet across, separates Kīlauea Iki from the main caldera floor on the other side, with Mauna Loa arching gracefully across the horizon in the background.

When Reverend Ellis visited in 1823, Kīlauea was around twice as deep as it is today, with a small pad of fresh lava at the bottom. Eruptions in 1832, 1868, and 1877 poured more lava into the crater, mostly pooling on the deep eastern floor, although

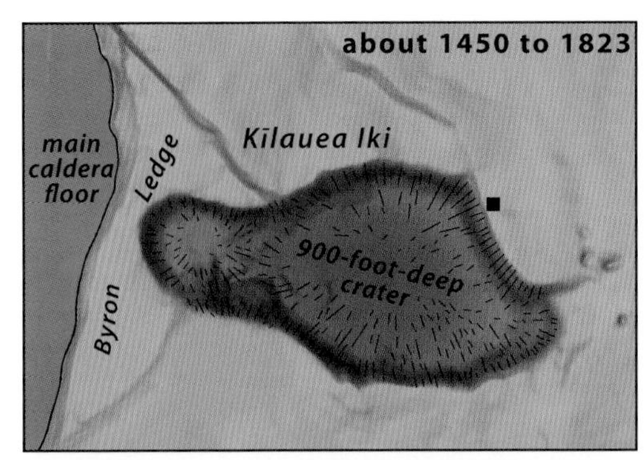

Eruptions have gradually refilled the pit of Kīlauea Iki since it was first observed by Western visitors in 1823.

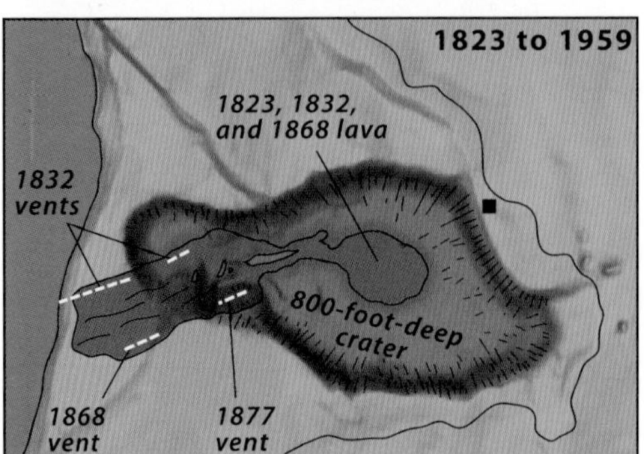

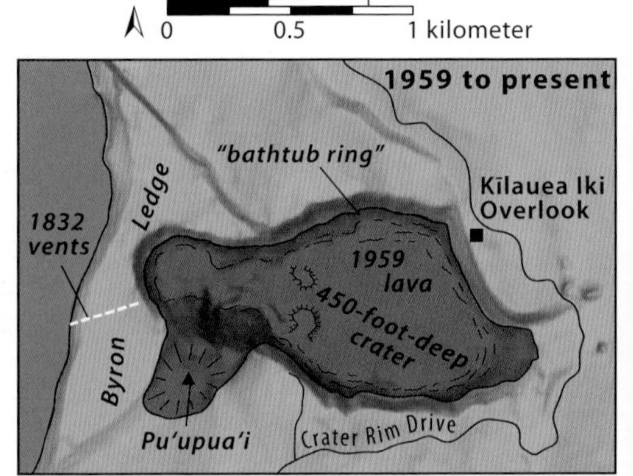

View into Kīlauea Iki Crater from Kīlauea Iki Overlook. Mauna Loa looms on the horizon. Inset is fountaining from Pu'upua'i into the lava lake in 1959. —Photo courtesy US Geological Survey

it remained about 400 feet deeper than it is now. The 1959 eruption—known for producing the highest lava fountains recorded in Hawai'i—broke out on the far southwestern wall of Kīlauea Iki. As lava streamed into the eastern pit directly beneath the overlook, fountaining and spattering concentrated around a single opening, gradually building up a mound of fresh tephra downwind. This grew into Pu'upua'i, the prominent 350-foot-tall cone perched on the rim. Large cracks and slumps developed on the steep, layered face, mostly sliding when the deposits were still partly molten. Hot gases and steam filtered through the tephra, giving the fallout deposits a pale orange-brown color. The central vent area is a deeper red color from interaction between lava and oxygen at higher temperatures.

The accumulating lava lake filled to a maximum depth of around 450 feet, burying forever the 1868 and older crater floors beneath. The eruption was episodic: towering lava fountains spewed up, filling the crater with molten rock until the rising lava level drowned the vent and choked off the fountains. After a short period of rebuilding pressure, the vent would erupt again—seventeen episodes in total. As activity finally waned, lava partly drained back into the vent, leaving a veneer of basalt plastered around the edges of the receding lake—a "bathtub ring" about 50 feet tall.

After the end of the 1959 eruption, scientists converged on the fresh lava lake to study it as it slowly cooled and hardened. Nothing like this had ever been studied before. Eight years after the eruption, the hard crust of lava across the lake had thickened to 90 feet, and by 1975 it was about 180 feet thick. It is fully solid today, though still hot enough inside to cause steaming of infiltrating groundwater in many places. Orange and white patches on the floor mark cracks where hot steam and gases oxidized the iron in the rocks and deposited gypsum. A walk to the bottom and back takes most people about an hour and a half. The lava is noteworthy for containing great numbers of large green olivine crystals—a rarity at Kīlauea's summit.

Past the Kīlauea Iki Overlook, Crater Rim Drive passes the entrance to Nāhuku (Thurston) Lava Tube. About 400 feet of the 1,500-foot-long cave passage is open. It is artificially lit, though a flashlight is helpful for seeing details, and the path is well graded. Entry into the cave is through the wall of a small, forested collapse pit—one of three hidden craters in the immediate vicinity. This lava tube formed from a lava river during the 'Ailā'au eruption about 500 years ago. The intense heat above the downcutting lava stream caused the tube walls to partly melt and bubble out, drip, or ooze down the walls. Near the entrance, a thin layer of this melt coated the walls—even puddling at the base—and dripped from the ceiling. The name Nāhuku ("the protuberances") probably refers to these stalactites and/or the small, hollow buds of partial melt protruding from the walls deeper in the cave. The exit climbs out a natural skylight, revealing the thickness of the tube roof.

Crater Rim Drive winds down past the Pu'upua'i turnoff, which leads to the eastern base of the young tephra cone on the southwest rim of Kīlauea Iki. Devastation Trail, which crosses the area of forest destroyed by falling tephra from the 1959 lava fountains, begins and ends on remnants of the old, now-buried Crater Rim Drive. Look for tree molds and spatter along the way, and pick up a handful of tephra to look for spongy, glassy pieces of pumice, Pele's tears, and loose olivine crystals. The gentle pathway, only a couple thousand feet long, leads to the Devastation Trail parking area. You can also reach the parking area by driving a short distance farther down Crater Rim Drive.

The entrance to Nāhuku (Thurston) Lava Tube. Note the texture of the walls where partial melt oozed down the walls while they were still hot.

From the Devastation Trail parking area, you can walk along a 2-mile round-trip segment of Crater Rim Drive, now open only to foot traffic, which leads to Keanakāko'i Crater. As the road approaches Keanakāko'i, you may see large cracks in the pavement related to caldera collapse in 2018. You also pass by a series of spatter ramparts, fissures, and lava flows that erupted along the southern caldera rim in July 1974. These shelly pāhoehoe flows thinly coat a landscape mostly covered in fallout debris from previous eruptions: larger blocks from explosions in 1790 and finer cinder and Pele's tears from the 1959 Kīlauea Iki high fountains directly upwind.

Overlooks north of the drive provide the best views of the chaotically dropped southern caldera area, where eruptions in 2020–2022 filled the newly collapsed Halema'uma'u pit with more than 1,000 feet of lava so far. Look for one block along the southern (left) edge of the deep inner pit with a straight band and lane divider crossing it. This part of the former Crater Rim Drive fell into the pit during the 2018 collapse. It now lies 400 feet below its original position. On the caldera wall immediately to the west of the overlooks, the yellow-crusted outcrops of South Sulphur Bank stand out brightly; you are standing almost directly above Kīlauea's magma reservoir here. The continuous release of sulfurous gases from the reservoir accounts for the buildup of sulfur.

Keanakāko'i Crater measures around 1,000 feet across and 200 feet deep. No one knows when it formed, but it is probably only a few hundred years old. Cascades of lava pouring over the crater wall during the day-and-a-half-long July 1974 eruption left dark stains on the far crater walls. The eruptive fissure opened in the forest to the south and partly crossed the crater floor at the base of the southern wall.

Keanakāko'i means "cave of the adzes" (an adze is an ax-like cutting tool). Inside the crater was a quarry prized by Hawaiian stone-carvers for its dense, smooth rock used to make chopping tools. Lava flows in possibly 1877 and in 1974 buried the quarry site.

Look for roadcuts just downslope and across the street from Keanakāko'i Overlook. These show layers of ash and gravel laid down during the mammoth 1790 eruption and preceding explosions. You may spot a block sag or two here—places where large, blasted rock fragments dropped into soft layers of ash, depressing them on impact. Later ash falls preserved the partly embedded rocks and their related sag structures.

CHAIN OF CRATERS ROAD

15 miles

This road guide describes some of the many interesting features you'll encounter as you drive along Chain of Craters Road from the summit to the sea. Zero your odometer at the intersection of Crater Rim Drive and Chain of Craters Road across the street from the Devastation Trail parking area. The first 3.2 miles pass one pit crater after another: Luamanu, Puhimau, Ko'oko'olau, Hi'iaka, and Pauahi. Still other craters mark another 6 miles of the east rift zone. Most of the others are accessible via the Nāpau Trail, which begins from the Mauna Ulu parking area around Mile 3.6.

Lava trees, where lava flowed around tree trunks and then receded, record the maximum height of the July 19, 1974, lava flows.

At mile 0.4, Chain of Craters Road crosses a branch of the lava flow erupted on July 19, 1974. The flow partly filled a shallow, 250-foot-wide pit crater named Luamanu, "Pit of the Bird," perhaps in reference to the white-tailed tropicbird, koaʻe kea, that often nests in crater walls. The rough surface of the 1974 flow may be explored on foot to discover fissures, spatter ramparts, lava trees, and tree molds left by the lava as it poured through the forest. In some tree molds, charcoal from burned vegetation still remains, and the bleached, partly charred wood of trees toppled by the lava still litters the flow surface in many places. A piece of stone wall and some rusty railing are all that remain of the old Luamanu overlook, buried as the last fissure ripped through the road. Amazingly, all the lava here erupted in less than 4 hours.

At mile 0.8, a short distance past Luamanu, the road ascends a short slope, which is the easternmost fault scarp related to Kīlauea caldera. It is largely obscured by dense forest. On cool, cloudy days the scarp can be seen steaming gently.

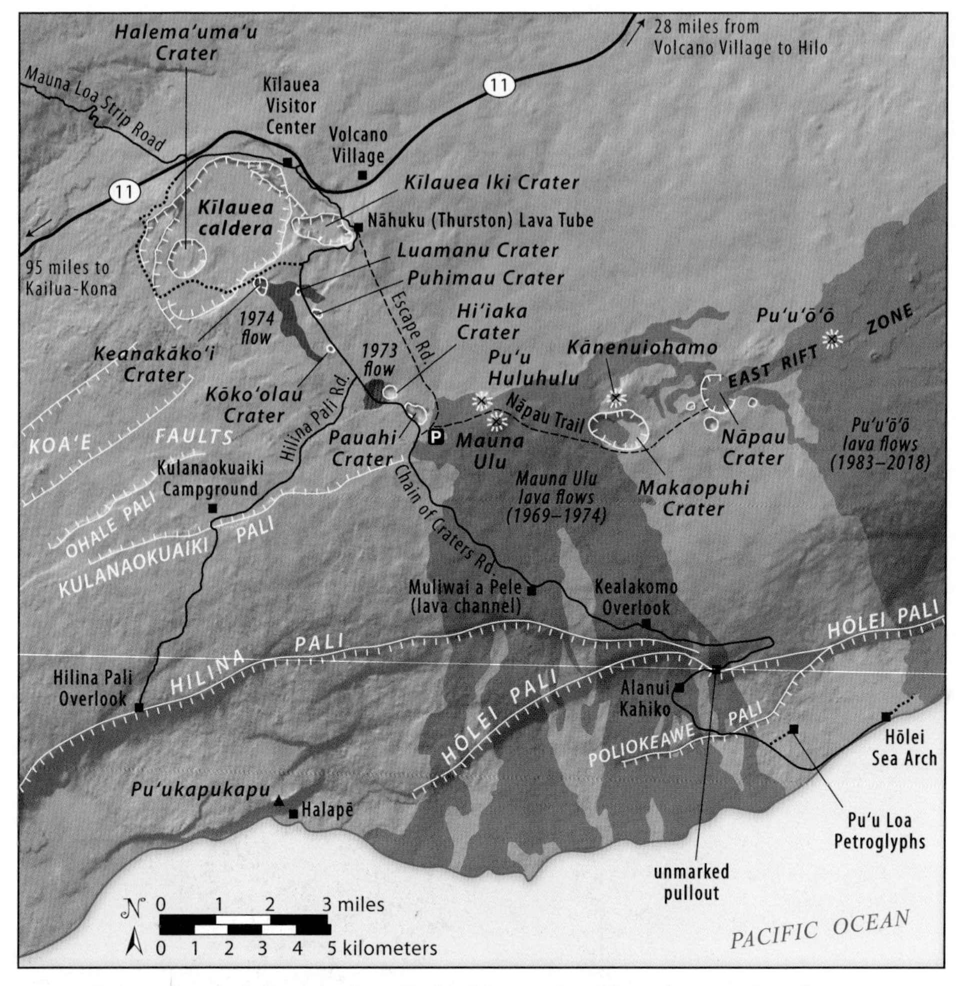

Points of geologic interest along Chain of Craters Road from the summit to the sea.

HILINA PALI ROAD

At mile 2.2 on Chain of Craters Road, Hilina Pali Road leads 8.25 miles to a lookout at the crest of Hilina Pali, one of the largest fault scarps on the slumping south flank of Kīlauea. Zero your odometer at the intersection with Chain of Craters Road. The road is well paved but narrow and winding, so it will take longer to explore than you might estimate from the distance on a map. Multiple trails from Hilina Pali Road lead to the south coast and into the Ka'ū Desert to the north, allowing greater exploration of the Kīlauea backcountry.

The first 3.6 miles cross a spectacular series of normal fault scarps related to the Koa'e fault system (see figure on page 194). Like the large coastal palis, the Koa'e faults are a product of the mass seaward movement of Kīlauea's flank but with a twist: they result from the prying open of the volcano in response to countless intrusions of magma in the rift zone to the east. The many faults in the system are splayed like a great, 10-mile-long fan on the map, opening to the east and converging or dying out to the west, where they merge with the Kamakai'a Hills, a remote chain of cinder cones and other vents adjoining the southwest rift zone.

The best place to explore the Koa'e faults is around Kulanaokuaiki Campground at mile 3.4. Just west of the campground, the road climbs 30 feet up and over Kulanaokuaiki Pali. The young face of the fault scarp includes toppled blocks, the tilted trees growing or barely rooted in them evidence of recent fault movements. Where the road crosses the pali, the scarp transitions into a monocline—an arching mass of lava broken by a thrust fault at its base. This structure suggests that the Kulanaokuaiki fault plane is listric; that is, it curves underground to the north and inclines at a gentler angle at depth. The highly localized thrusting results from shallow buckling of the crust above a curving shear plane. The name *Kulanaokuaiki* means "the shaking of a small spine."

Kulanaokuaiki Pali, a 30- to 50-foot-high fault scarp along Hilina Pali Road, is a (mostly) normal fault within the Koa'e fault system.

During the 2018 eruption, the Kulanaokuaiki fault slipped approximately 7 inches over a period of a few months and ruptured the road. A short walk north of the campground to the prominent tumulus surmounted by a metal surveying post takes you to the rim of another large fault scarp, Ohale Pali, which has also dropped the crust on its northern side. The youngest lavas in the area, all broken by ground cracks and fault scarps, are around 600 to 550 years old. But the Koa'e faults were certainly active before this time, too.

Beyond Kulanaokuaiki, Hilina Pali Road winds through the lush, 1,500- to 1,200-year-old savannah of Kīpukanēnē. Next you cross younger pāhoehoe flows erupted as recently as the fourteenth century, featuring numerous tumuli, ropy surfaces, and entrail forms. Around 6 miles in, the road enters another large kīpuka similar in age to Kīpukanēnē but with an even longer name—Kīpukakeanabihopa. The name means "Bishop's cave kīpuka," in reference to a nearby large lava tube cave undeveloped for visitor access.

At road's end, enjoy a breathtaking view from the rim of 1,150-foot-high Hilina Pali to the wide coastal plain below. You can easily see the South Point area of the Big Island from here on a clear day, together with the lower southwest rift zone of Mauna Loa. The oldest known lavas exposed at the surface of Kīlauea—20,000 to perhaps more than 100,000 years old—crop out in the face of Hilina Pali and the nearby fault knob, Pu'ukapukapu. The Koa'e faults, upslope, act as natural barriers sheltering this flank of the volcano from fresh coverings of lava.

Puhimau Crater, at mile 0.9, is considerably deeper than Luamanu—530 feet. The youngest rim-forming flow cut by the crater is 650 to 700 years old, indicating that collapse took place sometime later. There is no evidence that molten lava broke out when the crater formed. A large crack slices the far northeastern wall, and steam may occasionally be seen rising from this opening, probably heated by a more recent intrusion of magma. Puhimau tellingly means "ever smoking" in Hawaiian. At its greatest width, Puhimau is nearly 1,000 feet across.

Just past the Puhimau turnout at mile 1.2, Chain of Craters Road passes a large meadow of steam vents, the Puhimau thermal area. Prior to 1936, this area was simply thick woods similar to the surrounding forest. The soil and shallow groundwater grew hot as fresh magma intruded the area, killing the trees from their roots. The area of forest dieback and meadow growth has expanded ever since, finally crossing the road in 2000–2010. Wahinekapu, near Kīlauea Visitor Center, is a similar, though more stable, thermal area. Another hot spot like this also existed at the site of the 1969 eruption that led to the formation of Mauna Ulu lava shield. Some geologists believe that the Puhimau thermal area could one day erupt or collapse to form a new pit crater.

Ko'oko'olau Crater at mile 1.5, a pit now deeply wooded throughout, resulted from two coalescing collapses taking place sometime between 700 to 650 years ago and 350 to 300 years ago. It is only about 400 feet wide. The crater is named for a flowering native plant that early Hawaiians brewed into a tea.

At mile 2.4, Chain of Craters Road cuts through the side of an oxidized spatter cone marking the site of an eruption that took place here more than 500 to 600 years ago.

At mile 2.6, the road crosses a narrow pāhoehoe flow that erupted on May 5, 1973, from a 0.5-mile-long string of vents both to the west and east of the highway. The

Hourglass-shaped Pauahi Crater is floored by 1979 lava, but has a "bathtub ring" from the withdrawal of a 1973 lava lake. Small 1979 spatter vents on the crater floor are just visible at the bottom of the photo. The Mauna Ulu lava shield is at the top left.

lava spilled into 1,000-foot-wide Hi'iaka Crater, visible on the floor of the collapse basin through the low trees to the left (east) of the road. A day-long eruption in August 1968 also added lava to the floor of Hi'iaka, though most of this rapidly disappeared back into a fissure in its floor. In the Hawaiian tradition, Hi'iaka is the youngest sister of the fire goddess, Pele.

Pauahi Crater at mile 3.2, largest of the Chain of Craters accessible by road, is a composite double pit 1,650 feet long and 360 feet deep. In May and November of 1973 and again in November 1979, fissures opened and erupted lava on both the northern rim and at the bottom of Pauahi. A low November 1973 spatter rampart lies just to the left (west) of the short pathway from the parking area to the overlook. The associated vents can be explored easily by strolling around to the other side of the cones. Another 1973 fissure ascends the far northeastern wall of the crater. More impressive still is the 1979 fissure-spatter rampart to the west across the road from the parking lot. The restless activity in and around Pauahi underscores the meaning of its name, "destroyed by fire." This volcanic area is one of the most active on Kīlauea.

Each of the 1973 and 1979 eruptions essentially lasted for less than a day—brief spasms. The November 1973 eruption within Pauahi Crater, however, did linger sluggishly for almost a month within the pit directly below the overlook. The lava initially accumulated in two deep circulating pools, one on each side of Pauahi Crater's floor. The molten rock then drained back underground as activity waned. The remaining bathtub ring on the lower crater walls is the tallest presently displayed at Kīlauea. Rockfalls from subsequent earthquakes have partly covered the wide ring. Most of the visible floor is 1979 lava.

The presence of two prominent hills on the horizon, visible from the overlook, attests to the frequency of eruptions in this area. The rugged, tree-covered hill to the left is Pu'u Huluhulu, a cone composed of mostly welded spatter formed sometime between 600 and 400 years ago. To the right is a gently sloping, half-mile-wide, 300-foot-tall lava shield called Mauna Ulu, or "growing mountain." Mauna Ulu grew over a period of almost continuous lava fountaining and vent overflows between 1969 and 1974. Up until then no historically documented flank eruption of Kīlauea had lasted so long. But just a few years later an even longer-lasting outbreak started at Pu'u'ō'ō, 7 miles to the east. Lava from Mauna Ulu completely filled two pre-existing pit craters, Aloi and Alae, few traces of which remain today. These now-buried craters were the next ones in the Chain after Pauahi.

Mauna Ulu and Nāpau Trail

At mile 3.7, a short spur road leads east to the Mauna Ulu parking area. The spur road is a remnant of the original Chain of Craters Road from before 1969. You can walk to where the old pavement disappears beneath lava just past the parking lot—a stark reminder that come what may, natural forces ultimately prevail.

Fissure vents from 1969 show where lava drained back into the vent, surrounded by smooth-coated boulders that were once trees. In the background are large spatter mounds; the horizontal ledge in the upper right corner shows the maximum height of the shelly pāhoehoe flow that emanated from the fissures.

An easy, quarter-mile-long loop trail leading from near the road's end guides you over 1973 pāhoehoe flows from fissures to the north, past a 1974 ʻaʻā flow from Mauna Ulu, to a spatter rampart and fissure that formed in May 1969. Large lava trees—many topped with spatter into a mushroom shape—stand above the surface of the 1969 lava flow nearby. Shortly after the May eruption began, activity shifted eastward a half mile, focusing in the older thermal area between now-buried Aloi and Alae Craters. Here the new Mauna Ulu grew over the next five years. During shield growth, lava fountaining periodically became very strong, up to 1,770 feet high in September 1969. Falling pumice and Pele's hair and tears from the high lava spray carpeted the landscape for miles downwind. Fallout deposits cover much of the earlier 1969 flow as well, making it sound soft and crunchy underfoot.

Another trail leading from near road's end, the Nāpau Trail, leads to Puʻu Huluhulu and the base of Mauna Ulu. The 2-mile round-trip hike to the summit of Puʻu Huluhulu is certainly worth it, particularly on a clear day. Along the way you'll walk mostly on 1973 shelly pāhoehoe that erupted from fissures to the north and pass by some nice specimens of lava trees. Near the end, the trail ascends pāhoehoe overflows from Mauna Ulu, then up the steep, forested cone. From the top of Puʻu Huluhulu, you'll have an outstanding view of the summit area and east rift zone, including other lava shields, cones, pit craters, steaming areas, and flows extending for miles around. Lava streaming off the growing shield gathered in the saddle between Puʻu Huluhulu and Mauna Ulu. This lava pooled and developed levees to form a 500-foot-wide perched lava pond. The channel that fed fresh lava to this

A perched lava pond, one of the best examples of this type of feature in the world, formed as pāhoehoe pooled in the saddle between Mauna Ulu lava shield and Pu'u Huluhulu cinder cone, from which this picture was taken. (Don't confuse this locality with the other Pu'u Huluhulu in the Humu'ula Saddle area.)

extraordinary feature can be plainly seen stretching down from the summit of Mauna Ulu to the edge of the pond. Lava within the pond largely drained away through tubes penetrating the surrounding levees, dropping the level of the lava by as much as 10 feet as activity waned.

Past Pu'u Huluhulu, the Nāpau Trail continues 5.5 miles to the rim of 0.75-mile-wide Nāpau Crater, with its outstanding view of Pu'u'ō'ō lava shield another 2.5 miles to the east. Eruptions in 1965, 1968, 1972, 1983, and 1997 all added fresh lava to the broad, flat floor of Nāpau. Translation of the crater name is "the endings," perhaps because this is the last of the big pit craters in the Chain of Craters reaching downslope from the summit.

On the way to this remote destination, the trail crosses the summit of Alae lava shield, a parasitic vent related to Mauna Ulu where lava filled and circulated within former Alae pit crater. Other sights along the trail include Makaopuhi Crater, the largest pit crater in the east rift zone; the forested 550-year-old Kānenuiohamo lava shield; a dense 'ōhi'a tree-fern forest; and ancient spatter ramparts of unknown age. More information about this challenging but worthwhile day hike is available at Kīlauea Visitor Center. A primitive backcountry campground is set up near the rim of Nāpau Crater for adventuresome hikers who wish to spend more time savoring this wild volcanic wilderness.

Mauna Ulu to Coast

For those of you who drove to the Mauna Ulu parking area, reset your odometer to zero. If you did not turn off to visit the Mauna Ulu area, continue with the mileages in parentheses. The Chain of Craters Road leaves the craters and descends through the vast Mauna Ulu flow field that formed from 1969 to 1974. Pullouts along the way overlook gigantic fault scarps caused by seaward slumping of the volcano's southern flank. The road switchbacks across these scarps, which are draped by spectacular cascades of 'a'ā and pāhoehoe. Finally reaching the coast, the highway ends at a sea arch next to the Pu'u'ō'ō lava flow field, formed between 1983 and 2018.

Between mile 0.2 and 0.4 (3.9–4.1), Chain of Craters Road crosses the May 1969 lava flow from the opening vent of the Mauna Ulu eruption. Pillars and mounds of crumpled lava are lava trees formed where the molten rock swept around living trees and hardened before the rest of the lava drained away. A thin layer of windblown cinder from high fountain eruptions on Mauna Ulu carpets the flow. On the far side of the flow, the road passes through a large roadcut marking the trace of the up-thrown block of the southernmost fault in the Koa'e fault system, Kulanaokuaiki Pali.

Between miles 2.5 and 4.2 (6.2–7.9), Chain of Craters Road winds gently downhill and begins crossing Mauna Ulu flows near the Mau Loa o Mauna Ulu pullout. The vast expanse—known as a flow field—gives a good sense of just how much lava poured out of Mauna Ulu over five years. More than 10.5 square miles of forest and grassland vanished beneath fresh lava. Flows also buried 12 miles of the original Chain of Craters Road and reached as far as 7.5 miles, all the way to the ocean.

A pullout and viewpoint for Muliwai a Pele, "the river of Pele," is at mile 3.7 (7.4). A short walk from the parking area here takes you to the rim of a well-developed, 5-mile-long lava channel in a Mauna Ulu 'a'ā flow, formed toward the end of the eruption in 1974. The lava pouring through the channel was quite fluid at its peak. Surges of this fluid lava flushed large lava-coated blocks—breakdown rubble from channel walls upslope—out of the channel to leave them stranded on the enclosing levees. These are called accretionary lava balls, and they may travel for miles bobbing down a channel before being either disgorged as seen here or swept out across the distant terminal area of a flow. As the flow waned and the lava cooled and became stiffer, it took on a rubbly ʻaʻā form in the channel once again. The roadcut exposes thin layers of lava overflows that built the channel levees.

Kealakomo at mile 6.1 (9.8), a viewpoint picnic terrace, overlooks a tremendous landscape of Kīlauea's south flank fault scarps and the lava-covered coastal platform more than 1,000 feet below. Kealakomo, which means "the entrance path," lies just upslope from the top of Hōlei Pali, one of the largest and most active normal fault escarpments in the world. Gray, fresh-appearing Mauna Ulu lava flows contrast sharply with older grassy terrains in the landscape far below. ʻAʻā stands out as the darkest lava in the flow field, which is dominantly pāhoehoe. The narrow, scarp-bounded knob to the west is Puʻukapukapu, which looms 1,000 feet above Halapē, a seaside campground where campers were washed inland by a tsunami following a magnitude 7.5 earthquake in 1975. The name Halapē translates as "crushed" or "missing." In 1868, an even larger earthquake precipitated a tsunami that destroyed a number of Hawaiian settlements along this coast. During that event, the south flanks of both Mauna Loa and Kīlauea slipped tens of feet seaward.

Huge normal fault scarps—palis—dominate the south flank of Kīlauea. View from near Kealakomo toward South Point and Mauna Loa on the horizon. Lava flows pour over the palis and gradually build out the flat coastal plain.

Chain of Craters Road makes a big hairpin turn as it descends Hōlei Pali at mile 8.5 (12.2). The walls of the hairpin beautifully expose stacked pāhoehoe flows and small lava tubes in cross section. Many of the pāhoehoe lobes drained to form small pockets and voids in the lava as it hardened.

Chaotically mixed ʻaʻā and pāhoehoe cascading over Hōlei Pali in 1972 produced the entrail-form pāhoehoe well displayed at an unmarked pullout (mile 9.8 or 13.5) as the road descends to the coastal flatland. On steep slopes, gravity draws the toes of slowly extruding, active pāhoehoe into the long "entrails" so characteristic of this lava. These entrails cool independently and are unable to merge into larger lobes the way pāhoehoe toes would on flat ground. At higher volumes the lava forms ʻaʻā crust.

A pullout at mile 10.3 (14.0), signed Alanui Kahiko, provides one of the best views of the Mauna Ulu flows draping Hōlei Pali. Alanui Kahiko means "old highway," named for the patch of original Chain of Craters Road surrounded by lava just a short distance downslope. Look inland to see a spectacular mix of ʻaʻā and pāhoehoe on the face of Hōlei Pali, a result of multiple pulses of Mauna Ulu lava flows pouring down the steep slope. The steepness caused the larger, faster-moving flows to rip and tear into ʻaʻā. Smaller, slow-moving flows or the liquid cores of stagnated ʻaʻā flows oozed out in places to mantle rough, dark clinker with fresh, smooth pāhoehoe, most of it in entrail form. This complex interfingering of flow types contrasts strikingly with the lava at the foot of the pali where the lava is almost all pāhoehoe. Tearing forces no longer acted on the lava as it poured out across the gentle coastal plain. If the lighting is right, you may see some dark cracks in the slope near the crest of Hōlei Pali, results of the latest faulting to have occurred here, mostly within the past few decades.

The road descends the lowermost of the fault scarps, Paliuli, as it approaches the coast at mile 11.8 (15.5). A 1.8-mile round-trip walk not far from the base of Paliuli at mile 12.6 (16.3) leads to an archaeological site with over 23,000 petroglyphs. The artwork, at a place called Puʻu Loa (or Puʻuloa, "long hill"), is carved across a broad area of hummocky, prehistoric pāhoehoe, a short distance east of the Mauna Ulu flow field. Scattered large tumuli stand out in the flow surface along the trail to the petroglyphs. Also note the position of the steep slope in the lava directly uphill from the rupture marking Paliuli. This slope is an older fault scarp, buried by younger flows that may have broken in turn due to more recent faulting.

The road ends at Hōlei Sea Arch on the shoreline near the western edge of the big Puʻuʻōʻō lava field. A short walk from the road's end takes you to a viewpoint of the precariously thin 90-foot-high arch carved by countless storm surges and wave action, probably during the past century. The surface lava in the area erupted from the Kīlauea east rift zone sometime between 800 and 500 years ago.

The closed section of road leads to edge of the Puʻuʻōʻō lava. Like that of Mauna Ulu, this flow field is mostly pāhoehoe where it reaches the coast. Lateral spreading of lava as new land builds out from the shore accounts for the relative flatness of the coastal plain. Puʻuʻōʻō and its satellite vents were active from 1983 to 2018—the longest flank eruption in Kīlauea's recorded history.

Puʻuʻōʻō lava first reached the shore nearby in 1986. Over the next 32 years, these flows provided spectacular views of lava spilling into the ocean at various points along 11 miles of shoreline. But these flows took their toll, destroying hundreds

Rugged coastline around Hōlei Sea Arch. Lava flows from Pu'u'ō'ō are visible on the slopes in the distance.

of homes and burying the famous thirteenth-century temple, Waha'ula Heiau, and countless other archaeological sites, a park visitor center and popular campground, plus 4 miles of the historic Kalapana Trail. The source vents at Pu'u'ō'ō and nearby Kūpaianaha lie out of sight atop the pali, 6 miles away. At the end of this long eruption, the summit of Pu'u'ō'ō—a slowly growing, coalescing complex of cinder and spatter cones, lava flows, and lava shields—abruptly collapsed into a new pit crater 1,500 feet deep. Its magma supply drained rapidly down into the rift zone to feed the 2018 eruption in the lower Puna District.

GLOSSARY

ʻaʻā. A very rough, rubbly type of lava difficult to walk across. ʻAʻā flows have a massive interior sandwiched between layers of rubble.

accretionary lapilli (pisolites). Rounded pellets of hardened ash embedded in tuff layers. They range in size up to small marbles and form when ash falls through clouds of steam during volcanic eruptions. Static electricity in the eruption cloud may play a role in their formation.

alkalic basalt. Basalt with a high content of sodium (island areas such as Hawaiʻi) or potassium (continental areas). Sodium and potassium are the alkali elements in geology.

amphibole (hornblende). An iron-magnesium silicate mineral, usually black with an elongated or needle-like shape. Rare in Hawaiian rocks.

andisols. Fertile soils, excellent for agriculture, produced from the weathering of volcanic rocks.

ankaramite. A type of alkalic basalt rich in black pyroxene and green olivine crystals, sometimes with plagioclase feldspars.

ash (volcanic ash). Fine-grained particles of lava, the consistency of fine sand or even dust, explosively erupted with great force from a vent.

ash cone. Low, broad cone with a wide, flat floor, and a rim composed primarily of volcanic ash. A product of powerful steam blast eruptions generally nearshore or in an area of shallow groundwater. If the loose ash is hardened into rock, the feature is called a **tuff cone**.

basalt. Dark-gray lava rock with high concentrations of iron and magnesium and a low silica content compared to other volcanic rocks. Basalt is the principal rock type in the Hawaiian Islands.

beach cusps. Points and embayments of sand lined up along a beach at the shoreline. They are produced by the swash of waves repeatedly running up onto the beach at an angle, then washing directly back out again.

beach rock. Ancient beach sands, generally calcareous, hardened into solid rock.

benmoreite. A rare type of alkalic basalt, generally light gray in color and containing elongate, black amphibole (hornblende) crystals.

blocks. Jagged pieces of rock thrown out in explosive volcanic eruptions. The pieces are cold and hard when ejected. They may be many feet in diameter but generally are no bigger than a football.

blowhole. A geyser of seawater blown through a hole in the surface of a terrace when a wave rushes into an opening in the rock at waterline, driving air ahead of it. The compressed air forces eruption of the geyser as it escapes through the nearby hole atop the terrace.

bombs (volcanic bombs). Large blobs of molten or semimolten lava thrown from a vent during more explosive eruptions. They may be smooth sided, show stretching features, and contain widely scattered vesicles.

calcareous. Material composed of calcium carbonate ($CaCO_3$). Examples include coral or beach sand derived from the erosion of coral reefs, and limestone.

caldera. A large volcanic crater exceeding about 1 mile in diameter. It forms by collapse above a magma chamber generally accompanied by a large volcanic eruption.

cinder (scoria). The fine bubbly pieces of lava thrown out by cinder cones during explosive eruptions. The pieces rarely exceed golf-ball size, have numerous vesicles, and may be oxidized, commonly giving a cinder cone an overall reddish-brown coloration. Cinder fragments are usually already solid when they hit the ground, so they won't weld together the way spatter does.

cinder cone. A steep-sided volcanic cone with a cup-shaped crater formed during an explosive basaltic eruption. A cinder cone is typically only a few hundred feet high and forms where a vent opens on the flank of a much larger volcano, though many cinder cones (mostly in continental areas) may be stand-alone features. Strong wind blowing during an eruption may pile up the cinders to one side of a vent, giving the resulting cone a horseshoe shape, viewed from the air, with the high side downwind. Lava flowing from a vent can produce a similar U-shaped rim because cinder that would ordinarily build a rim is carried away on the back of the flow instead.

clinkers. Jagged lava rubble. The rubbly pieces making up the surface of an ʻaʻā flow are called clinkers.

coarse. A coarse-grained rock has crystals that are mostly more than 5 millimeters (0.2 inch) in diameter.

columnar joints (cooling fractures). Evenly spaced, smooth-sided, roughly straight joints (fractures) in a lava that form column-like structures within the massive cores of flows. They form during cooling of the lava, developing at right angles to the surfaces through which the flow cooled.

corestones. Spherical or rounded stones embedded in soil, or fractured soil. They are all that remain of the original rock from which the soil weathered.

cross bedding. Layering in which one set of layers, usually the upper one seen in an outcrop, slices across another set, usually below. Typically observed in beach rock and ancient, hardened sand dunes.

debris avalanche. A type of energetic landslide in which a large mass of intact crust breaks loose and disintegrates on its way downslope, producing randomly scattered pieces of debris embedded in much finer material. Debris avalanches typically travel great distances across landscapes or under water.

dike. A sheetlike body of igneous rock, which may have the appearance of lava and show columnar jointing, inserted within and across preexisting rock layers. A dike forms when magma fills a crack in the crust and hardens into place.

dome (volcanic dome). A pasty mass of highly viscous lava filling a vent like a plug. Buildup of pressure beneath the dome as gases continue trying to escape can lead to its explosive destruction, only to be replaced again by a new dome after the gases have been released. Volcanic domes in Hawaiʻi typically form only during the alkalic postshield and rejuvenated stage of volcanism.

drowned valley. A valley where the floor sinks below sea level, allowing sediment to fill in the deeper, submerged portion of the valley. A new, broad valley floor develops atop the accumulating sediment, close to current sea level.

entrail pāhoehoe. Pāhoehoe on steep slopes that features toes and lobes drawn into elongate forms downhill by the pull of gravity as the lava flows.

fault. A crack in the Earth's crust along which slippage takes place.

fault scarp. A cliff or ledge formed where a fault ruptures the surface.

fine. A fine-grained rock or sand has crystals that are mostly less than 1 millimeter (around $^1/_{32}$ of an inch) in size.

fishpond. A walled coastal pond, built by native Hawaiians to cultivate edible fish. Many had grates to the sea that allowed small fish to enter but kept larger fish from leaving.

fissure (eruptive fissure). A large ground crack through which lava may erupt. Fissures form as magma flowing in a dike approaches the surface at the start of an eruption and forces open a crack. Fissures break in straight segments, though they may sidestep as they advance.

flow field. A broad area of lava flows erupted over an extended period but all from the same eruption.

gabbro. A dark or speckled, coarse-grained rock of pyroxene, plagioclase, and olivine formed where a magma body that could potentially erupt as basalt stays in place and crystallizes underground instead.

graben. An area of crust dropped down between two closely spaced faults. In younger landscapes, a graben appears as a narrow trench-like valley with steep sides.

Hawaiian Arch. An upwarp in the lithosphere of the seafloor surrounding the main Hawaiian Islands. It is caused by flexing of the lithosphere from the great weight of the islands the arch surrounds.

Hawaiian Deep. A downwarp in the seafloor right next to the heaviest Hawaiian Islands, accompanying the Hawaiian Arch, defined above.

headscarp. A cliff formed at the upslope end of a landslide, from where it tore away as sliding began.

hyaloclastite. Glassy, fragmental debris with the general consistency of sand formed where lava is blasted to pieces by steam as it enters the ocean or a deep pool. Hyaloclastites typically include larger chunks of torn lava, some of which may have flowed underwater before breaking apart.

inflated pāhoehoe. Pāhoehoe that deforms like rising bread dough when it flows onto flat ground. Molten lava continues to feed into the flow beneath a hardened surface crust, swelling it. Tumuli and other distinctive surface features commonly result.

iron oxide. Much of the reddish coloration of rocks and soils in Hawai'i comes from oxidation (rusting) of the iron in the rocks.

kīpuka. An area of older landscape surrounded by younger lava flows.

laccolith. A highly swollen sill, generally with an arching upper contact where it forced overlying rock layers upward during intrusion.

lag deposits. Deposits of cobble, boulders, and other rubble lying atop a surface of finer soil, in many places lateritic. They represent the heavier material in a soil bed that has eroded, but which could not be removed by wind and water. They have "lagged" behind as the surrounding soil has been washed or blown away.

laterites (oxisols). Highly leached alumina-rich tropical soils that may show a deep-red coloration owing to iron oxide staining. They are generally infertile.

late-shield stage. Toward the end of the shield stage, lava flows may become less frequent and more alkalic as the volcano moves off the center of the hot spot. Stored magma may cool and crystallize, producing viscous lava.

lava balls (accretionary lava balls). Rounded lumps of lava often found sitting atop the levees enclosing lava channels, or spread out across the surface of a flow downslope. They form when chunks of solid lava broken from the wall of a vent or channel are swept in flowing lava downslope, rolling and tumbling to acquire ("accrete") a coat, much like rolled snowballs.

lava cave. A cave formed in a lava flow, most often by partial drainage or channel deepening of a lava tube during an eruption. Caves in lava can also form at shorelines by wave action eroding a flow along fractures completely unrelated to volcanic activity.

lava shield. A gently sloping mound or hill of lava surrounding a vent, often capped with a deep, narrow pit crater following eruption. Lava shields are typically no more than a few hundred feet high and a few thousand feet across. They form as lava repeatedly overflows from a vent, spreading across the surrounding landscape for months or even years. Most lava shields, like cinder cones, are not stand-alone volcanoes in Hawai'i, but secondary features fed by the magma chambers of much larger host volcanoes.

lava trees. Pillars of lava left where a flow has enclosed living trees and formed hardened cases around them before they burn away. The surrounding molten lava later drains, leaving the quenched pillars behind as tree monuments. Many lava trees are hollow, preserving fossil molds of the original trees in their centers.

lava tube. A channel inside an active lava flow that conducts molten lava to a flow front downslope. The channelway may later drain or melt down its bed to form a lava cave. The term *lava tube* is often applied to indicate caves, too, long after the end of an eruption.

limestone. A light-colored calcareous rock generally formed from precipitation of calcium carbonate in seawater or via evaporation. Corals produce reefs made of limestone biologically.

lithosphere. Earth's crust and the rigid, uppermost layer of mantle that together form a tectonic plate.

longshore current. A nearshore current in the ocean that flows basically parallel to the shore. Longshore currents can be quite strong and represent a potential threat to swimmers in deeper water.

magma. Underground molten rock.

magma chamber. An underground pool or reservoir of molten rock that commonly (though not necessarily) feeds eruptions.

mantle. The thick layer inside the Earth that lies between the molten core and the crust.

massive. There are two meanings for this word, depending on the context of its use. Something massive is simply very large—the ordinary meaning of the word—or, geologically speaking, a massive rock is a rock with unbroken, homogeneous texture showing no distinctive features, like a smooth board or blank sheet of paper. Whether the rock is especially big or not is irrelevant in this geological meaning.

obsidian. A rock composed entirely of volcanic glass without crystals or bubbles; usually high in silica.

olivine. A light- to dark-green, dense, glassy mineral composed of magnesium, iron, and silica.

pāhoehoe. Basalt lava with a smooth, bulbous, or rippled surface texture.

paleosol. An ancient soil, now covered by younger layers, that appears in an outcrop or erosional cut.

pali. The Hawaiian word for cliff or steep slope, and widely applied as a common geographical term in Hawai'i.

Pele's hair. Thin threads of golden basaltic glass that form as droplets of lava fall from high lava fountains or blow from the surface of flowing lava with the escape of gases. The droplets are stretched and spun into the glassy threads.

Pele's tears. Small glassy beads that form as droplets of lava fall from high lava fountains. They may be nearly spherical to teardrop-shaped and some may have a strand of Pele's hair attached.

peridotite. The greenish rock, rich in olivine and pyroxene crystals, that constitutes Earth's mantle.

pillow lava. Pāhoehoe that flows underwater may form pillow or sack-like lobes as it quickly quenches. The pillows may break off or accumulate in a thick stack as part of a pillow lava flow.

pit crater. A crater less than 1 mile wide with vertical walls that forms from collapse due to drainage of an underlying, shallow body of magma. Eruption need not accompany collapse. Many pit craters have flat, lava-covered floors simply because they partly filled with lava erupted at a later time.

plagioclase feldspar. A light-gray to white mineral with an elongated blocky or needle-like shape. Composed of calcium, sodium, aluminum, and silica.

plate (tectonic plate). See **lithosphere**.

postshield stage. Growth stage where eruptions become infrequent and alkalic in composition as the volcano moves well off the hot spot. Calderas and shields are buried under a cap of alkalic cinder cones and more viscous lava flows.

pumice. Highly vesicular, frothy glassy fragments thrown from a vent during explosive eruption. Both basaltic and trachytic pumices are found in Hawai'i.

pyroclastic cone. A cone of fragmental volcanic debris formed around a vent during an explosive eruption. Depending upon the size and character of fragments, it may be classified as an ash, pumice, or cinder cone, with or without embedded volcanic bombs and blocks.

pyroxene. A group of minerals composed of iron, magnesium, calcium, and silica. Most are blocky in shape and black or dark green in color.

reef. Reefs are structures of calcium carbonate ($CaCO_3$) built by colonies of coral polyps in warm, shallow waters. Reefs are diverse ecosystems that support a great deal of marine life and help protect coastlines from storms.

rejuvenated stage. A return of volcanic activity, often taking the form of widely scattered, infrequent, small eruptions, long after a host volcano has become extinct. Eruptions are highly alkalic, indicating much deeper mantle melting sources than those of the main shield-building stage of volcanic activity in Hawai'i. Not all Hawaiian volcanoes experience volcanic rejuvenation.

rift zones. Zones of weakness and fracturing in the flank of a shield volcano, frequently injected by dikes fed from a central magma reservoir underlying the summit. Rift zones may be marked by numerous cinder and spatter cones, fissures, pit craters, and other volcanic features.

riptide. A strong current related to tidal movement in inlets and narrow bays or water with submerged channels close to shore. A riptide can pose a hazard to swimmers, even in shallow water.

shield stage. The main growth stage of Hawaiian volcanoes as they move over the center of the hot spot, characterized by frequent eruptions of fluid tholeiitic basalt that build a massive shield volcano. Calderas are common due to the presence of large, shallow magma chambers.

shield volcano. A gigantic volcano with gentle slopes built of countless thin lava flows piled atop one another, fed from eruptions taking place at a summit and in many instances from rift zones and vents radiating from the summit. A large, continuously active magma chamber underlies a shield volcano. Smaller **lava shield** vents may form on the larger shield volcanoes.

silica. SiO_2 (silicon dioxide) is an important constituent of most rocks and minerals in Hawai‘i, usually making minerals or volcanic glass. Basalt is about 50 percent silica. Amorphous or microscopic types of silica (quartz)—sometimes called chalcedony, agate, or opal—can form from hot fluids circulating through rocks.

sills. Bodies of crystallized or hardened magma injected between rock layers. The magma wedged the layers apart as it intruded.

slump. A type of landslide in which a mass of material breaks into slices that do not internally disintegrate but instead slowly shift downslope maintaining some structural coherency within them. Individual slumps may remain active for many thousands of years, in contrast to the far more catastrophic debris avalanches.

spatter cones. Steep-sided cones formed around a vent where spatter (blobs of lava) have piled up and welded together at the base of a long-playing lava fountain.

spatter ramparts. Low, steep-sided ridges and heaps of lava spatter piled up along the edge of an eruptive fissure.

tafoni. Pockets left in shoreline basalt by corrosion and crystallization of salts from sea sprays.

terrace (marine terrace). A flat bench of rock or tuff near waterline or a short elevation above sea level, formed by wave erosion when sea level stood higher relative to the present-day shoreline.

tholeiite (tholeiitic basalt). A type of basalt with a shallow melting source that is rich in iron and magnesium and poor in alkali elements (sodium and/or potassium). It is the most common type of rock on Earth.

trachyte. An alkalic volcanic rock with a larger amount of silica relative to other alkalic rocks, giving it a light-gray color. Eruption of trachyte tends to begin explosively, producing abundant pumice. The trachyte lava flows that later erupt also are much thicker than typical basalt flows and do not travel very far because of their high viscosity.

tree molds. Cylindrical holes or molds in the ground marking the locations of tree trunks surrounded by lava. The trees burned or rotted away to leave the holes.

tsunami. A giant sea wave, or series of waves, coming off the ocean, generated by undersea earthquakes or great landslides in the water that have occurred in some cases thousands of miles away.

tuff. Ash and other fine-grained explosive volcanic debris that has been hardened into rock.

tuff cone. See **ash cone**.

tumulus (plural is **tumuli**). A steep-sided mound of fractured lava, sometimes showing evidence of releasing molten lava from the fractures. Tumuli form within inflated pāhoehoe, as blocks of solid, thickening crust are slowly heaved upward by the pressure of the still-molten lava inside the flow. Tumuli may be 5 to 20 feet high and wide, or even larger.

vent. An opening in the ground through which molten lava erupts.

vesicles. Rounded pockets or holes where gas bubbles were left in lava as it cooled and hardened.

vog. A persistent haze (“volcanic smog”) created by sulfur gases and particles released by an erupting volcano.

xenolith. A rock from a different source embedded within a host rock. **Mantle xenoliths**, for instance, originate in Earth’s mantle and can be found embedded in Hawaiian tuffs and alkalic lavas. Magma incorporates the solid xenoliths from the older rock enclosing the magma conduit as it rises to erupt.

RECOMMENDED FURTHER READING

Blay, C. 2004. *Kauai's Geologic History*. Self-published, available online.

Bolt, B. A. 2004. *Earthquakes*, 5th ed. New York: Freeman and Co.

Clark, J. R. K. 2002. *Hawai'i Place Names: Shores, Beaches, and Surf Sites*. Honolulu: University of Hawai'i Press.

Culliney, J. L. 2006. *Islands in a Far Sea: The Fate of Nature in Hawai'i*. Honolulu: University of Hawai'i Press.

Cumming, C. F. G. 2016. *Fire Fountains: The Kingdom of Hawaii, Its Volcanoes and the History of Its Missions*. Sydney, New South Wales: Wentworth Press.

Dvorak, J. 2015. *The Last Volcano: A Man, a Romance, and the Quest to Understand Nature's Most Magnificent Fury*. New York: Pegasus Books.

Farndon, J. 2018. *The Illustrated Guide to Rocks and Minerals; How to Identify and Collect the World's Most Fascinating Specimens; With Over 800 Detailed Photographs and Illustrations*. Dayton, OH: Lorenz Publishing.

Goff, J., and W. Dudley. 2021. *Tsunami: The World's Greatest Waves*. Oxford: Oxford University Press.

Grigg, R. W. 2013. *Archipelago: The Origin and Discovery of the Hawaiian Islands*. Honolulu: Island Heritage.

Juvik, S. P., and J. O. Juvik. 1998. *Atlas of Hawai'i,* 3rd ed. Honolulu: University of Hawai'i Press.

Lockwood, J. P., R. W. Hazlett, and S. de la Cruz. 2022. *Volcanoes: A Global Perspective*. Hoboken, NJ: Wiley-Blackwell Publishing.

Lopes, R. 2012. *Volcanoes: A Beginner's Guide*. London: Oneworld Publications.

Poland, M. P., T. J. Takahashi, and C. M. Landoswki. 2014. *Characteristics of Hawaiian Volcanoes*. US Geological Survey Professional Paper 1801.

Thompson, C. 2019. *Sea People: The Puzzle of Polynesia*. New York: HarperCollins.

Scheuer, J. L., and B. K. Isaki. 2021. *Water and Power in West Maui*. Lahaina, HI: North Beach–West Maui Benefit Fund.

Wyban, C. A. 2020. *Tide and Current: Fishponds of Hawaii*. Honolulu: University of Hawai'i Press.

Ziegler, A. C. 2002. *Hawaiian Natural History, Ecology, and Evolution*. Honolulu: University of Hawai'i Press.

REFERENCES CITED IN TEXT WITH GRATITUDE FOR WORK ACCOMPLISHED

Clague, D., and D. Sherrod. 2014. Growth and degradation of Hawaiian volcanoes. In *Characteristics of Hawaiian Volcanoes*, ed. M. P. Poland, T. J. Takahashi, and C. M. Landowski, US Geological Survey Professional Paper 1801, pp. 97–148.

Langeheim, V. A. M., and D. A. Clague. 1987. The Hawaiian-Emperor Chain Part II: Stratigraphic framework of volcanic rocks of the Hawaiian Islands. In *Volcanism in Hawai'i*, ed. R.W. Decker, T. L. Wright, and P. H. Stauffer, US Geological Survey Professional Paper 1350, volume 1, pp. 55–84.

McMurtry, G. M., J. F. Campbell, G. J. Fryer, and J. Fietzke. 2010. Uplift of Hawai'i during the past 500 ky as recorded by elevated reef deposits. *Geology* 38: 27–30.

Neal, C., and J. P. Lockwood. 2003. *Geologic Map of the Summit Region of Kīlauea Volcano, Hawai'i.* US Geological Survey Geologic Investigations Series I-2759.

Nullet, D. No date. *The Natural Environment.* (Online textbook for physical geography with emphasis on the Hawaiian Islands). https://laulima.hawaii.edu/access/content/group/dbd544e4-dcdd-4631-b8ad-3304985e1be2/book/toc/toc.htm.

Sherrod, D. R., J. M Sinton, S. E. Watkins, and K. M Brunt. 2021. *Geologic Map of Hawai'i.* US Geological Survey Scientific Investigations Map 3143. Scales 1:100,000 and 1:250,000.

INDEX

Pages in boldface include photographs.

ABOUT THE AUTHORS

Rick Hazlett received his PhD in geology from the University of Southern California and his master's degree from Dartmouth College. His graduate work included volcanic hazards evaluation around San Cristobal Volcano in Nicaragua and mapping of the Mopah Range volcanic field in California's Mojave Desert. Subsequent research included studies in Italy, the Aleutian Islands, and Hawai'i. He taught for nearly thirty years at Pomona College in Claremont, California, where he helped found the Environmental Analy-sis Program. He is coauthor of *Volcanoes: A Global Perspective* (Wiley-Blackwell Press) and is presently retired in Hilo on the Big Island of Hawai'i.

Cheryl Gansecki teaches geology courses at the University of Hawai'i–Hilo and intensive volcanology field camps in Hawai'i for the Black Hills Field Station. She holds a BA in earth science and archaeology from Wesleyan University and a PhD in geology from Stanford University. She has researched volcanoes in Greece, Yellowstone, and Hawai'i. She works with the Hawaiian Volcano Observatory to monitor Kīlauea Volcano, both in the field and in the lab studying the chemistry of erupted lava. She also filmed and produced many educational videos—and a card game—about volcanoes along with spouse and fellow volcanologist, Ken Hon.

Steve Lundblad experienced firsthand the eruption of Mt. St. Helens because it covered his hometown with ash during his high school years. This encounter, coupled with his interest in the outdoors, fostered his interest in volcanoes and geology. Steve received his BA from Harvard University and earned a master's degree at the University of Wisconsin and a PhD at the University of North Carolina. He is currently a professor of Geology at the University of Hawai'i–Hilo and is the recipient of the 2018 UH Board of Regents Excellence in Teaching Award. Along with his students, Steve monitors ground deformation of the Kīlauea Volcano.